Managing Environments for Leisure and Recreation

Most people live for their leisure, recreation and travel. *Managing Environments for Leisure and Recreation* seeks to bring together the different strands of thought that inform the management of settings for leisure and recreation. The text shows how a positive difference can be made to such activities, by taking a balanced approach in managing for the environment and for people, both now and into the future.

The text begins by examining the scope of leisure, recreation and tourism, including a brief historical survey and a discussion of the web of players involved. The book then looks at the benefits, the costs, the settings and the trends within leisure and recreation. It also looks at how we attach or measure value and examines some of the issues in environmental economics, as well as the need to consider value from different angles. The final part deals with management techniques, from preparing to manage, the research and planning, to actual management for the environment (the physical and often natural setting) and for people (the socio-economic setting). *Managing Environments for Leisure and Recreation* then explores possible future scenarios, and discusses what skills managers will need to deploy.

Managing Environments for Leisure and Recreation urges managers to balance the need for a systematic approach, with the need for a systemic development of an environmental management ethic. It is highly illustrated with over fifty line drawings and includes many case studies from around the world.

Richard Broadhurst is the Policy Officer for the Forestry Commission's National Office for Scotland.

Routledge Environmental Management Series

This important series presents a comprehensive introduction to the principles and practices of environmental management across a wide range of fields. Introducing the theories and practices fundamental to modern environmental management, the series features a number of focused volumes to examine applications in specific environments and topics, all offering a wealth of real-life examples and practical guidance.

MANAGING ENVIRONMENTAL POLLUTION
Andrew Farmer

COASTAL AND ESTUARINE MANAGEMENT
Peter W. French

ENVIRONMENTAL ASSESSMENT IN PRACTICE
D. Owen Harrop and J. Ashley Nixon

ENVIRONMENTAL MANAGEMENT
Principles and practice
C. J. Barrow

MANAGING ENVIRONMENTS FOR LEISURE AND RECREATION
Richard Broadhurst

Forthcoming title:

WATER RESOURCE MANAGEMENT
N. Watson

Managing Environments for Leisure and Recreation

Richard Broadhurst

London and New York

First published 2001
by Routledge
11 New Fetter Lane, London EC4P 4EE

Simultaneously published in the USA and Canada
by Routledge
29 West 35th Street, New York, NY 10001

Routledge is an imprint of the Taylor & Francis Group

Typeset in Ehrhardt by RefineCatch Limited, Bungay, Suffolk
Printed and bound in Great Britain by
T.J. International Ltd, Padstow, Cornwall

British Library Cataloguing in Publication Data
A catalogue record for this book is available from the British Library

Library of Congress Cataloging in Publication Data
Broadhurst, Richard
   Managing environments for leisure and recreation/Richard Broadhurst.
      p.   cm. – (Routledge environmental management series)
   Includes bibliographical references and index
   1. Environmental management. 2. Ecotourism. I. Title. II. Series.
GE300.B76   2001
333.78–dc21   2001018064

ISBN 0–415–20098–9 (hbk)
ISBN 0–415–20099–7 (pbk)

Contents

v

CONTENTS

Tables, figures and boxes

Tables

Figures

Boxes

Preface

I am convinced that (in Britain, at least) we grossly underestimate the contribution that leisure, recreation and tourism can make to our health, our well-being, our development and our environment. As we rush towards the twenty-four-hour society, we seem to be heading for almost certain gridlock, not just of traffic in our cities, but in the superhighways, in information and communications technology and our consumption of information. Think your way through the day: the radio programmes we listen to, the paper we buy in the morning, the papers we take to meetings, the telephone calls, internet, the e-mails, the post, television, the books and of course the conversations. Our computer systems crash, jammed up through the sheer mass of daily communication (never mind the actions of viruses). The same communication systems that bring so much good so swiftly can also be used by others to spread crime and other antisocial practices. The control systems, the societal norms, seem not to be working adequately to cope with the speed at which we now live. Our human brains have limits too. We have let technology run away with us, and we now need to pay more attention to some other facets of human development, and quick. I believe leisure, recreation and tourism have a major part to play in our rehabilitation.

More than a quarter of a century ago Dower and Downing (1975) wrote that the focus was more on the ecology of leisure activity than on sociology or psychology. The emphasis still seems to be more on the natural or physical aspects of the environment than on socio-economic setting. In a similar vein, there is still more attention focused on facilities than services, and too much compartmentalisation. We seem to forget how diverse people are and how their chosen forms of leisure, recreation and tourism vary. We focus on the detail and lose the big picture. We have an intense interest in the immediate, in terms of space and time, the here and now, egotistical in the extreme. Our interest (or time preference) drops off too swiftly for our own good. If we are not careful, we risk long-term losses for short-term gains.

To the rescue (we hope) comes Environmental Management, an emerging discipline, which implies a much richer approach than we have used in the past. It seeks to take into account the whole range of effects: the costs and the benefits on our total environment, and that of our neighbours; taking account of the socio-economic and biophysical world. It is an inclusive approach that in many cases is still seeking to devise and incorporate new methodologies to allow very different kinds of costs and benefits to be considered. It not only allows appraisal before decisions are taken, through impact assessments, but crucially encourages reflection and continual development. Perfectly applied, it will lead to sustainable development.

What this book does is outline how leisure and recreation managers can borrow from, and contribute to, this developing approach. The best managers have been using these techniques for years, intuitively; simultaneously operating at different scales, thinking long-term, considering the effects of actions on communities, as well as on environments, without the benefit of buzz words. Management in this field is not all about systems and methods, it is also about feelings and emotions, and about common sense. Smile at the next person who comes your way, and the smile will ripple out throughout all the people in the system. This ripple effect works for other outputs too, whether economic, environmental or social. Ripples of pleasure and enlightenment will lead to ripples of action. We need a systemic approach, more than a systematic one, but if we do it right, managing should lead to more opportunities now, and more opportunities for future generations too. The values embedded in this systemic approach will also give us the grace to recognise when not to manage, or more properly when good management means giving people space and time, to manage their own leisure and recreation environments.

Acknowledgements

First, I must thank those who gave me the idea for this text, especially the fourth-year students and staff involved with the Recreation and the Leisure Studies BA courses, in what is now part of Edinburgh University. A number of people also gave wise counsel while I was shaping the proposal especially Alan Barber, Kevin Bishop, Derek Casey, Allan Patmore and Elaine Thompson (and received precious little thanks!). This text is very different from what was envisaged then, but that input was crucial.

Second, I must thank those who have supplied material and helped along the way. I cannot mention them all by name, there is not room, but I remain indebted and only hope I can give the same help to others when they need it. Thank you to colleagues Graham Cullen, Simon Gillam, Steve Gregory, Rob Guest, Wilma Harper, Paddy Harrop, Alastair Johnson, Emily Ramsay and James Swabey for looking over parts of the text, for your humour and your advice. Thank you to colleagues in other agencies, and in the Countryside Recreation Network, for all the help over the ages.

Thank you to those who, amid busy schedules, looked out materials and gave advice, especially to Rob Green and the Countryside Agency for helping to locate Table 5.3, Chris Barrow for the ideas that led to Table 7.3 and Celia for the idea which led to Figure 9.3. Thank you too, to my reviewers: Gill Day, Chris Barrow and Allan Watt, who had a major bearing on the development of the text, and crucially to those at Routledge: to Sarah Lloyd, Sarah Carty, Andrew Mould, Ann Michael; also Ray Offord, Belinda Dearbergh and others behind the scenes.

Third, the publishers and I would also like to thank the following for granting permission to reproduce material in this work: Professor John Adams, University College London, for the figure 'Travel by Britons: cycle, bus, train, car and plane, 1952–2025' in 'Hypermobility: too much of a good thing', *Countryside Recreation*, 8 (1), 2000 (Figure 9.2); Dr Ian J. Bateman, Centre for Social and Economic Research on the Global Environment, University of East Anglia, Norwich, for a figure from his chapter with F. Bryan, 'Recent advances in monetary evaluation of environmental preferences' in *Environmental Economics, Sustainable Management and the Countryside*, ed. R. Wood, Countryside Recreation Network, Cardiff, 1994 (Figure 5.4); Neil Bayfield for the figure 'Limits of Acceptable Change: annual cycle at Aonach Mor' from his chapter with G.C. McGowan, 'Monitoring and managing the impacts of ski development: a case study of Aonach Mor resort 1989–1995' in *Landscape Ecology Theory and Application*, ed. G.H. Griffiths, IALE, Aberdeen, 1995 (Figure 8.1); the Canadian Association of Geographers, Montreal, for the figure 'Conceptualisation of the relationships amongst leisure, recreation and tourism' from 'Some notes on the geography of tourism' by Z.T. Mieczkowski in the *Canadian Geographer*, 25 (189), 1981 (Figure 1.4); City of Edinburgh Council Recreation Department and Pentland Hills Regional Park for 'The Pentland Cycle of Wildlife and Farming' from the calendar of the Pentland Hills Regional Park by Duncan Monteith (Figure 8.9); CABI Publishing, Wallingford, for figure 11.1 in the chapter by R. Broadhurst and P. Harrop 'Forest tourism' in *Forest Tourism and Recreation*, ed. X. Font and J. Tribe, 2000 (Figure 5.2); the Council for the Protection of Rural England, London, and the then Countryside Commission for the map 'Tranquil areas, south east region', Tranquil Area Maps, CPRE and Countryside Commission, October 1995 (Figure 7.3); the Countryside Agency, Cheltenham, for table 2.1 in 'The non-timber benefits of trees and woodland', report by Environmental Resources Management and the Countryside Agency, 1998 (Table 5.3); the Countryside Agency for the map of the Countryside Agency's Designated and Defined Interests, 2000, Countryside Agency copyright, based on the Ordnance Survey map (Figure 5.11); Forestry Commission for 'Day visits in the United Kingdom, 1998' in Social and Community Planning Research, *Leisure Day Visits: Summary of the 1998 UK Day Visits Survey*, 1999 (Figure 5.7); International Thomson Publishing Services Ltd and Mike Stabler for figure 2.1 in his chapter 'The concept of opportunity sets as a methodological framework for the analysis for selling tourism places: the industry view' in *Marketing Tourism Places*, ed. G. Ashworth and B. Goodall, International Thomson Business Press, 1990 (Figure 3.7); International Thomson Publishing Services Ltd and Brian Hay for figure 10.1 in the chapter by A. Seaton and B. Hay 'The marketing of Scotland as a tourist destination, 1985–96' in *Tourism in Scotland*, ed. R. McLellan and R. Smith, International Thomson Business Press, 1998 (Figure 6.2); the New Economics Foundation, London, for the figure 'UK index of sustainable economic welfare *per capita*, 1950–96', p. 3 in *More isn't always Better*, ed. E. Mayo, A. MacGillivray and D. McLaren, 1997 (Figure 5.5); Jonathan Gershuny and Kimberly Fisher, University of Essex, for a figure from their chapter 'Leisure' in the third edition of *Twentieth Century British Social Trends*, ed. A.H. Halsey with Josephine Webb, pp. 620–49, Macmillan Press, 2000 (Figure 1.1); the Policy Studies Institute, London, for data from the article by A. Feist 'Comparing the performing arts in Britain, the United States and Germany' in *Cultural Trends 31*, 1998 (Figures 3.4–5); A.P. Watt Ltd, London, on behalf of the National Trust for Places of Historic Interest or Natural Beauty, for the four lines of verse from *Just So Stories*, by Rudyard Kipling, which open Chapter 1; John Wiley and Sons Ltd, Chichester, for a figure from the chapter by R. Broadhurst 'The search for new funds' in *Heritage Interpretation* II, *The Visitor*

Experience, ed. D. Uzzell, Belhaven Press, London, 1989, pp. 29–43 (Figure 6.4). Every effort has been made to contact copyright holders for permission to reprint material. The publishers would be grateful to hear from any copyright holder who is not acknowledged and will undertake to rectify any errors or omissions from future editions of the book.

Fourth, I must thank those who ensured doors remained open, which includes all the teachers, lecturers and students I ever had. In first place is R.D. Harris and closely thereafter all the librarians (especially those who smiled), other people who lent or even gave papers and books and Ian Ricketts, who tamed a very wild computer (for a second time!). I should also thank long-suffering colleagues in the Forestry Commission and my friends and family, whom I have sorely neglected. The biggest thanks and acknowledgements go to Celia, Marie-Louise, Henry, Katie and Annabelle, who have between them given me all the most memorable and delicious slices of real leisure and recreation, and to whom I owe the biggest debt.

How to use this book

This book is written very much with students and managers in mind, mostly in leisure and recreation, also for those in other disciplines who may wish to explore the boundaries or interactions with leisure and recreation, or use it to develop their own strands of thinking. Students in many disciplines have an interest in leisure and recreation; psychology and sociology, geography and economics, education, environmental sciences and business management amongst them. The book's remit is unashamedly broad, and therefore often skims across the surface. As Driver (1974) has pointed out, 'The gap between what we profess and what we do must be narrowed. Only the discriminating critic can bring this about, and he needs to be a generalist, if not a universalist.' This particular generalist is likely to be heavily criticised, but the intent is there.

There are fewer diagrams than I would like, but the student and manager is urged to take up the suggestions for developing mind mapping and other visual techniques, which can be extremely helpful in applying whole brain solutions. Many readers are likely to feel shy about drawing diagrams, but they should persist, and enjoy it. After all, there are good precedents. St Thomas Aquinas, patron saint of schools, colleges and universities, is attributed with saying 'Man cannot understand without images' (Hughes 1997: xv). And it was said (de Santillana 1956: 68) of Leonardo da Vinci that he refused to think only in words. (A good job too, you might think, if you had to read it all using a mirror!)

The structure of the book is simple: beginning (chapters 1–3), middle (4–5) and end (6–9). The beginning gives an idea of what it is that we are managing. The first chapter looks at the scope of leisure and recreation as we view it now, before looking, in chapter 2, at some of the changes down the years. The entire book, but particularly the chapter focusing on history, is written very much from a Western point of view, but recognising that there is much to be learned from other cultures. The intention with the history chapter is to sense the breadth and depth of the different activities in which people have engaged, and of course still do. Different cultures have adopted different practices at different times. Chapter 3 describes the web of people and organisations typically involved (using Britain as an example), and at the flows of information which hold the web together.

The middle describes many of the different benefits and costs associated with leisure, recreation and tourism; the settings and trends. It explores how we can measure or attach values to these.

The end focuses on the act of managing: preparing to manage, managing the physical setting, managing the people and managing our future. Chapter 6 focuses on how we can prepare to manage, in the research, planning and design stages. In practice, these techniques and actions are jumbled up with our interventions, or management activities, but here we have the luxury of treating them separately. Chapter 7 focuses on managing for the environment, with a focus on the long term and ensuring choices remain open for generations to come, whether by managing the resource itself, or the behaviour of visitors. Chapter 8 then focuses on managing the socio-economic setting, with more of a concern for effects today, and equity. The final chapter takes a look at what we may need to be considering in the years to come, to make sure that we use leisure and recreation to best effect.

At the beginning of each chapter there is an overview, to give a brief summary of what follows. Each chapter closes by suggesting some ideas for further study or work (through five questions) before pointing to some authors who may add something special in the area covered. Some other sources are listed in the appendices: web sites, and a bibliography with references. The extensive use of figures and boxes may help to encourage students and managers to immerse themselves in the real joy of the subject.

Chapter 1

The scope of leisure and recreation

Overview

> I keep six honest serving-men
> (They taught me all I knew);
> Their names are What and Why and When
> And How and Where and Who.
>
> (Rudyard Kipling, 'Just so Stories for Little Children', 1902)

Kipling's six honest serving men serve as the framework for introducing the subject. After exploring some difficulties in defining the scope of leisure and recreation, we settle on some pragmatic definitions to answer the question, 'What?'. Leisure is more than just the disposable time available after work and other obligations, and the 'non-productive consumption of time' (Veblen 1899). It also reflects a state of mind, an optimal balance between the desires of an individual and the obligations felt to others. Each individual calibrates obligations on his, or her, own scale. Personal development and assertiveness training are just two of the techniques people are using to recalibrate their sense of obligation. Recreation is described as the activity (or lack of it) that people choose to engage in when at leisure, and it may be multi-faceted, comprising physical, cognitive, emotional and social components. It may manifest itself as watching or taking part in sport, the arts, outdoor recreation, and a realm of other activities. Music and reading, as examples of recreation, are briefly explored. Management is described as arranging resources to achieve particular ends, in bio-physical and socio-economic environments.

Theories of motivation abound, and among them Maslow's hierarchy of needs is helpful in exploring what it is that people are seeking from leisure and recreation. Behaviour is dependent on personality and environment (both bio-physical and socio-economic) and it is in these settings the manager can intervene. What distinguishes leisure from work is the freedom to choose, the activity, the place, the company, the intensity of physical, mental and emotional effort that is deployed. The end may be happiness, a search for meaning, or the journey itself. This is a matter for the participant. What we choose to do depends in part on the life stage we are involved in, as an individual, whether independent, dependent youngster, or with young or old dependants. Similarly, the amount of time we have available will reveal different opportunities. Our desires and the opportunities afforded will often have a relationship with natural periodicities or cycles, whether during the day, lunar month or year, even though technologically we can have a twenty-four-hour day.

Recreation is often a blend of physical, mental and emotional activity, most often enjoyed in company. We can study at different scales of resolution, and at each, we should be able to identify five stages: anticipation, journey, event, journey back,

and recall (Clawson and Knetsch 1971). We can study an even
itself, the holiday or the life of the participant. Each event
middle and an end. Play is important in communicating an
for life.

Home, school and the workplace will claim a large
Other settings are important too: the spaces within our neig'
inside public buildings, important neutral spaces. Travellir
with a moving backcloth, and holidays provide opportuni
with the family, or other referent group. We tend to shar
with others, even when in pursuit of wilderness experienc..,
playmates, colleagues from work and fellow club members. The distribution of
opportunities for recreation and leisure is not, though, equitable. Typically, and certainly in Britain, women, children, older people, people with disabilities and ethnic
minorities tend to have fewer opportunities (Bramham *et al.* 1989).

How we approach leisure will depend on numerous influences. Genetic make-
up, personality, the influence of the family, teachers and peer group will all play a part.
Science and the manager's analytical and creative thinking will help to address the
issues, which are awaiting intervention. Among the prominent issues are: balancing
obligations with leisure, travel, the motor car, sustainable development and management of leisure, equity, the roles of the state, private and voluntary sectors, and the
role of management. Managers need to determine where to intervene to best effect,
towards useful policies and programmes and in areas where the intervention will
make a real difference.

What?

We could define 'leisure' simply as the time available that we can spend as we choose.
Simple though this definition is, we immediately find a complication. In its purest
state, it is probably as difficult to find as the Holy Grail. The problem is that each of
us feels very differently about our obligations, whether to our colleagues at study or at
work, our families, our friends, our communities or to society at large. How we feel
will be the product of many influences and many of these will be specific to our own
culture. The time we are describing here is the time left, after taking away the hours
we spend on:

◆   work;
◆   study (for pupils and students only);
◆   sleep;
◆   caring for dependants;
◆   housework.

Many use the term 'leisure' to describe the time we have left after we have
carried out essential functions, but that is unhelpful. This book will develop the
argument that leisure itself is an essential component of our lives, and not a residual.

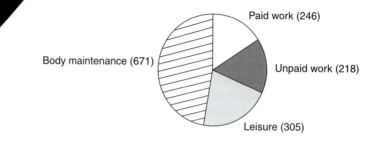

FIGURE 1.1 Work or leisure? Proportions of the day (minutes)
Source: Gershuny and Fisher (1999)

There are many alternative views, and the reader in search of definitions will not be disappointed (Dumazedier 1967; Godbey 1994; Goodale and Godbey 1988; Glyptis 1993; Haworth 1997; Kelly 1993; Neulinger 1974; Parker 1976; Rapoport and Rapoport 1975; Roberts 1978; Seabrook 1989). It has been said that in the past the focus of study has been on the impacts of leisure and recreation on the physical environment rather than on the socio–economic environment (Dower and Downing 1975). Some consider leisure as a state of mind, an approach to life, a part of our lives set aside from other obligations. Often the terms 'leisure' and 'recreation' are used interchangeably, but not in this book. Here the term 'recreation' is reserved to describe what we do with, or at, our 'leisure'. It is the chosen activity (or lack of activity) in which we take part, when we have the choice as to how to spend our remaining or disposable time.

Think for a few moments about the recreations you enjoy. Chances are that they will include different kinds of activity, from which you may choose according to the time of day or year, the time available, the money available, your mood and the preferences of others close to you. If you made a list, physical activity and exercise are sure to be there. The list would include more cerebral activity as well (even emotional or spiritual activity), and would often include activities you did not ordinarily, or otherwise, do. There will be times when you choose rather less intense activity and wish to be reflective or just reactive.

You might wish to exercise your senses, be concerned with a more passive role, or have a desire to consume goods, services and (famously) food and drink. Or you might want to listen to music. Commentators have found that there is a reasonable consistency across people, about what activities they engage in, which is neatly explained by the *core plus balance* model, described by Kelly (1999: 143–4). The *core* consists of activities that are easily accessible, and therefore also low–cost. Typically this core can be engaged in around the house, or neighbourhood, and among the daily pattern of events. It may include: reading, walking, listening to the radio, watching television, chatting with friends and family informally, conversation generally and engaging in sexual activity. The *balance* of leisure is more likely to be planned, and could change or develop during the life course. Managers are more usually focused on this area, the balance. To be effective managers we should consider how our

TABLE 1.1 Favourite recreational activities for periods of up to five minutes

Mind-centred	Body-centred	With others
Write a poem	Run (a mile for some!)	Smile
Write a song	Run up steps	Laugh
Compose a tune	Run on the spot	Meet someone
Sketch	Do 100 press-ups	Share some space
Mental arithmetic	Kick a football	Share a song
games	Keepie-uppie	Meet someone new
Memory games	Juggle	Kick a ball
Invent	Dance on your own	Play catch
Draw a cartoon	Walk for 2.5 minutes and	Throw a frisbee
Doodle	return	Phone a friend
Take a photograph	Explore for 2.5 minutes	Talk to a friend
Draw a mind map	and return	Listen to a friend
Make a list	T'ai Chi	Walk and talk, in the park
Remember an event	Jump	Share a favourite place
Look at a picture and	Touch the tree	Dance with a friend
savour it	Taste the water	Kiss a friend
Look for a photograph	Eat an apple	Share coffee with friends
Open a book	Take a drink	Share a snack
Look for quotes	Sit in the sun	Learn a trick, a game
Listen to music	Sit in the shade	Teach a trick, a game
Listen to birds	Sit in the wind	Share a joke
Watch some flowing	Sit in the calm	Share a prayer
water	Totally relax	Share thoughts
Meditate	Catnap	
Day dream	Sleep	

interventions will help facilitate (or at least not impede) core leisure too, by providing and maintaining appropriate settings.

As well as enjoying watching others, listening or otherwise consuming, recreation can also take the form of creative, expressive or productive activity. We all have a need to turn our innermost thoughts into some more tangible form which we can share with others, in the hope that they too will share something of what we felt when we first had that idea, or perhaps to provoke thought and a different reaction. For whatever reason, some of us spend a good proportion of our lives engaged in such activity that spans the complete range of arts and crafts. There is nothing more enchanting than to receive a gift, whether sketch, piece of pottery, poem or even letter, created or crafted by someone close to you.

Sport is more than simple physical activity (Cashmore 1990). Usually each sport has a set of rules that ensure fair play and encourage full enjoyment and

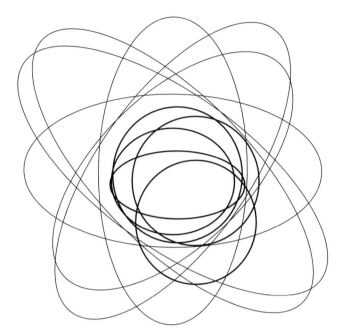

FIGURE 1.2 Core plus balance. For simplicity, each individual's interests are represented by two ellipses. The first, inner, ellipse describes core activities and interests, such as conversation, reading, listening to the radio – generally easily accessible, and no great effort is required. The second, outer, ellipse describes the personal blend of activities and interests that provides the balance, often requiring greater planning or effort. The opportunities for management intervention are quite different in each case but focus on managing settings, physical and social

development of potential within some convenient boundary. Sport may involve competition, co-operation, risk, physical exertion, stamina, concentration, hand–eye motor co-ordination of self, co-ordination, rhythm and timing across a whole team and crucially, fun. Some sports encourage cardiovascular activity, which among those in an increasingly sedentary technological world is a welcome break, from driving a desk or computer, or an escape (Segrave 2000). Physical activity of this kind releases endorphins, chemical messengers (hormones) that increase our sense of pleasure (Greenfield 1997).

The arts encompass every conceivable branch of practical activity guided by a set of principles (Chambers 1972): visual arts like drawing and painting; literary arts like poetry and prose; music; performing arts like drama, dance, ballet, song and opera; plastic arts like sculpture and modelling; radio, television, film and video.

More practical skills include the further application of the arts, such as through carpentry and metalwork, decorating the inside of our houses. For some, activities such as car maintenance are recreational and an enormous industry has

BOX 1.1 'Music, the greatest good that mortals know'
(Joseph Addison, 1672–1719; from Cohen and Cohen 1960: 1)

One of the most ubiquitous art forms enjoyed by all ages and cultures is music. It starts with the heartbeat. People have been aware of the restorative and mood-changing effects of music for thousands of years (Plato, in Abraham 1979). It can be used to put people into the right frame of mind (or mood) for a particular purpose (Argyle 1996: 203–5), for social activity, for relaxation, for expressing emotions (Csikszentmihalyi 1992). It can be used to marshal courage when troops are going into battle, or to galvanise football fans into a single crowd. Musicians and composers are well aware of the characteristics of music that can cause different effects (Meyer 1956).

There are descriptions from biblical times of the effects of trumpets, cymbals and other instruments. Rameau wrote his *Traité de l'harmonie* in 1722 (Rameau 1971) describing the effects of different treatments, cadences and fugue, with an indication of the resultant effects. Recently P. Robertson (leader of the Medici String Quartet), the psychiatrist and writer A. Storr and colleagues have set up the Music Research Institute (www.mri.ac.uk) to advance 'appreciation of the beneficial role that music can play in the enhancement of human experience'. Some of the work in which they have been involved has demonstrated the scientific basis of what many music lovers intuitively know. Music can make us feel very good indeed. Using magnetic resonance imaging, it is possible to plot and record brain activity responding to music. This brain activity releases chemical neurotransmitters, which in turn, depending on the type of transmitter, makes us feel different types of 'good'. Scientific evidence is required to persuade people of the value of music, as many Western societies are spending less time and money on music education as they try to squeeze ever more 'important' subjects into the curriculum.

For some people, music is essential to life. A survey (reported in Storr 1997: 123) revealed that most of us spend at least one or two hours producing music each day, quite apart from the time spent in listening. We should not be surprised that music can be so potent in therapy, in treating autistic children and children traumatised by war, and in pain reduction. The different moods and effects which can be induced are so wide-ranging that song writers and musicians will never run out of surprises, harmonies and melodies with which to please us, whether pop or classical, rap or jazz.

FIGURE 1.3 Different kinds of recreation: mind-centred, body-centred and with others

built up in some cultures supplying every gadget and material which you could imagine useful (and some rather less so) in the pursuit of Do-It-Yourself. Most of us also spend some time decorating or managing the spaces outside our houses, our gardens, becoming immersed in the earth and feeling some connection with the real world.

Another group of recreations takes us outside our immediate home setting, where we are more concerned with contact with the elements and the natural and physical world. Experiences such as these, whether through white water canoeing, sailing across oceans, walking among the mountains or along coast, watching birds or other animal and plant life, provide people with opportunities to marvel at the natural world. Inevitably, the more we marvel the more the experience becomes an emotional and often spiritual one, as we feel more a part of the complex world around us. These experiences can be passed on with eloquence, using the written word (Shackleton 1999; Simpson 1997; Slocum 1949).

Enough of recreation and leisure; what is management? In this book, 'management' is used to describe the activities of managers, directing and arranging resources to achieve particular ends. A manager is anyone engaged in that activity. The book describes some of the processes which managers can use in respect of the environments in which we enjoy our leisure and recreation. The classical factors of production – land, labour, and capital – represent one simple classification of those we have available. Another approach is to view our world as a number of interacting processes. If we understand these processes we can begin to work with them, to direct and arrange resources to shape and develop the environments we want, as landscape ecologists do (Diaz and Apostol 1993). In this model, it becomes important to understand the existing dynamic background. When managing impacts on the physical environment, we should recognise that there are many natural cycles and processes operating, which we need to take into account. In just the same way, when we intervene in the socio-economic environment, we need to recognise the other political, economic and social processes at work. We need to recognise the differences between those we have to work with, and those we can barely shape.

We choose the environments in which we spend our leisure. Our urge to explore our environment, and seek out new worlds, new settings and new people, is given almost totally free rein through tourism. Much of tourism is very closely related to leisure and recreation. Mieczkowski's elegant conceptualisation (Figure 1.4) is useful when considering the scope of interest.

The freedom to choose the place where we enjoy our own time is fundamental. In olden times we might have been constrained by choosing between environments or settings that existed; within the woodland, open savannah, beside the river or at the coast. For some thousands of years, man has been shaping the environment, inside and outside, and we have a sophisticated understanding of what effects different treatments may generate.

Just as we choose the physical setting, so too do we choose with whom we spend

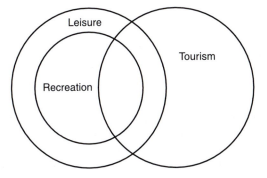

FIGURE 1.4 Leisure, recreation and tourism: Mieczkowski's conceptualisation
Source: Mieczkowski (1981)

our time. This is a fundamental freedom associated with leisure and recreation. We intuitively manage the social and economic settings within which we operate. The social setting is just as important in determining the outcomes for leisure as the bio-physical environment. The person I choose to play squash with may not be the same person I would choose to accompany to a concert.

BOX 1.2 Reading and writing for pleasure

If you have ever tried your hand at writing with a quill pen, you will be able to imagine the hours it would take to write a book by hand. The Venerable Bede, working in the north of England, completed his *Ecclesiastica Gentis Anglorum* in 731, by when (aged 58) he had written nearly forty books (Harvey 1967: 75). Spare a thought, then, for the dutiful monks setting down on paper, or vellum, stories which would be seen by only a handful of people each month or year. The Book of Kells, produced in the eighth and ninth centuries (held at Trinity College, Dublin), demonstrates the endless hours of effort and concentration applied by these devoted craftsmen.

In 1476 William Caxton (Harvey 1967) assembled a printing press in Westminster. In other cultures similar developments ensured that the written word could be reproduced and increasing numbers of books circulated. Reading was mostly a matter of learning, and was available only to the highest classes or castes in society. Imagine what a monk would make of the explosion in reading today. In 1998, in the general book market in the United Kingdom alone, some 72,567,951 books were sold, with a value of £597,105,093 (*The Bookseller*).

Why?

Theories of motivation abound, and it is instructive to review such theories (Gross 1996). The simple view that behaviour is dependent on personality (inner drives, or push) and the physical and social environment (or pull) (Mannell and Kleiber 1997: 187–9) is helpful in emphasising where the manager can intervene to best effect. Why people engage in recreation must in part depend on what it is that they perceive recreation to be, and whether we study the person in physical, social, psychological or even spiritual terms; in reductionist or holistic terms (Bullock and Mahon 1997: 11–12). Motivations have been categorised in clusters (Crandall 1980: 45–54) and psycho-logical benefits grouped in different ways, 'but the literature on human needs is fraught with disagreements about how many human needs exist and their importance' (Driver *et al.* 1991: 283). Others have sought to describe these in terms of related influences: characteristics, basic life goals, recreation orientation and preferences (Chubb and Chubb 1981: 236–7). The subject is so multi-faceted that we gain more by looking from a number of different angles. Maslow's hierarchy of needs (1954) is one way of separating out different realms of needs or drives. In essence, this

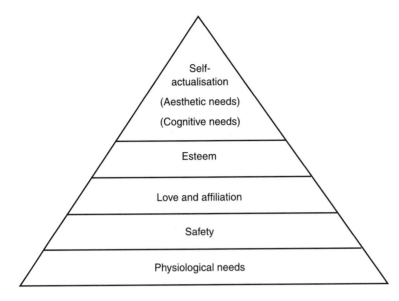

FIGURE 1.5 Needs
Source: Adapted from Gross (1996), after Maslow

postulates that certain basic biological needs are likely to predominate before there is great motivation to consider more social, psychological or spiritual needs. Interestingly, veterinarians have identified five freedoms in the welfare of livestock, the last of which is the freedom to express normal behaviour. (The others are concerned with avoidance of physical and mental distress.)

What do we learn from this? As managers we could build on such a framework, to help meet the needs people have (to express what they consider normal behaviour) in leisure and recreation. Fundamental to what distinguishes our usual construct of leisure from our usual construct of work, is the freedom to choose. The freedom to choose implies other freedoms – of movement, action, expression, speech – as well as the freedom to associate with whomever we choose, in whatever setting we choose. For a chosen few there is little to distinguish between work and leisure, as these lucky people feel they have freedom in choosing what they are doing at any time, whether at work, or leisure. There are also some for whom choice can appear to be easier to exercise at work than at home.

The choice of activities and settings within which to enjoy them is vast. Beyond choice of activity, participants choose the quantity of physical, mental, emotional, spiritual or any other characterisation of activity, and also the intensity, focus or concentration of effort. The end is not clear either. For some it will be arousal, and others relaxation, or perhaps a pattern of alternation at a particular frequency. For some it will be happiness, and for others meaning. Leisure provides the crucial space in which individuals can express themselves and develop personally. It also provides conditions for the development of intimacy and social relation (Kelly 1993: 23).

Earlier we described leisure as being time left to us, after work and other obligations are met. We have also emphasised that it is a state of mind in which time is defined as free time, free from obligations, when we have reached some sort of balance, when we are at peace with the world and ourselves. At this point, we are free to indulge in recreational activity, in tourism and travel or something altogether more sedentary. It is this freedom of choice, to indulge in whatever activity we wish, with whomever we wish, at whatever time we wish, that we all crave. Physical, and mental, health and well-being are the fundamental goals. In the end, perhaps all the different drives can be reduced to a desire to seek or to escape (Iso–Ahola, in Mannell and Kleiber 1997: 198). As has been pointed out, this pattern of behaviour is something that we share with even the simplest single-cell animals, such as amoebae (Campbell 1973: 61).

What we do at leisure is also manifestly important because it has an effect on others now and into the future, through its effects on our environment (ecological, economic and social). Some of the effects may appear simple, predictable, short-lived and reversible, but all too often we know very little about what the long-term effects may be, or how extensively they may be felt. The focus of impacts is usually related to the natural and cultural environment, because of fear of action that will result in irreversible changes with unknown effects. This is the basis of the precautionary principle (see pp. 201–2). Quite often we do know quite a lot about the effects, but we still choose to, for example, travel more and more. The socio–economic environment is generally considered to be more robust, but given the arrow of time, it is clear that anything which has been done cannot be undone, and any effects may ripple far and wide. As the two realms are actually integrated, we can (and must) develop mechanisms to encourage behaviours that maximise benefits and minimise costs.

When?

What we choose to do changes over time as we develop and pass through the different stages of life. The family life-cycle is a useful tool to explore how our needs change. The concept is often used in studies of consumer behaviour (Engel and Blackwell 1982). For consumer scientists the interest is in the money spent. The version developed by Murphy-Staples begins with the bachelor stage when the individual is characterised as having relatively low earnings. Our interest extends far beyond this. Leisure is time and consequently our leisure family life-cycle starts with the very young, who have no earnings at all! The time we have for leisure is the obverse of that which we allocate (or have allocated) to our obligations (Hook 1967). These often follow a series of rhythms at different periodicities. Each of us will have a different profile that can be explored over different periods, by means of time–use surveys. These can be aggregated to a give a national picture. All, obviously, bear a strong resemblance to natural rhythms concerning the movement of the earth about the sun, the rotating earth, the movement of the moon about the earth and all that follows. The requirement for leisure has often been institutionalised into festivals, rituals, religious and cultural festivals.

As we have developed technology we no longer need (technologically) to be tied to these natural rhythms. Our ancestors would have been restricted to enjoying summer at one time of the year. Now we merely transport ourselves to another place on the globe, enjoying the season we wish to experience. In the same way, we use floodlighting to enable us to play or watch outdoor sports in the long evenings (in Scotland, in midwinter, the sun sets before 3.30 p.m.). For many of us, the working day becomes more flexible (Future Foundation 1998). Information technology allows access to money (twenty-four hours a day), and our use of energy allows us to select whichever part of the day or night we choose. Shops and services develop to meet the needs of people, such as shift workers who live in a different world from most of us, and the difference between the day and night continues to be eroded.

In a similar way, weekends and weekdays are more similar than they once were. A succession of celebrations, holy days and holidays throughout the year remind us of seasons and times that had great significance when we were more dependent (or realised more fully the nature of our dependence) on natural cycles. Typically these include festivals which celebrate the coming of light; the flush of spring when plant and animal life appear to wake from the sleep of winter; midsummer; harvest; and midwinter. Some become more clearly linked with our use of the natural environment, such as harvest festivals. Others relate more to social needs such as the social excesses allowed at specified times of the year, such as Carnival in Venice or Mardi Gras. Many have been adopted by successive cultures, so that previously pagan customs are recognised now as holy days. Public holidays have in many areas lost their spiritual significance but remain of enormous social and economic importance. The festivities at Christmas and the New Year in the United Kingdom bring a peak in alcohol sales, which during the course of 1997 amounted to £29,200 million or more than £500 per head for each man, woman and child. This amounts to more than 5 per cent of the disposable income. Because of the tax on alcohol some £10,300 million finds its way into the government's coffers, which amounts to a reimbursement of well over £175 per head (Brewers' & Licensed Retailers' Association 1999).

How?

Recreation as an activity is usually a blend of physical, mental, emotional and spiritual components. It is unhelpful to try to define it too tightly. We think of most activities taking place in a specific location or area of space and over a finite period of time. This is only for convenience. The effects of recreation are often more spread out, and perhaps more far reaching, than we usually think, playing a part in our subsequent development, and that of others. At the very least we can think of recreation as comprising five stages: anticipation, travel to activity, activity, travel back, and reflection (adapted from Clawson and Knetsch 1971). We can use this model at different levels of resolution, so that we can analyse a meal out on holiday, the day of the holiday (including the meal), the whole holiday (including the day), or the contribution of the holiday to the development of an individual.

We are adaptive creatures, and what we learn from our recreation activity is fed back into the system and affects how we move forward. Play is a major element that contributes to the development of children. In the same way, we can be profoundly moved by recreational activities, after which we may never be able to see things in quite the same light. Sometimes we forget the power of play (Huizinga 1949). Recently a prodigious industry has built up around the notion of using play to develop qualities of leadership, under the heading of management development.

Some people spend every spare moment of their own time acquiring know-ledge, learning skills and in search of meaning. Such activities are educational but, carried out at leisure, also lie within the realm of recreation.

Searching for meaning depends on more than purely rational mental activity and soon enters the world of spiritual activity. For some people this is tied up with attempting to understand our relation to others, and to the animal and plant world, or to the universe. Contact with the natural world or escape from the usual social environment (providing escape from more everyday human concerns) can be extremely powerful in reshaping how people approach their lives. Coming into contact with the elements, perhaps standing at the top of a hill, being blasted by the wind, or alternatively basking in the warm sea, looking at reflections of the sun, provides a context outside everyday trivia.

BOX 1.3 Eastern approaches

Yoga and the martial arts share a common heritage. Having emerged from the East, both these forms of activity have in the last thirty years or so become enormously popular in the West. They are an antidote to stress, providing a measured discipline that seeks to give practitioners more control over their bodies and their minds. Through these activities, people can learn to relax completely (physically and mentally) and then learn to focus their energies. There are many variants of yoga, each having been developed by an individual and grown as classes grow into movements. China gave rise to a string of martial arts, from T'ai Chi, judo, and karate, and Japan to others like Kendo (Csikszentmihalyi 1992). There are few towns in Britain where the eager novice would not be welcomed into a class.

Few people can foresee what the outcomes of any particular recreational activity might be. As with many other systems and activities, the outcome is extremely sensitive to initial conditions. Because the activity feeds into the development of self and of others, the ramifications soon approach infinity. Even so, there are usually expected inputs and outputs, as well as unexpected inputs and outputs. Much depends on the scale at which we study the activity, over time and space.

Where?

The home, school and the workplace between them capture a good deal of our lives. We may not voluntarily choose to spend more time than necessary at work or school, or if we do, we see it as leisure. There is, though, a great deal of leisure and recreation in both. Keeping a diary or time log will reveal just how much. Job satisfaction is complex, but for many people the social interaction at work is, if not central, then extremely important. Humour, the exchange of jokes and conversation, is often only very vaguely connected with the direct purposes of work, although this interaction is probably essential for the health of the workplace.

The home is a very flexible set of environments where we often spend time at the pleasurable end of obligations, or in settings over which we have a good deal of control. Although the home has been transformed with the development of in home electronic entertainment – radio, television, compact discs, home computers and video – most people still resort to public buildings or centres in large numbers, at certain times. In the United Kingdom the public house or 'pub' is a venerable institution, where people meet friends, and like-minded souls. Typically, people adopt a particular public house as a 'local', which provides a neutral venue for meetings and provides a means of escaping the usual pressures (Smith 1995). Locals or regulars may visit the pub at the end of the working week, or working day, to unwind or exchange experiences. Their popularity has never been in doubt. Each year, in Britain,

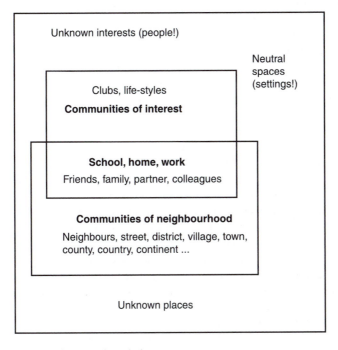

FIGURE 1.6 The setting: people and places

consumers spend about £17 billion on alcoholic drinks in 60,000 pubs. Other countries have their equivalent: for example, in France the 'café'; and in America the 'bar'. Differences in cultural values are reflected in the nature and way these premises work. Cinemas, theatres, libraries and sports centres are all centres that provide for different recreations. Increasingly they have been developed to provide for extended visits, with facilities for meals and drinks, and meetings.

For youngsters living in towns and cities, the street is a place of great excitement. Each visit wins graduated degrees of freedom (from the discipline of family life) for a period of time. Here a different set of rules applies. The motor car has invaded the street and in many areas reduced radically the role streets can play in safe leisure. The motor car does not win everywhere, though. Planners have once again been encouraged to design car free spaces where people can meet, share a picnic, talk, play games. Some of the space has been won back, providing areas where concerts, street music and theatre are once again made possible. These spaces once again conform to a human scale, and banish the noisy machines which otherwise invade almost every corner of our world. The most civilised of our cities include a portfolio of spaces of different kinds: hard, soft; paved, grassed; intimate, open; with and without trees; formal, informal; with or without water or sculptures. With care these spaces can be linked to provide access to the city and all its amenities, by walkways and links with public transport (Buchanan 1964). If these spaces are linked together with greenways, canals and other pleasant linear routes the value increases exponentially.

Gardens and parks provide lungs and spaces for people to come into contact with something more natural. While parks such as Central Park in New York may

---

BOX 1.4 Jardin du Luxembourg, Paris

Like many of the most beautiful green spaces or parks in the charming old cities of Europe, this was once the grounds of a royal palace, the Palais du Luxembourg. The palace was completed in 1631, and remained a royal palace until the revolution. Since then it has had something of a chequered career, which has included operating as a prison and a wartime headquaarters of the Luftwaffe. The building now houses the French Senate (Bailey 1994). The gardens follow the typically formal pattern found in France, with geometrical paths linking sections of the gardens, and providing many different spaces or settings. Parisians (and visitors) can relax, sit in the shade, sit in the sun, sunbathe, play tennis, play with their children, sail model yachts on the pond, admire the beautiful statues (many from the eighteenth century), take a snack or just talk with friends. There is surprisingly little grass, and what there is is well used. The various facilities are framed if not hidden by the trees. In this way, on a hot summer's day, the relatively small gardens can play host to several thousand visitors at any one time. A crucial element, and focal point, is the rectangular pond beneath the façade of the palace. The gardens provide a haven of peace, away from the traffic and bustle of the city, for everyone to enjoy.

---

play host to thousands of visitors each day, smaller local places, even temporary green spaces in our cities, can provide important oases to give a little more space for leisure and for play.

The car may be seen as the villain in our cities, but it also provides us with the personal mobility that we crave. With the car, we can go anywhere, any time and escape. In the Britain of the 1960s, the growth in car ownership allied to other factors led some authors to predict massive growth in leisure visits to the countryside (Dower 1965). Beyond its ability to take us (actually to be driven!) anywhere, as a travel capsule, the car itself provides a setting for conversation and social interaction among friends or family. The car is not the only means of personal transport. In many parts of the world, the cycle provides the key to independence. Much of the value is in the freedom of movement which ownership bestows. There are now more cycles sold each year in Britain than cars, but many spend hours, days, and even weeks locked away, out of use. In other parts of the world, the boat takes the place of the car.

The journey itself can provide the leisure experience we seek, whether a long-distance walk, canoe trip, car trip with the family, a sightseeing tour on a bus or a Mediterranean cruise on a ship with the population of a town or city (with amenities to match). Travel exposes us to novelty and the excitement, if not the shock, of the new. The desire to see what is round the corner is, to many, irresistible. As earnings rise, people spend an increasing proportion on travel, and on experiencing other cultures. Between 1983 and 1993 the number of passenger kilometres flown between ICAO countries increased by some 83 per cent (Friends of the Earth 1999). One segment of the market appears to be increasing the intensity of travel experience, by increasing the number of holidays (including holidays overseas) taken each year. There are others for whom more extensive and authentic travel is the aim.

Travel to some place away from home is often the fundamental building block of a family holiday, by definition taken outside the school term. The new setting provides novelty and provides a chance for parents, children and parents together with children to relax and enjoy each other's company. Each member of the group will have activities and pursuits that they as individuals wish to pursue, and there will be other games and activities in which several if not all members take part. Children will

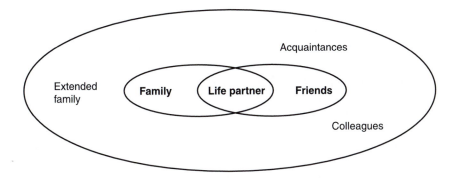

FIGURE 1.7 Recreation, and who with

learn skills of all kinds; relating to playing games, and perhaps later and more memorably for developing social skills. The learning of new skills is often high on the list of priorities for parents too. Prolonged contact with the holiday environment may lead to the development of a special attachment to a particular place, as the relationship develops through a deeper understanding of the character of the place and its people, through repeated visits.

Because holidays are likely to be among the most concentrated periods of family social interaction in the year, the results are not always predictable. The preparation for a holiday can itself be a stressful time. Parents try to finish tasks at work, prepare the home for an absence, pay bills, pack, remember all the details required: passports, tickets, currency required, vaccinations, appropriate clothing. The journey then begins with driver and navigator tired, and children constrained and uncomfortably compressed into the back of the car. How things develop from such a beginning can be extremely sensitive to initial conditions and to what would otherwise be minor events. For some this holiday represents too much time (or too concentrated a period of time) spent in the company of some family member(s). Holidays can be surprisingly stressful, with people really needing another holiday when they return home.

Holidays allow us a little more time to be close to our natural environment, taking part in activities that require access to land and water, particularly natural resources and the elements. Enjoying the rich diversity of natural and cultural environments converts awareness into understanding and a more considered approach as to how we interact with our surroundings. Understanding can lead to action to safeguard our natural and cultural heritage. In this way, recreation and conservation become mutually supportive. It is important to remember that holidays allow us to choose our leisure, and each of us has our very own picture of what leisure is. We should avoid stereotyping, and instead look at what people do. Activities that seem mundane to some people will be at the top of the list for others. Not everyone will always seek peace and quiet; some will wish to go shopping, head for the centres of our busiest towns and cities, or look for culture.

Who takes part?

As individuals, we can choose which activities to take part in. We can explore our environment in any way we choose, and at our own pace. Some will wish to take risks in adventure, to set new challenges and to compete. Others will wish to explore the self, go on an inner journey, exploring thought processes, and reactions to the world around them. Personal development requires a level of independence that such activity promotes. However, even in exploring wilderness areas, we tend to travel and enjoy activities in the company of others. The family group is the first, and natural, group in which we enjoy leisure. The extended family (or other group) may provide some additional leisure and recreation opportunities, in a way which allows freer rein. Experience in such groups will inevitably contain a dynamic (if not explosive) mix of leisure and obligation, varying over time and between individuals. We cannot choose

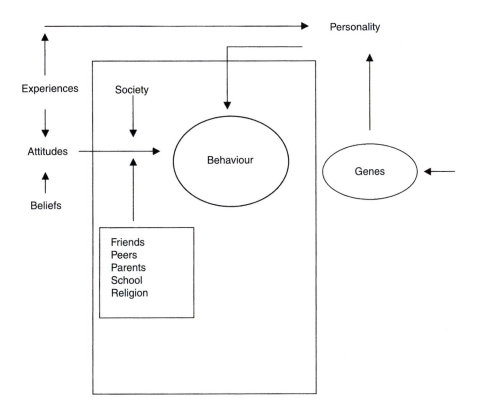

FIGURE 1.8 Influences on behaviour

relatives, but we can choose our friends. Friends provide us with the essential social component to allow experiences to be shared, expressed, amplified and reflected upon, without quite the same risk of being judged (or feeling that it matters). Often relatives are friends also, but friends who share similar circumstances, age view, gender view and interests remain crucial to our leisure, and indeed work.

Where interests cut across different groups, activities are often organised. Participants in such activities may then form or join a club, a team, a governing body of sport, a society, a walking club or any number of organised groups. Such groups provide their members with a range of social settings within which to share experiences, and the appropriate level of pooled skills, resources, knowledge and interest to allow the meaningful pursuit of some aim. People who enjoy their leisure together are attracted to like-minded souls. The characteristics that define the people in these groups can be very useful in defining how such people live. Leisure, work (or school) and the home are the principal areas in which people live. People who play together will often share a certain lifestyle and have a similar cluster of affiliations. This is of interest to marketeers, who like to identify people with similar lifestyles and (of course) ways of reaching them.

If leisure is the time available once other obligations have been met, everyone should have some leisure. In Britain, though it is a relatively prosperous country, there are some people, for whom society does not provide so well. Typically, these groups include:

- *Women*, because of constraints (real or culturally imposed) associated with roles commonly adopted by women (Goode *et al.* 1998; Kay 1996; Warrington and White 1986) or threats perceived (Burgess 1995; Scraton and Watson 1998).
- *Children*, especially teenagers (Gordon and Grant 1997; Wooley and Noor-ul-Amin 1999).
- *Older people*, typically (by younger people) classed as one group from the age of 60 upwards.
- *Ethnic minorities* (Floyd 1998; Khan 1976).
- *People with disabilities*, whose needs are as varied as those of the rest of the population (Pomeroy 1964), for whom recreation may provide special benefits (Shivers and Fait 1975), and who need to have the information available to make informed choices about what recreation to take part in (Woolmore 1995).

There are also some people who take their obligations to society, to the workplace or to the family rather more seriously, or feel them more deeply, so that discretionary leisure time is something of a luxury. Mothers are a special group for whom this sense of obligation or protection of the family is particularly strong. This may, in part, help to explain the difference which still exists between the amount of unpaid work carried out in the home by father and mother.

Issues

The meaning of leisure will vary across cultures, and among individuals: the time available which we can spend as we choose. We each of us have a different sense of obligation. In early times when we lived in extended families or tribes, obligations to others would have been so strong that choices would have been very differently formulated. Our choices now are made in a context that recognises the contribution of others, whether family or friends, teacher or pupil. The values that shape the choices we make are formed by a combination of forces. Personality, genetic make up, the influence of family, teachers and peer group will all have a part to play. The link between normative influences, attitudes, intentions and behaviour has been explored by psychologists and consumer behaviourists, and is helpful in understanding how people approach leisure (Engel and Blackwell 1982; Tuck 1976).

In recent years there has often been concern about the impact on the bio-physical environment of recreation activities (Department of Employment 1991; House of Commons Environment Committee 1995). Ancient cathedrals and monuments are being worn away by the passage of feet, the stonework being eroded

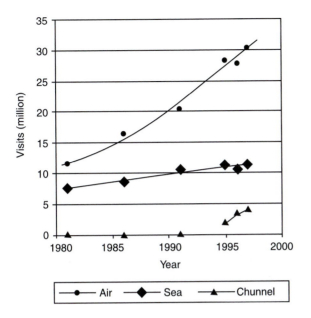

FIGURE 1.9 Visits abroad by UK residents, by air, sea and Channel tunnel, 1981–97
Source: Office of National Statistics, International Passenger Survey

by a combination of chemicals, exacerbated by a build-up of exhaust gases. And, in the countryside, people have traditionally been concerned with trampling and disturbance of wildlife (especially nesting birds, in Britain). These are the local impacts. Arguably, the indirect (or even global) effects of travel associated with much of recreation are of greater importance. Even so, this is subjected to rather less study. The effects of recreation on wildlife pale into insignificance when compared with the effects of different changes in land use (Sidaway and Thompson 1991). In just the same way, travel associated with commerce and trade has a very much greater impact altogether. Until recently, those of us brought up with influences like the Protestant work ethic would not even have considered the impacts in the same realm. We would always have subjugated the needs of leisure to the needs of work. Of course, any impacts connected with work are legitimate and those connected with leisure are less so, if you consider leisure to be a frivolous use of time. Attitudes are changing, though. All our actions are measured now against the yardstick of sustainable development. We should view the impacts, environmental and social, short-term and long-term, in context. How do we determine which impacts are significant? This is the crux of the Environmental Management approach, determining what the effects are, and for whom – in social, economic and environmental terms. Having reached this far, it is easier then to determine which are the most significant impacts (and for whom), and to focus on the required management interventions to bring about the desired results.

Among the many developments that have shaped the way we enjoy leisure, the motor car must surely have pride of place. The personal freedom that this form of transport affords has a social impact that is truly outstanding. Depending on your viewpoint this may be associated with enormous costs or enormous benefits. The truth is that there are both costs and benefits. Day by day, we become more aware of the need to use energy wisely. There is no doubt that we can make a major contribution by encouraging environment-friendly means of transport, and that the result can often be less stressful on psychological and social, as well as biophysical environments. Policies in developed countries, where the enormous growth in car ownership has resulted in congested cities and queues from city to coast every public holiday, need to provide for a measure of car-free leisure. In times past a car was often required to gain access to leisure facilities, although some have always striven to show the benefits of providing for people on foot (Buchanan 1965). We need to rethink how we use energy for recreation, and how we deal with this hypermobility (Adams 2000).

Einstein is alleged to have remarked that it is often when we are at leisure that we have time to think. Recreation, particularly out of doors, can make us think deeply about how we relate to the rest of our environment. Wonder and awe, whether for natural or cultural heritage, awaken the spirit. Exposure to the elements, intense experience, awareness and understanding will in most of us lead to a desire to conserve what we have, whether this is to share the experience with valued others now or for future generations. The Earth Summit in Rio (Children of the world 1994) encouraged us to think in terms of sustainable development rather than conservation. Some would argue that development is a rather awkward word here. In conjunction with 'sustainable', it suggests that we can for ever improve our quality of life, in a materialistic way. Actually, what goes on is a constant reorganisation of the resources we have at our disposal. If development is viewed in crude economic or financial terms then it is surely a nonsense. Some suggest that recreation and tourism cannot ultimately be sustainable. Most would agree that pursuit of swifter and swifter travel, and experience, cannot be developed sustainably, but we will return to this later.

In the best transactions that accompany recreation and leisure, everybody wins. In many transactions, though, some win and some lose. The thinking we have been developing suggests that we need to take every step to make sure that our children and their children's children should not be losers, that we should not squander their inheritance before they even see it. We have perhaps tended to conveniently ignore other inequalities, within our respective societies or between them. There are enormous ethical problems associated with some post-industrial countries asserting measures and standards by which all (including developing) countries must abide. Tourism is an area of legitimate concern.

Recognising differences and inequalities is surely the first step in providing for a more equitable distribution of costs, and more especially benefits. There is great room for lateral thinking here and for a comparative approach, recognising how other cultures and societies have solved a problem or seized an opportunity that we too might identify.

We advance by making choices. These choices are shaped by the environment within which we live; that is, the political, economic, sociological, technical and

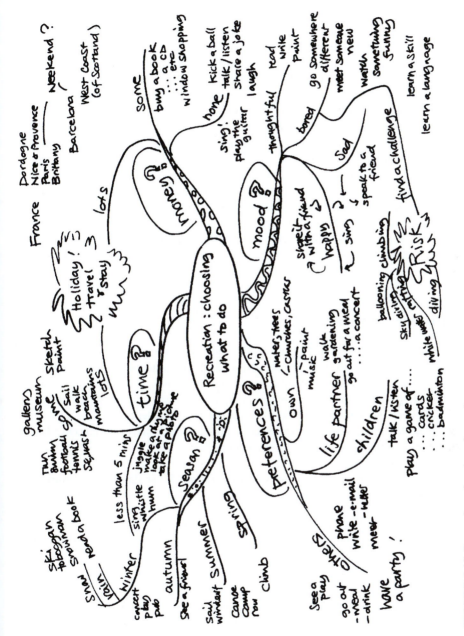

FIGURE 1.10 Recreation: choosing what to do

natural environment. We have considerable choices about the kinds of leisure and recreation we can enjoy, share or provide for others. Some of these in turn depend on our attitudes to time, to work, to life and to society. What kind of a life do you want to lead? What kind of speed do you wish to travel at? What kind of intensity of experience do you seek? With whom do you wish to share these experiences?

We cannot act alone. We are connected with everything and everybody else, and yet recently Western societies have given increasing focus to the needs of the individual. But where do you draw the line that separates the individual from its environment? Is there a clear line? The more you think about it the less obvious the answers become. Indeed, lines are not always easy to draw between leisure and work.

To live we have to simplify. In this book, there are many aspects of leisure and recreation that are barely touched on. Few would deny that sex is one of the most pleasurable activities indulged in at leisure, but this is not a topic dealt with in this book. Also, if the proportion of pages allocated to subjects reflected the time we choose to spend on different activities, then the art of conversation would surely take up all but one page of the book. Choices have to be made, and this book seeks to focus on those areas where managers would choose to intervene, and on those areas where intervention might be most beneficial. We should, though, remember how intervention may affect core areas, as much as the balance of leisure with which we are usually concerned.

Ideas for further study or work

1   What different personal leisure needs might you have as (a) a youngster; (b) a working adult; and (c) after retiring from work? You can produce a more refined analysis (e.g. for intermediate stages) if you have time.
2   What sense of obligation do you have, and to whom, that sets up tensions between the time which you can and cannot allocate to leisure?
3   Are the tensions between leisure and non-leisure necessarily destructive? Can they interact or be managed to good effect?
4   Why do we play?
5   Who are the managers of leisure and recreation? What should their priorities be?

Reading

Psychology

Argyle, Csikszentmihalyi and Neulinger provide an insight into the psychology of leisure and recreation, and Gross gives a more general introduction to psychology, and understanding attitudes, intentions and behaviour. Much of the really detailed work on decision making has been carried out in the field of consumer behaviour. Engel and Kollat is a valuable source.

Activity

For insights into different realms of activity, the choice is vast. Classics in the arts include Gombrich, and Storr. For insights into outdoor recreation and exploration, the reader is truly spoilt, but start with Shackleton, Simpson and Slocum. For a more gentle or lighter approach to recreation and travel try Stevenson, Newby and, in a rather different mould, Bryson.

Recreation and leisure

For more serious study, try Kelly, Roberts and Godbey. Glyptis reviews definitions of leisure; Chubb and Chubb look at behaviour, and Jackson and Burton draw together many expert views in the one volume. Some older texts are well worth revisiting, such as Opie and Opie, reporting on street games and play. For a more philosophical approach touching on some of the inequalities, and seeking to uncover the real motivations, read Zeldin.

Management

Books on management abound, and to keep the mind alert, keep abreast of recent advances in marketing and management, and look for reviews by skimming through management journals and magazines, of which *Management Today* and the *Leisure Manager* are examples. There are also some classic books by Handy, de Bono, Peters and Waterman, Kotler, Buzan, Kosko and Straker which will help to keep lateral thinking alive. Within the field of leisure and recreation, there are also some classics such as Torkildsen, and Godbey.

Histories of recreation

Overview

People in the earliest days will not have had the same concepts of work and leisure that we now do. This chapter provides the swiftest of tours down the ages, from prehistory to today. There are surviving artefacts from ancient cultures that clearly show ornamentation on a scale that signifies that expression was an important component of living, and was perhaps bound up with the rudiments of religion. Although early artwork, such as the cave paintings and rock art (Chippendale and Taçon 1998), survives, we have only fragments of evidence of what people did with their leisure. We have to deduce from what people left behind, and how others behave, what they may have done.

By the time of the ancient Greeks and the Romans, the development of civilisation, of knowledge, of written language, architecture and the arts generally ensures that we have a much fuller picture of leisure. Music was a broader concept than our restricted use of the word now suggests, and was an important component of a man's education (certainly among the elite), and this was balanced by education in gymnastics and sport. We know more about what such men did than about poorer people or, for that matter, women. It was not all about high-minded ideals though: excesses of food, drink and sex are all well recorded.

The movement of people across the globe, the various conquests and emigrations, spread ideas (good as well as bad), and allowed trade and the spread of diseases too. The search for novelty, and riches, was bound up with adventure. Introducing new ideas was sometimes merely part of the colonial process, and sometimes seemed to be a form of succession. Viewed from our vantage point, there are some very uncomfortable passages of history. These are balanced by some glittering moments such as the confirmation of happiness as a right in the US constitution, and festivals and carnivals to brighten different times of the year, or provide lighter interludes in hard working lives.

The industrial revolution threw into sharper relief the difference between work and leisure (including disposable time). The development of steam, and later of the motor car, speeded up the rate at which we all live, work and play. Many social developments were accelerated by the turmoil across Europe, and the rest of the globe, during the World Wars. This frenetic activity has been mirrored by technological developments concerning communications: radio, television, videos, mobile telephones and personal computers among them. The sedentary lifestyle has recently encouraged a spate of fashions to encourage more (and more meaningful) use of leisure. Just as in the times of the Greeks and the Romans, drugs (including tobacco and alcohol), betting and fantasy still have a part to play. These sit alongside the higher ideals of cultured self-improvement through leisure. Some of life's pleasures remain simple indeed. Speed, movement and song are among the body and mind centred pleasures that seem to be universal down the ages. Mostly, though, we take our pleasure in company with others.

Meanings through time

The twin concepts of work and leisure will have had little meaning in the earliest times. The first (and often only) goal of every individual would have been survival. If only for reasons of survival, people would have felt strong obligations to family or tribal groups. Free time of the kind we now enjoy is unlikely to have been available. Even so, a visit to any museum of ethnography, archaeology or ancient history will reveal that the earliest of peoples put considerable effort into decorating artefacts and tools and making ornaments. Often this craftwork seems to have had a religious significance. The significance was probably related to a belief that pleasing various deities would produce more benefits: a better hunt, more bountiful harvest or stronger partner. The making of sounds (or music) and of images, through rock art and cave paintings, were early preoccupations (Cole 1965). Some of the images persist, but for an idea of the music we have to resort to experimental archaeology (Coles 1973: 158–67).

Different parts of the world developed at different times. Fireworks and paper were known in China long before people in Britain and northern Europe started carving runes on stone (*Encyclopaedia Britannica* 1999). Some ideas take thousands of years to travel across the world. The word 'leisure' appears first in the fourteenth century, and refers to freedom from occupation, and hence free time (Onions 1966: 523). Recreation derives from the Latin *creare*, to create. In the fourteenth century it appears to have meant 'refreshment by pleasant occupation', and by the fifteenth 'restore to a good or wholesome condition'. Various definitions of work, suggest a purposeful effort towards some end, the making of something or what one is employed in doing (*Chambers' Dictionary* 1969: 1571). Some of the early artefacts made during leisure would surely also have qualified under the heading of work.

Accident of birth, in certain cultures of old, could have resulted in a life as a slave, or a noble. In most forms of social organisation, certain groups have access to more resources, whether time or rewards from employment, than others. With the industrial revolution came the development of large concentrations of people in towns and cities. Following this, a clearer distinction between work and leisure emerged. Certainly everyone drew the distinction between their own time and that of an employer.

Prehistory

Recreation is a fundamental component of life. How far among the animal kingdom play extends is not easy to discern, but there is ample evidence that mammals (Thorne 1992: 76–8) use play as a tool for learning and sharing behaviour patterns which are useful, just indeed as humans do. Many of the antics of birds too must also fall into this category. We do not know when humans first realised that they were indulging in recreation. The idea of any sudden realisation brings forward images of Adam and Eve! Certainly both the eating of the apple and what followed are activities in which all humans and a good many animals (Young 1971: 520) remain very interested.

Social activities would have involved elements of behaviour that fall outside our normal description of work. Prehistoric hunters would surely have enjoyed the thrill of the chase with the high levels of adrenalin being produced. Recreational activities would have been used to develop fighting skills: speed on foot, strength, and accuracy in throwing stones and spears. Hunters moved round the country following their quarry, and this would have brought them into contact with other groups. There were many possible outcomes: trade, love and war among them.

As hunter-gatherers settled down to an agrarian way of life, there will have been more time to spend on social activities. The roles of individuals would have become more specialised. Aptitudes, skills and strengths were utilised to the full. The natural order required females to rear their young. Menfolk went hunting. Gender differences will have given rise to disagreements then, just as they do now.

From very early times (Mesolithic onwards) man has developed a close relationship with animals. Initially, this will probably have been concerned with making hunting, killing and eating these animals simpler. Through experiment a wider range of uses emerged – keeping animals for their products, whether goods or services. Soon our ancestors were using wolves and canids (dogs) as an aid in hunting, or to protect or warn against incursions by others. The range of animals which have been domesticated is very wide, from honey bees to elephants, and including silkworms, cormorants, geese, hens, dogs, cats, sheep, goats, cattle, camels, donkeys, horses, water buffalo and rabbits, to name some of the obvious ones. The behaviour of animals kept in domestication is changed by their being kept in a human social environment (away from their own kind), and by selective breeding. The development of domestic animals as pets was probably gradual.

In these ancient times, meals would have been an important time of gathering, imbued with ritual. The discovery of fermentation, and then the production of alcoholic drinks, would have played an important part in the development of the meal as a focus of social activity. According to some accounts (Johnson 1971: 12), the cultivation of the vine possibly started in Caucasia or Mesopotamia in 6000 BC before passing to Egypt and Phoenicia in 3000 BC. Decorations on artefacts from this time demonstrate the high value placed on wine.

Craft and culture

Stone and flint are so durable that our record of what kind of lives our ancestors lived is deduced in large measure from stone and flint remains. Embellishing flint tools for artistic purposes may seem something of a luxury. Making a flint tool at all is a skilled business, although even a simple flake (often a by-product) can be used as a tool. Anyone who has tried to fashion a flint tool will appreciate that an axe or an arrowhead takes a different order of skill (and patience) to make (British Museum 1968: 82–103).

Possession of such a tool may give an abstract pleasure in the anticipation of what it might achieve. Later carving of stone, bone and wood could give expression to the activities and uses to which these tools were put. Some of the craftwork appears beautiful to us now, but in all likelihood the beauty had a purpose (Gombrich 1995: 40–53; Young 1971: 534–7). It is entirely likely that many of the markings were associated with the supernatural powers that would then be in the gift of the owner of the tool.

Celebration and beliefs

Analogous situations in 'primitive' tribes may give an insight into the reasons for which art and craft were employed. Detailed ornamentation and craft were employed in the manufacture of masks and costumes worn on ceremonial occasions. It seems likely that the associated singing, dancing, feasting, drinking and storytelling have an even longer pedigree. After all, it is not just humans who use dance to communicate (von Frisch 1970: 113). Rhythmic music leads people toward the same mood, and is a basic form of communication, while rhythm itself is also a common component of play (Stillman 1966: 3). Rhythm and patterns of sound (music) not only enable expression, but they can intensify emotion (Robertson 1994; Storr 1997: 67–76). Many of the dances and stories are linked with past events, and with beliefs about what forces shape our lives. Some will encourage a good harvest, a fruitful marriage or a successful hunt. Celebrations are linked with cyclical events, notably (in the temperate parts of the world) the beginning of spring and the different stages of the moon. Others will be linked with rites of passage through life: birth, adulthood, marriage and ultimately death. In many parts of the world, these events are still celebrated with gusto. Where the significance has been lost, people are working to retrieve that meaning (Gill and Fox 1996) through the use of popular art.

Cave paintings

The limestone caves of the south-west of France and the Pyrenees have yielded paintings created thousands of years ago (10,000–15,000 BC). There are representations of particular animals, and scenes of events witnessed, or hoped for. Some interpretations (Gombrich 1995: 42) suggest that the pictures were created by hunters

---

BOX 2.1 Gardening

The origin of gardening belongs to Neolithic peoples, who were as interested in the domestication of plants as they were of animals. Initially this interest will have been stimulated by the collection of edible plants. This must have been accompanied by some horrific errors! Gathering edible products of the forests, whether fruit, berries, nuts or fungi, is still very important to people. It may be a recreational activity in some parts of the world (Mabey 1972) but in

---

others, gathering food products from the wild still forms an essential part of the household economy. The ancient Greeks and Romans had well developed gardens producing foodstuffs, and the practice has been passed on from civilisation to civilisation (Grenfell 1975: 7). Gardening, for many, is as much about settings as it is about food. Garden designs have been traced back as far as 1400 BC, in Thebes, for a garden of an Egyptian court official. It included a pergola with vines, walled enclosures, formal gardens and ponds (*Encyclopaedia Britannica* 1999). From these ancient times, different cultures have developed their own particular kinds of preferred gardens. The development of garden landscapes is a study in its own right.

In Western countries, gardening seems to have really taken off in the early nineteenth century (Argyle 1996: 216), perhaps as small parcels of land in urban or suburban settings, adjacent or close to houses, came to serve as a reminder of a rural past. Plant collectors, accompanying expeditions, brought back exotic plants from all corners of the world, and this novelty spurred interest. Today, no less, people garden to develop settings that they find pleasing. These may be for use as outdoor spaces over which they have control (although they may have to battle somewhat with indigenous plants), and in which they can establish a feeling of being close to the earth, connecting with nature. They may be interested in particular groups of exotic plants, in maintaining colours or scents, in keeping a pond, or in gardening to provide food plants for wildlife (Baines 1985). The popularity of gardening is incontestable. Recently much of the interest has been directed at creating gardens at breakneck speed, literally over a day, rather than the painstaking work of selective breeding that would have been the stock in trade of gardeners of old.

In other countries, Japan for example, different traditions have been established, and gardens have an almost religious significance, creating appropriate settings for reflection. And, as space is at a premium, gardening in window boxes and the keeping of houseplants (Wright 1974) provide some of the enjoyment of gardening. There has also been a resurgence of interest in raising vegetables and fruit under organic conditions, and for self-sufficiency (Seymour 1978). This is fuelled by increasing concern over the effects of pesticides, herbicides and other chemicals, as well as a fear of genetically modified foods.

Some garden escapes, like any introduced species, have spread very quickly and caused considerable damage. Rhododendron has spread in certain areas of the west of Britain (Mabey 1996: 158), where its systematic removal provides opportunities for another recreation activity, conservation projects carried out by enthusiasts working in teams, under the auspices of the British Trust for Conservation Volunteers and Scottish Conservation Projects. Japanese knotweed, introduced by a Dutch horticulturalist in the early nineteenth century, has spread rapidly in the twentieth, extending its range from Middlesex in 1900 to the north of Scotland by 1960 (Kelly and Whitaker 2000).

before the chase, hoping that by capturing the image, power would be exerted over the animal in reality. The images are painted using natural pigments made from the earth. These same pigments are still important to artists today.

Classical and early

Ancient Greece

Plato considered that leisure was the real goal of life. Leisure was seen as resting on a rather higher and more intellectual plane than most of us would recognise today. This civilisation promoted two quite different streams of thought, *logos* 'reasoning', and *mythos* 'myth'. Using reasoning allowed great mathematicians like Pythagoras and Euclid to lay down laws that are still of great value today. Using myths enabled Plato to describe things as intangible as love. He suggested that the Creator had originally made humans as having four legs and four arms, but being then scared at how powerful they would be, split them in half to reduce their power. This then created two halves, each of which would have to search for the other. This is the nature of love. In Greek mythology (Grimal 1965: 110), Mnemosyne (as reflected in 'mnemonics', and signifying memory without which we could make no sense of the world) gives birth to the nine muses. These goddesses, who inspire learning and the arts, each looked after a different branch. Music was a much broader concept to the Greek then than it is to us now. Three years' tuition in music was considered reasonable for the ordinary young man. Education in the arts was balanced by education in gymnastics. Drama, poetry, literature and music had a higher profile. This education in youth provided the adult with the intellect to spend time thinking – both reasoning and daydreaming or dealing in myth. School and leisure were closely associated (Bramham *et al.* 1989: 116–17; Pieper 1965: 25–6).

Sport was a crucial part of any man's education. The skills learnt from sport would prepare the individual for war. Throwing the javelin, putting the weight, throwing the discus, and running, would all be useful skills. After the Athenians defeated the Persians at the battle of 492 BC on the plains of Marathon, a soldier ran to Athens to carry news of the victory. This has been commemorated in the marathon race (26 miles 385 yards) ever since 1896 in the Olympic Games, and now the world over.

Romans and 'civilisation'

The Roman Empire stretched across the Mediterranean, and enveloped many different cultures, customs, languages and approaches to leisure. Romans were particularly adept at organising social events, and city life. Even in parts of Britain, on the edge of the empire, the influence of Roman culture has had an impact on our way of life: from the architecture of our cities (Gombrich 1995: 117–31) to our laws, and no doubt on our recreation and leisure. Emperors took leisure seriously. Attention often seemed to focus on sex and on food. The emperors of Rome excelled in this respect (Suetonius 1957). Suetonius gives quite sufficient details of the excesses that were taken (if not permitted) by those of noble birth. Drink was a major component, particularly among the less well off (Toner 1995: 77). Feasts were important as celebrations of victories, and as occasions for thanks. The emperor often paid for the event out of the public coffers, but sometimes contributed himself. There were feasts associated with changing seasons. Saturnalia, when momentarily slaves and masters exchanged clothes, lasted several days. Plays would be held in different parts of the city, and in different languages to suit each quarter, and games and races in the amphitheatres.

With our views about animal welfare today, we would have enormous difficulty in comprehending the cruelty in some of the entertainments. Women were banned from attending the games, or admitted only at certain times. Between the races, there would be other attractions: wild animal hunts, panther baiting and the like, even boxing bouts between fighters from different provinces. Gladiators were pitted against one another, or against animals. Caligula himself took part in many shows, as a Thracian gladiator (using the round shield), as a singer, a dancer and in supporting actors. Nero became obsessive in his playing of the lyre and in singing, taking part in the games in Olympia. Some of the excesses are hard to imagine: a stable made of marble and the stall of ivory for Incinatus (one of Caligula's horses), or the mock sea battles that took place on special lakes created for the purpose. An emperor himself tried out a chariot pulled by ten horses (and fell out!). There were other, less savoury practices, with Christians put through unspeakable acts of torture.

Some public shows may have been arranged to keep social order, to control and moderate the behaviour of soldiers. Away from these public events, we have rather less account. The survival of so many Roman baths makes plain their importance for enabling conversation and recreation, exercise and relaxation. We know from archaeological finds that there must have been board games. The knuckle bones (astragals) of sheep provided crude dice, and these were later made from pottery and ivory. One of the Caesars even wrote a text on the subject of dice. Love of poetry and literature would have marked Romans out as a race apart from their European neighbours of the day, and certainly those in the north.

The Greek column and the Roman arch have been crucial to the development of architecture in the Western world. Walking around our cities you see echoes of these forms even in new houses being built today (Gombrich 1995). So strong an influence have these civilisations had on Western development that we are apt to overlook the contributions made by other civilisations. Influences percolated from one civilisation to the next, as people travelled, traded, intermarried and settled.

First nations

In America the peoples seem to have developed without quite the same degree of intermingling, until relatively late on. Until the eighteenth century, the first nations were rarely disturbed in quite the same manner as the cultures of the Mediterranean, by such large-scale change. However, in the South, the Incas and the Aztecs had developed settlements to rival Roman towns, and they were to suffer early incursions by Europeans. There were also nomadic peoples following the game herds and keeping to natural rhythms, particularly the cycles of the moon. This lasted until the coming of the Europeans, and the railroad. We know that these peoples had a rich culture, from the survival of artefacts, and the living history provided by the flourishing arts and crafts, stories and songs passed down from generation to generation, and still in evidence today.

BOX 2.2 Head-smashed-in-buffalo-jump

This is the name given to a place in Alberta, used for thousands of years by native Americans to drive buffalo over a cliff to their certain death. Whilst the buffalo had kept the first nations supplied with hides, meat and all manner of materials, for thousands of years, the invading Europeans felled the magnificent bison with rifles and took their hides (whilst leaving carcasses, much valued by the first nations, rotting). It became ever harder to hunt down sufficient beasts to keep the peoples going. Added to that, the US cavalry was well armed. Under these circumstances, a fast and hardy horse was all-important. Peoples of the first nations took great delight in racing, and other sports designed to develop horsemanship to aid survival. To this day, many nations (e.g. the Sarsee nation in Alberta) hold games or festivals that include races and rodeos.

Gaelic culture and the environment

In a quite different setting on the wild western shores of the British Isles, Gaelic developed late as a written language, a preserve of the learned. Traditions had been passed down orally, in rhyme and in song. From translations you can glimpse the relationship which Gaelic peoples had with the environment, as in this Gaelic blessing:

> Deep peace of the running wave to you
> Deep peace of the flowing air to you
> Deep peace of the quiet earth to you
> Deep peace of the shining stars to you
> Deep peace of the gentle night to you
> Moon and stars pour their healing light on you
> Deep peace of Christ, the light of the world, to you.
>
> (Based on an old Gaelic rune. John Rutter composed music to make this blessing into an anthem. Hear it on *Gloria – the sacred music of John Rutter*, Collegium Records)

Eastern civilisations

Language used in the Western world derives from the Indo-European root, and it would be surprising if there were no other aspects of culture which had not been passed across at the same time. Unravelling influences in the development of leisure is as difficult as unravelling the development of culture as a whole. Some Eastern cultures have retained their integrity over very long periods. Chinese calligraphy, and its use of characters, have ensured the retention of a quite separate way of life for thousands of years.

From Vikings to United States

Ships and sagas

Voyages can be long and boring, but enriched by surprise and serendipity. Voyagers have always sought diversions and found ways of blending recreation with work. It would be impossible to view a trip in an open canoe, or in a sailing ship, as wholly work. Raising sail, pulling up the anchor, turning winches, all demanded great strength. Despite ingenious pulley systems, the only way to achieve some of these tasks was to ensure that everyone pulled together. The solution was rhythm, and the mechanism was song. The result is a legacy of sea shanties. Different songs would be used, depending on the nature of the task. The rhythm of the song would be matched to the length of the pull: in halyard shanties, capstan shanties, short drag shanties and inevitably forecastle shanties for singing at leisure (Boni 1947: 129–31). Sailors on such voyages would also make souvenirs and works from anything that they could find, including scrimshaw (carvings on the teeth of sperm whales) and in knotting (Ashley 1947: 1–3).

Sailors in port found other diversions, just as they do today. The lonely life at sea was generally compensated for when sailors reached shore. The Vikings had a reputation and they were great storytellers. Thanks to the tradition of passing down stories by word of mouth these sagas are available to us still, including those that tell of the 'discovery' of America (Magnusson and Palsson 1965) well in advance of Columbus, albeit after St Brendan's voyage (Severin 1978).

Saxons, Normans and hunting forests

Ine, King of Wessex, created forest laws as early as AD 693. Saxons recognised the value of protecting the animals of the forest. Royal forests were created in areas close to where the King might take up residence or establish his court. Forest laws gave certain rights to the Crown, even on privately owned land. The Normans were keen on forests, for providing food for feasts, and for the thrill of hunting. In 1097 King William the Conqueror established a new royal forest, not being content with the Saxon royal forests, which the Normans fell heir to. His Nova Forestis, the New

Forest, provided great scope for hunting and served his court at Winchester. By the time of Henry II there were some eighty royal forests or so, and their chief role was in supplying venison and game for the King's table, or for gifts. For his Christmas dinner in 1251 King Henry II had 430 red deer, 200 fallow deer, 200 roe deer, let alone all the wild boar, swans, cranes, salmon and lampreys. The list also included 2,100 partridge and 7,000 hens (Rackham 1986: 119). The taking of deer was a crime, punishable by death. There were other penalties for lesser crimes such as the loss of soft organs, which meant the removal of eyes, and other soft parts.

Exploration: Columbus, Conquistadors and Cook

Columbus, Conquistadors and Cook had much in common. They were all seeking novelty, exploring the unknown, and hoping in the process to make personal gain. Columbus had read Marco Polo's account of his travels. He made marginal notes in his copy (McKee 1990: 25), and underlined words, including 'pearls', 'precious stones' and 'spices'. When he had inadvertently reached America in 1492, he believed he had discovered the route to the Indies, and the seat of all riches. Those who joined these explorers were spurred on by a mixture of motives. Reflecting on our motivations for travel today, some individuals may have been as interested in escaping, just as much as they were in arriving. Columbus had originally found the natives friendly. Despite this exhibition of friendliness, he kidnapped ten of them to take back as curiosities; a grisly souvenir (Brown 1970: 2). In South America, the Conquistadors searched for silver and gold. In the process, they exhibited every kind of ruthless savagery known. They introduced to the first nations of America the practice of scalping. Captain Cook, and his entourage, in their travels (1768–1840) through the South Seas, were responsible for spreading diseases of many kinds (including venereal diseases) which were to cause havoc (Moorehead 1968: 56). The emergence of a trade in arms fuelled a succession of mini dictatorships, a social disease just as deadly as the other diseases.

The Grand Tour

For centuries the young of aristocratic European families were encouraged to travel among the capitals of Europe. During the eighteenth century, a journey of a year or so (the Grand Tour) to take in the riches of France and Italy, and possibly the Netherlands, became an almost essential part of the education of an English gentleman. There seems broad agreement that it was this tour that gave rise to the term 'tourist', and as has been pointed out, what tourists were encouraged to do then is not so very different from how people touring European capital cities behave today. They would sketch views, and collect pictures, particularly by Italian masters, such as Canaletto. They would climb local hills, or steeples, to enjoy fine views, and ensure that they saw the main sites. They would be encouraged to use the local language and engage in conversation with local people rather than keeping within their groups. As has been pointed out, many of the behaviours of these first tourists are quite recognisable in the rather more time pressed visits of today. Today's tourists visiting the capital cities of Europe are more likely to take photographs and collect postcards; go by funicular or elevator to the tops of hills, or communication towers for views; and sample local food and drink (Butler 1999: 100).

Drama and festivals: Venice Carnival and Mardi Gras

The Venetian government arranged street games and shows, for much the same reasons as the Romans. Carnival began on Boxing Day, 26 December, and lasted through until Lent! During this period, many of the usual strict procedures and laws were set aside. The wearing of masks allowed people of different social standing to take part. Balls and parties were arranged and people danced in the squares. The regatta, which started through unofficial races between teams for amusement, was given official blessing. They provided a means for oarsmen to develop their skills, strength and stamina, necessary in times of war. The first was held about 1300, and they continue to this day (Roberts 1993: 50–4). Nowadays the Carnival in Venice is restricted to February. In France, and parts of Germany, Mardi Gras (the last day before Lent) was (and is) accompanied by a good deal of celebration. Mardi Gras means Fat Tuesday. The day was associated with the using up of foodstuffs which would not last the duration of Lent, and therefore had to be eaten, or wasted. In Britain the day is known as Shrove Tuesday, often referred to as Pancake Day. In Cologne, during the celebrations, all sorts of misdemeanours were forgiven.

The Age of Steam

Towns and specialisation

In 1776 the United States declared their independence. The Declaration of Independence, drafted by Thomas Jefferson, and passed by Congress in 1776, confirmed happiness as a right.

> We hold these truths to sacred and undeniable: that all men are created equal and independent, that from that equal creation they derive rights inherent and inalienable, among which are the preservation of life, and liberty, and the pursuit of happiness.
>
> (*Oxford Dictionary of Quotations* 1953)

President Abraham Lincoln, in a speech on 16 October 1854, reminded Americans of these core values (MacArthur 1996: 344–7).

In that same year, Adam Smith's *The Wealth of Nations* was published. Smith was a student (aged 14) at Glasgow University before moving to Oxford. At the time Scotland was a great centre of learning and civilisation and enjoying the period of

Enlightenment. Smith's contribution fuelled the industrial revolution already under way. He was convinced that the key to wealth was the division of labour. He illustrated his point by reference to the manufacture of pins. If each man carried out all the operations required in making a pin – drawing of the wire, fitting of the head, sharpening, etc. – it would take much longer than if individuals specialised on a single task. He went so far as to suggest that production would increase a hundredfold with specialisation.

Already, in 1769, another influential Scot, James Watt, had patented his steam engine. After some development it was applied to locomotives on rails, in 1804, and to European shipping in 1812 (Morgan 1984: 428). Steam represented a major development for transport, and for leisure. Britain was at the forefront of an industrial revolution. This induced a massive drift to towns and cities, with their dark satanic mills. In New Lanark in Scotland, beside the beautiful Falls of Clyde, there is the remnant of a more benign system, where Owen and Dale sought to put into practice their social ideals. New Lanark is now cared for as part of our industrial heritage. The crowding became severe in some of our towns and there was a much crisper division between work and leisure. The distribution of wealth was rather more skewed than it is now. If you were not well off, you were unlikely to have much leisure to yourself, other than holy days (which became holidays). In 1801, 30 per cent of people in Britain lived in towns, and by 1901, 80 per cent. Rising agricultural production, and a reduction in the spread of epidemics, was in part responsible for the increasing population. The strength of Britain's industrial revolution, allied to the increase in trade and military might, ensured that growth would continue.

Trains

The development of railways has a lineage stretching back to Roman times. The coming of steam heralded a major leap forward. Much of the original development in Britain was driven by the increasing demand for haulage of coal and steel. Lines such as the Stockton & Darlington and Liverpool & Manchester railways were crucial. George Stephenson's enthusiasm, organisation, and locomotives such as his son's *Rocket* played a large part. Visits to the seaside had been made popular by the enthusiasm of royalty. The development of the railways made resorts accessible for extended excursions. By 1850 an extensive network of railways covered Britain. As early as 1829 orders had been placed for locomotives from Britain, to start up the railroads in the United States. Soon railroads would be instrumental in pushing exploration across the continent, heralding the rush to the west.

Ships and cruising

Wooden built paddle steamers appeared in the early nineteenth century. The earliest commercial boat to carry passengers was the 43 ft *Comet*, built by an amateur, Henry Bell, for the Clyde in Scotland. In the north-east of North America many such ships were developed, and a sailing across the Atlantic from Canada to England occurred in 1833. Ten years later, the first iron hulled steamship, the SS *Great Britain*, driven by a screw propeller, took to the water. Towards the end of the nineteenth century the passenger liner had truly arrived, although the clippers still worked certain routes until the advent of the Second World War (Newby 1956).

Sail, raft and canoe were still preferred by some. Robert Louis Stevenson and his family travelled by steam and sail in the South Seas, in the 1880s and 1890s, before settling in Vailima. Early tourists had to entrust themselves to the means of travel employed by the residents of the countries they were exploring.

Charabancs and tours

In Britain, buses within towns were chiefly operated by municipal authorities or through the setting up of special companies, or boards. Services operating in the country between urban areas were mostly operated by private companies. Companies sought to augment their income by putting on excursions to beauty spots or to sporting events. Many of these trips would be signed up by individuals, and there was also something of a tradition of works outings. Buses retained their popularity until the 1950s, by which time car ownership was within reach of many. Cycles were popular too, giving access to low cost personal mobility. In the late nineteenth century the Cyclists' Touring Club was formed and other organisations followed, as champions of the recreational use of the cycle.

Ballooning and early aviation

From the earliest times (BC), the Chinese used kites to lift people into the air (Pelham 1976), mostly for military purposes, but sometimes for pleasure. Other attempts to conquer the air have included gliders, balloons, fixed wing aircraft, helicopters, autogyros and hovercraft. Early attempts at balloon flight were restricted to wealthy aristocrats and sponsored flights, often in competition. Only very recently has travel by aircraft been widely available.

Health: resorts, spas and seaside towns

In much of urban Britain, until the mid-nineteenth century, the water was simply undrinkable. This partly explained the popularity of ale and spirits. Drinking water had to be boiled first. The taste was still often so foul that it required flavouring. Tea from China and latterly India, cocoa from Mexico and coffee from Africa (and much later South America) were expensive but popular options across Europe (Gleeson 1998: 79). Epidemics were common, and pollution was rife. Typhoid was one of the most feared diseases. Prince Albert, Queen Victoria's consort, himself fell victim. This gave an impetus in Britain to the reinvigoration of health spas. Water in different spas will differ in taste. The quality (and taste) of water varies according to the profile of trace elements and minerals present. Some will be good for drinking, and others for bathing. Nobody who has bathed in the natural hot water of Sulphur Mountain, Cave and Basin National Park in Alberta would be tempted to drink it!

Parks: promenades, fountains, bandstands

Increasing urbanisation, concerns for health, and increasing wealth from manufacture and trade, gave rise to the creation of many parks and recreation grounds, as well as promenades, piers and winter gardens (Walton 1983: 158–85). Within public spaces ornament was often added, sometimes gifted in memory of loved ones: bandstands, fountains and statues, adding character and giving a sense of identity. Communities would hold sporting fixtures with adjacent communities in these parks and grounds. Football and cricket continued to grow as spectator sports through these times. Fairs, circuses and concerts would also be held.

Collecting: an obsession?

At some stage, most people seek to collect something – whether seashells or coins, stamps or postcards, bottle tops or pictures. Mankind is naturally acquisitive, and is driven to make sense of the world by discerning patterns. Often the first stage is deriving a classification system to make sense of the world, and thereafter by finding representatives of each group. It is a little difficult to see how collecting bottle tops or cards is likely to make a significant difference to the progress of society, but collectors live in a world of their own, and the activity is surrounded by all kinds of symbolism (Hughes 1963). People collect, not just to amass the material, but to enter the social world attached to the hobby and for the pleasure of discovery, for ever extending the collection (Argyle 1996: 220), and to live in the world of 'stamps' or 'playing cards' or any other object of collection.

Street games

Children have always found the doorstep and the street favoured play areas. There is a range of street games the world over. Collectors (Opie and Opie 1951, 1969, 1997) have eagerly captured and recorded variants of games that involve and stimulate motor skills, cognitive skills and above all social skills. The games often involve running, catching each other, singing, throwing skills, and memory. Some of the associated rhymes have a long pedigree. The nursery rhyme 'Ring a ring of roses, a pocketful of poses, atishoo, atishoo, we all fall down' allegedly reflects the time of the plague, the Black Death, when a sneeze was the sign that the individual had caught the disease, and was doomed.

Shows and the circus

Collecting animals and plants has been an irresistible urge for those with sufficient resources. Of greater interest still have been freaks, midgets, giants and individuals from hitherto unknown races. The circus is a piece of living history, with the elements remaining largely unchanged for hundreds of years: wild and trained animals, clowns, acrobats, fortune tellers, and sideshows. It is the balance that changes to reflect our values at any one time.

Sporting estates – Scotland

Queen Victoria was only the second monarch to visit Scotland since the battle of Culloden (1745), in which so many Highlanders died at the hands of one of Victoria's ancestors. Nevertheless, the Queen's first visit to Scotland in 1842 was swiftly followed by others, and she grew particularly fond of the Highlands (Duff 1983). Prince Albert shared her enthusiasm. By 1848 they had a lease on Balmoral Castle, on Deeside, and not long afterwards started to buy land. It soon became very fashionable to have a sporting estate in the Highlands, and industrialists from England started buying up large tracts of the Highlands, for stalking deer, shooting grouse (well, just about anything that moved!) and fishing for salmon. Scotland, to this day, has the most concentrated form of land ownership in western Europe (Cramb 1996; McGrath 1981; Wightman 1996), although the Scottish Executive has identified land reform as an early priority for the new Scottish Parliament.

Twentieth century

Radio and cinema

When Marconi set up his experiment to transmit radio waves, little could he have known how far reaching the effects would be. The 'wireless' soon became a source of reliable news, of music and entertainment, around which families would sit to share in each other's pleasure. In Britain the British Broadcasting Corporation was set up in 1922 and has had a special place in our affections ever since. Photographs became more popular and were used increasingly in printed matter. The moving image was a breakthrough that created a special art form of entertainment. Originally films were silent, and were accompanied by people playing pianos or organs. In time, these were replaced by talkies.

Wars and freedom

The turmoil across the world and especially in Europe after the First World War accelerated social change in many countries. The desire of people to wander at will across (what they thought of as) their countryside intensified. The membership of rambling clubs grew. Although the event has probably been exaggerated over time, the mass trespass at Kinder Scout (within what was to become England's first national park, the Peak National Park) has gained a special place in the collective memory as a crucial point in the campaign for access. In the fracas that occurred among the walkers (trespassers), gamekeeper and policemen, six arrests were made. Five of the ringleaders were given sentences of between two and six months (Rickwood 1982). The residual bitterness undoubtedly helped to fuel support for access legislation.

After the Second World War came a period of intense activity to safeguard peace, with the Universal Declaration of Human Rights in 1948. Articles with particular relevance here include:

Article 24. Everyone has the right to rest and leisure, including reasonable limitation of working hours and periodic holidays with pay.
Article 27. (1). Everyone has the right freely to participate in the cultural life of the community, to enjoy the arts and to share in scientific advancement and its benefits.

Health: exercise, food fads

After the Second World War people in Britain had to contend with rationing for some nine years. Diets were remarkably healthy, not least because people were urged to grow their own vegetables, and there was a good deal of advice available. Refined foods (of the kind that are believed to have subsequently caused problems) were simply not available in quantities sufficient to do harm.

Exercise was popular too. Youngsters wished to grow strong and the growing power of the mass media ensured there were plenty of role models to follow. People were also joining groups to exercise together. Various food products grew reputations associated with health, such as milk and vegetables, particularly green vegetables. Little children preferred sugar to the more wholesome foods. In America, cartoonists were commissioned to persuade children otherwise. Spinach, rich in iron, was adjudged especially good for health, but considered bitter by most children. Popeye the Sailor, who grew stronger as he ate spinach, was the result. There were other characters too, such as Olive (Oil) to help encourage a healthy diet.

The tastes of people (in Britain, for example) have developed over the ages as new foods have been introduced from other cultures. Feasts, such as those at Christmas, or for Americans at Thanksgiving, or for Chinese at their New Year, have always been important. Social patterns have changed. Meals are no longer always cooked at home. More and more people take meals out, especially for celebrations. Similarly, whereas many women used to stay at home and look after the house and growing family, now more and more women in the Western world take on different roles. Many work part time, or full time. This together with the increasing technology (in the kitchen) has encouraged the development of ready prepared meals, with recipes from all corners of the world. Air transport has ensured the delivery of 'fresh' foods and vegetables (otherwise out of season) at any time of the year.

Mass holidays (peace dividends)

In the 1950s and 1960s increasing wealth enabled more people to take holidays. Holiday camps, caravan sites (and later cabin sites too) grew up around the coasts of Britain.

As larger numbers of people began to travel, so each year another range of destinations, previously accessible only to intrepid explorers, was added to the library of places to which travel agents would take parties. Deals were struck with hotels, or with companies building hotels, and couriers would smooth out the wrinkles of a holiday taken in unfamiliar circumstances. The development of such integrated or package holidays, became a kind of colonialism, with ideas, customs and habits exported.

The car

The increase in car ownership, which accompanied automation, allowed personal mobility to take off. Travel by car allowed decisions about destinations to be left until the last moment. The car could be taken into the countryside, and frequently driven off road, right on to a picnic site. People tended not to stray far from their cars for most of the time, attached as if by some automobile umbilical cord. The car also served as a social time capsule. Interesting dynamics could operate, with different roles taken and played out among the occupants with the journey often as important as the destination. Trips out were of two kinds: either to a specific destination or for a specific time/distance.

Sun, sea and sex

During the summer months, millions of Europeans move farther south in a mass migration heading for the sun, with consequent pressures on the environment (Lewey 1996: 94). In a bid for freedom, people flock to the coast and contact with the elements, with sun, wind and sea. Under the law of the United Kingdom people have a right to 'besport' themselves on the foreshore, that area between high and low tides, whether looking in rock pools, building sandcastles, paddling or swimming. Such holidays allow people to meet on neutral ground and form friendships, away from the normal social tensions, whether for the very young playing on the beach or for those seeking the chance to admire members of the opposite sex. The sense of freedom associated with occasional skinny dipping is taken to greater lengths by naturists with varying strengths of belief (and codes for) walking about and living (at least for a while) naked.

From records to multimedia

Making and listening to music have always been popular, but the development of multiple recordings, cylinders, discs, vinyl records and compact discs has given rise to a major industry. Since the Swinging Sixties, pop stars have had their music played all over the globe, and become rich. Groups such as the Beatles, the Rolling Stones, Abba and the Spice Girls have each come to represent a particular era. Mixing and the electronic development of recording have established new rules. In Britain classical music on radio has been successfully repackaged for our short-term, high-speed world in the form of a commercial radio station which plays short excerpts or movements. The listeners to Classic FM had grown to about 6 million by 1999.

Visitor centres and museums

Museums went through something of a revolution. In Scandinavia, folk museums developed where examples of buildings from different areas would be brought to the one site and rebuilt. The American approach in developing visitor centres, which included the telling of stories, through sophisticated audio-visual programmes, was transferred to other parts of the world with varying success. There is now an increasing use of living history, of using the arts and (otherwise out-of-work) actors in character.

Fantasy, Disney

Visitors can now be delivered into a world of fantasy and make-believe. Simulator rides have developed which convincingly take people racing round circuits, on roller-coaster rides, into the air and under water. IMAX screens have been developed to allow the projection of films that give a wrap-round view. These have been used to show some of the wonders of mountaineering (notably on Everest) and space travel. Walt Disney has developed the world of fantasy originally portrayed in cartoons and film, into Disneyworlds and Disneylands in different countries.

Television, video and computer games

The range of television programmes aimed at people of different age groups and social groups is now very carefully developed and programmed in harmony with the marketing people concerned with the effect of their advertising. Programming is determined by many factors, and competition and negotiation among nations is an important element. Children in Britain spend more than 150 minutes a day watching television, although they prefer the idea of being out of the house, seeing friends, playing sport or going to the cinema. Fears for the safety of children are prompting parents to keep children indoors (Williams and Buncombe 1999). There are concerns over their future health and of the possible desensitisation to and stimulation of aggression (Gunse and McAleer 1997: 92–3, 102–5). In addition to watching television programmes, the computer monitor can be used as a games machine. The development of video recordings allows films to be played, or home videos to be made and viewed. It has been estimated that the average US child has seen '32,000 murders, 40,000 attempted murders and 250,000 acts of violence on TV, before they have reached the age of 18' (Parker and Stimpson 1999).

Recreational drugs: mostly alcohol?

Mood-changing substances and practices have always been used, in all societies. A variety of drugs remain in common use in the Western world. Despite court cases that have startled the tobacco industry, smoking is still popular. Links with disease have been shown to be strong. Despite this, the world sales of cigarettes reached 5,370 billion in 1997 (*Britannica* 1999) or not far short of 1,000 cigarettes for each man, woman and child on the planet. In the United Kingdom, where smoking has been generally declining, there has been a worrying rise in smoking among teenage girls, where it can be seen as a fashion accessory (Wearing and Wearing 2000: 45–58), and to a degree as a substitute for food. Figures for 1999 show lung cancer occurring with much greater frequency among women (Scottish Health Service 1999). Alcohol consumption varies enormously. The French may consume marginally less wine than the Italians (147 litres *per capita* of the drinking-age population compared with 153 litres), but by the time you add the beer and the spirits they consume more alcohol, and three times the figure in Britain (*Britannica* 1999). There are some inevitable consequences, with alcoholism and other diseases induced or affected by heavy drinking. Increasing consumption among younger people in many countries is worrying. Recreational drug use of illegal substances is growing, for example, with cannabis, cocaine, ecstasy and heroin. Unfortunately, some of these drugs are highly addictive.

Betting: dogs, horses, football pools, bingo, lotteries

The excitement and risk associated with chance, and our natural avarice, stimulate gambling (Bergler 1974; Cotte 1997: 380–406; Downes *et al.* 1976), and there may be an explanation for the apparent delusion of chance, lost in the search for escape into tension, pleasure and fun (Jong *et al.* 2000: 230–8). People wager on the outcome of sporting events, races, the likelihood of a white Christmas, and on every conceivable thing you can imagine. Many societies have sought to regulate gambling, rather than risk a repeat of the problems of the prohibition era when gambling was criminalised. Government then knows what is happening, and takes a handsome slice of the profit. Often a proportion is distributed to charitable organisations. In Hong Kong the Jockey Club puts money into projects with partners, as with the development of Hong Kong Park. Official lotteries operate in many European countries, and Britain is a recent convert. By 2001 the national lottery is expected to have raised some £10.85 billion for the good causes, whilst the proceeds prior to this division are distributed among the state, the operator and charitable causes. The lottery is sometimes referred to as an extra tax on the poor, as they take up proportionately more tickets. On the other hand, many of the facilities and services that are helped by the lottery in Britain are the preserve of those who are rather better off.

Surfing on air, sea and land

We love the exhilaration that comes from speed. The development of man-made fibres and materials has enabled strength to weight ratio of spars and masts and of sails to be increased. The equipment has become more portable, and (arguably) more affordable. A windsurfer can transport his board on a roof rack, and seek out appropriate favourable locations, according to conditions and preferences. The designers then create a fashion or style, and take advantage of what emerges through chance to promote a world with which aficionadas can identify. There are more people who wear the relevant clothes for a particular sport than ever take part in it. A great deal of money is tied up in these fashions, which may then spill over into mainstream fashion for some sub-group of society. As snowboarding depends on the presence of snow, whether natural or artificial, the skateboard has seen a resurgence as an alternative for snow-free times or places. Skateboarding has seen several waves of development, with skateboard parks being created by local authorities in many parts of Britain. The 1990s saw the development of the activity well away from dedicated facilities. Groups of skateboarders typically assemble in open public spaces and squares where there are ramps, steps and railings or other obstructions and challenges. Skis have been developed for use on snow, on water and (as roller skis) on grass and tarmac.

Personal motorised mobility, off road

Many machines have been developed to allow individuals greater freedom to travel at speed, whether motor bikes on roads, skidoos on snow, quad bikes across the open hill or personal water craft. In the air, the development of powered hang gliders has brought personal flight within reach of more people. Often developed by individuals, opportunities are soon seized by entrepreneurs to make the activity available to greater numbers. Although electric power units can be relatively quiet, motorised recreation to date has been inherently noisy. This has given rise to some tension.

Health again

The developments in medicine have pinpointed the need for exercise to improve cardiovascular health. Against the background growth in sedentary lifestyles, there has been a substantial rise in activities like swimming, cycling and the use of fitness centres, within certain groups. Others are pursuing dieting to achieve their ends. There is some evidence that these activities are stimulated more by a concern with physical appearance (or more properly, body image) than with health. Interest is growing in the use of leisure to promote mental and emotional health. At a personal level, this includes using various techniques for stress reduction and meditation. At a group level, it extends to persuading executives that good time management will provide them with more time with their families, which will be to everyone's good.

BOX 2.3 Excitement and adventure: bungee jumping

In the Pentecost Islands in the South Seas there is a rite of passage into adulthood for young males, which consists of jumping from a platform high off the ground. The young men make graceful dives into the air, their ankles are tied by lianas so that they are brought up gracefully, just before hitting the earth. The free fall and subsequent halt are accompanied by a release (understatement) of adrenalin. This activity has developed into bungee jumping, one example of the white knuckle sports, like rafting and parachute jumping.

New skills and learning: the boom in specialist holidays

People are choosing to spend more and more of their leisure time learning new skills. Painting, green woodworking, coracle making and a hundred and one other skills can be learnt on holiday, and these are being eagerly taken up. In these days, when flexibility is all important to securing work, an ever-expanding portfolio of skills is extremely useful. Pursuit of new skills may unlock doors for people; doors to opportunities never considered before, which can provide the basis of a career development.

Shopping for leisure?

Shopping, in many Western countries, has become (like so many other operations) a relatively high-speed affair. In times gone by, shopping would have involved many trips to specialist shops, such as the butcher, the baker, the greengrocer, the grocer, the tobacconist, the wine merchant, the bookshop and the newsagent. At each shop, news and gossip would have been exchanged, and the round of shopping was perhaps as much about communication as about gathering together the necessities (or luxuries) of life. Nowadays, news and gossip are national and reach us by mass media, and all these purchases can be made in the one place: the supermarket, *hypermarché* and shopping mall. Of course, many small family-run specialist shops find it difficult to compete on price, or convenience (as we head towards the twenty-four-hour society and expect service every hour of every day of every week), and they go out of business. With shopping available on Sundays as well, some friends and families engage in shopping for leisure, whether for leisure goods or as a form of leisure activity in itself, at home or on holiday (Jansen-Verbeke 1990: 128–37). Choices are often jointly made, and the role of the family and other referent groups has been shown to be hugely important (Engel and Blackwell 1982: 143–86). As with so many other aspects of life, the market has become very fragmented. Lunt and Livingstone, in a study carried out in 1992 (Lury 1996: 233), revealed five categories of shopper:

Alternative shoppers	12 per cent
Routine	31 per cent
Leisure	24 per cent
Careful	15 per cent
Thrifty	18 per cent

However, even the routine or thrifty shoppers were no doubt spending a reasonable proportion of their shopping money on goods which closely relate to leisure, of one form or another. When you next return from a shopping trip to a supermarket, analyse your own trolley load, and consider to what degree each article will contribute to leisure. For the leisure shoppers, the process of shopping – seeking out alternatives, evaluating and making comparisons, and engaging others in decision making – all becomes part of the pleasure. This is so, particularly where the shopping requires high-involvement, and conscious, decision making (Engel and Blackwell 1982: 23–34), e.g. for white goods, and in a different way for window shopping and impulse purchases. Some cultures keep goods for much longer than others.

Lee has suggested that there has been a transition in interest from material to experiential commodity (Lury 1996: 62), to an interest in time-based rather than substance-based commodities. However, the marketeer often seeks to associate a range of experiential opportunities, with particular products. The opportunities may extend beyond simple practical use. Holidaymakers often bring back a piece of artwork or souvenir, to remind them of their travels and extend the value of the holiday. The effects of such purchases of tourist art can be far reaching (Cohen 1992: 3–32) for the host community, even promoting a reorganisation of the way local economies and communities work. Most pertinent to the arguments presented in this book is the observation that every act of consumption has a cultural as well as an economic effect (Lury 1996: 51).

Communications and transport (speeding up time), internet and home PCs

Journeys that took our ancestors weeks to complete, or even years, are now accomplished with ease and at much less (personal) risk in a matter of hours. In the early development of horseless carriages (or automobiles) in Britain someone had to walk (and rarely run!) ahead of the carriage waving a red flag. Production models of some cars now travel in excess of 150 mph, and family cars comfortably travel above legally permissible speeds on motorways, no matter what the country. Some of the cheaper flights across the world now cost a few hundred pounds. We are constantly searching for new ways of speeding up travel and communications. We seem to be insatiable in our appetite for novelty, always looking to see what lies round the corner, and yet seeking to keep in constant contact with friends and colleagues. Mobile phones have enabled us to develop virtual communities linked by radio waves, and to keep friends (so long as we remember the time differences) across the globe. The increasing use of personal computers at home and rising connections to the internet will ensure that we keep pursuing the real-time communication if not time travel. In 1997 connections to the internet doubled to 1 million in Britain (Smalley 1998), and in 2000 it was reported that one in three households in Britain were connected.

Connection

Whilst some sectors have embraced these changes, there are people who have rejected these values, and others who have explored forms of living and travelling which are more gentle on our environment, such as soft, or green, tourism (Krippendorf 1987; Lane 1994). Similarly, the Earth Summit (United Nations Conference on Environment and Development, Rio de Janeiro, 1992) has been something of a watershed in developing a more integrated way of thinking among politicians and other opinion formers. However, as the preamble of Agenda 21 puts it, 'We are confronted with a perpetuation of disparities between and within nations' (UNEP 2000). The marketeers would say that within society (certainly in Western economies) people are continuing to develop specialised tastes.

And despite all the striving for communications and hi-tech solutions, one of the most common forms of leisure keeping us in touch with the earth (if not nature) is gardening. In the United Kingdom the increasing proportion of empty-nesters, and an upturn in the economy generally, are believed to be behind the increasing sums of money spent, estimated to reach £3.02 billion a year by 2000 (Howitt 1997) or roughly £50 for each man, woman and child. People are also increasingly interested in helping to care for the countryside, through planting trees and creating ponds, and through contributing to the care of environments far distant from where they live. One theme that has caught the imagination, not least because of the scale of change, is the tropical rain forests, from which an area the size of Switzerland disappears each year (http://www.greenpeace.org/~forests/main.html). There are many activities and schemes for conservationists to become engaged in to help combat destruction of the tropical rain forests of the world (see websites of FoE, Greenpeace International and WWF-Scotland). The interest in plants is maintained in offices, flats and workplaces through pot plants and window boxes. People are realising once again how everything is connected, and that some of the developing problems in medicine, such as the increasing incidence of asthma, and allergic responses, may be in part due to our love affair with the motor car. The 11 million working days lost though back pain in Britain each year (costing employers £5 billion) could be reduced, and the pain avoided or relieved, if we encouraged more physical exercise and other preventative measures, in our increasingly sedentary lifestyles (Campbell 2000).

Ideas for further study or work

1   Take an area of the world and map the different approaches to the concepts of work and leisure, as best you can at a particular time in history. (Beware, not in every language will there be – or will there have been – exact synonyms, and this will be among the clearest signs that the approaches are different.) What other factors define these zones?

2   Investigate the history of some particular form of leisure or recreation, within a single culture, or (if you have more time) across cultures – for example, the history of a particular sport, the development of a particular local festival.

3   Explore what records have been left behind to show what a particular people did, at a specific time, when at leisure, constructing a time diary to demonstrate how this changed through the day, week, month, year or lifetime. If you have more time, work on comparisons across time or in different parts of the world.

4   Pick a period of history and explore the different approaches to work and leisure among different groups, of your choosing, for example: men, women, children, older people, wealthy and not so.

5   Explore some form of leisure that has been interlaced with work, such as singing or dance, and chart the relationships: such as the derivation of the different sea shanties, or the gumboot dancing of South Africa.

Reading

First-hand accounts, artefacts and culture

To read a first-hand account of what people did in times past creates in all senses a very direct link. For those of us who do not shine at Latin it is easier to read a translation, for example, of Suetonius' work. Similarly much can be gleaned from visiting caves, old sites and monuments, castles and churches where tell-tale signs are still available for (best) discovery and inspection, or which have been interpreted by archaeologists and field historians.

General histories

There are many general histories, which will no doubt reveal what (some) people did at leisure, or for their recreation. Even works as broad as Morgan will provide useful sources from which to start, and provide reference trails. Often such reading is rather a circuitous way in. For example, while there is no entry for recreation or leisure in the index of Stenton's *Anglo-Saxon England*, Jones's work on the Vikings or Brown's work on the American west, or Miller's collected accounts, nevertheless much can be gleaned from such works. But it takes time. It is much easier to rely on those who have been there before. There are occasional references to recreation in works such as Rackham, Hoskins and Mabey, which all deal with the (British) countryside. For more general countryside recreation, read Patmore.

Specific histories of leisure and recreation

There are more general works such as Shivers and deLisle, Toner (ancient Rome), Cross (from 1660), and Walton and Walvin (Britain, 1780–1939). Of necessity, most focus on a particular period, or across particular lands or peoples. The references here are necessarily partial, but indicative of what you may find. There are also histories of specific subjects: Gombrich on art, or Abraham on music. Few subjects are without a recorded history, for example Opie and Opie on street games, Simpson on skiing in Scotland, Hargraves on playing cards, Gillmeister on tennis and even Carmichael on hymns and incantations collected on his travels.

Incidental

Plays, (historical) novels and stories all have a place in such study. Moorehead gives a great account of early exploration (tourism?), and McGrath gives a telling account of Scotland's history and the effects of the development of the sporting estate.

Chapter 3

The recreation web

Overview

Process links the actors together in the recreation web. At the simplest level of resolution, the web can be simplified into a chain. By mapping the process, we can understand what is going on and, importantly for us, when and where to intervene to achieve our management objectives. Simple examples are given, in different realms, of recreation chains consisting of the resource holder, agents (who may affect the rate or nature of the process) and principal participant. The part played by free markets is discussed as well as the need for intervention and regulation by government and its agencies. This provides the background for understanding the web that we will capture on the map, in order to plot the distribution of benefits and costs.

Actors and their roles are described within the context of the private, voluntary and public sectors. These sectors and their importance will vary according to the political environment and the recreational activity under consideration. The kind of government will be particular to the country, and for the purposes of this book, examples from Britain illustrate the approach. The detail is unimportant, as structures (and especially names) are continually changing (even within the process of producing this book). The host community and environment are larger players today than has been the case in the past, and by means of example, possible roles and interactions are described. To make any choices and decisions we need information. Sources of information can be very important in providing the array of leisure opportunities. They will also help in the development of attitudes, some of which will be deep-seated. Communication has two components – meaning and emotion – and both are important in leisure and recreation. Family, friends, school, work, experts and the media can all provide recommendations. The most influential source of all is feedback from experience.

To start the mapping process a simple stepwise approach is suggested, with simple questions. These help to clarify why we are intervening, who the participants are, who holds the resources, who can affect the process, and who stands to gain or lose. Types of diagrams which can be helpful in mapping and understanding the process are described, and strong encouragement is given to have the first map on a single sheet of paper (whether a wallpaper roll or the back of the envelope). Once the overall process is mapped to our satisfaction, different areas can be explored further at another scale, where it is warranted.

A web of actors and interactions

To manage leisure and recreation, we need to understand the process. What are the factors that determine what happens? What activity is undertaken, and by whom, over what time, and in which space? What happens in one part of the web will have

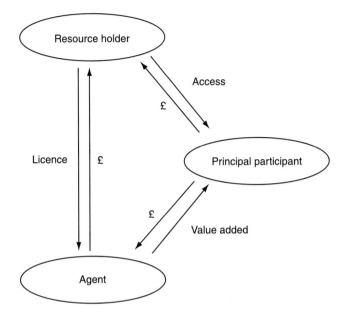

FIGURE 3.1 The principals: resource holder, agent and participant. The diagram shows the expected exchange of money for services but you could add in the many other exchanges. It is not all about money

consequences for people elsewhere in the web. All the actors are linked, and to help us explore and understand this we need to have an image, a map. In this chapter a variety of different charting techniques are used to help understand the interactions between the players, and the reader may enjoy (or benefit from) using more than one approach to reveal different facets of the problem.

Some components, actors and interactions will be key to this understanding. In the simplest case, it may be enough to identify the recreation chain: from resource holder, through agents of one kind or another, to the principal participant, analogous to a marketing channel (Kotler 1984: 538–63). In each case we need to determine the scale of resolution we will be operating at. How much detail is required to explain how the web functions? How far should the web extend? What time period are we concerned with? What is the appropriate geographical boundary? What scale should we focus upon? The answer must depend on the nature of the problem we wish to solve, or the opportunity we wish to seize. The leisure and recreation can begin only when the participant is in location. If the participant has not the means to travel independently, the provider of transport obviously has a crucial role.

Visitors, participants, audiences, users, customers and consumers: this list of synonyms for principal participants is prompted by the different forms of (and attitudes to) recreation. Those who think in market terms will recognise terms such as 'consumer' and 'customer' and be more concerned with aspects of service. Those who think of recreation as something rather more spiritual or social will use other terms. In

TABLE 3.1 The recreation chain

	Outdoor recreation – sailing	Indoor sport – squash	The arts – reading	Leisure – gardening	Tourism trip to Hong Kong
Resource holder	Boat owner (or owner of the water, or of launch site)	Court owner (club/local authority)	Author	Owner of garden	China, (Hong Kong) SAR
Agents	Boat builder Sailmaker Chandlers Harbour master Club Coastguard Weathermen Sailing magazines Relatives of skipper and crew	Manager Staff Clubs Equipment suppliers	Publisher Printer Bookseller Critic Reviews	The elements! Soil Climate Seedsmen Nurseries Garden centres Gardening pro- grammes and magazines	Travel agents Airlines Ferry operators Banks Hotels Restaurants Shops Hosts and host com- munity
Principal participants	Skipper and crew	Players	Reader	Gardener	Partners

mapping the recreation web, first capture the principal actors who gain direct benefit from the recreation, as individuals and as groups, and thereafter include those who indirectly benefit. For most recreation there will be a number of individuals who gain direct benefit from the event under consideration. Even for people visiting wilderness areas, most visits are conducted in the company of others. In the arts, expression and appreciation belong to the individual. Reading is an activity that is very much in this category, where it is the reader who gains the principal benefit. Even here, though, much of the pleasure comes from sharing some of the quotations or the highlights with companions or recommending books to friends.

In each event, individuals will obtain personal experiences, and benefits from their chosen recreation. These personal experiences each contribute something to the development of the individual. Just how, will depend in large measure on the nature, extent and intensity of the event; the individual's background, genetic make up, education, previous experiences, attitudes, motivations and intentions. At the first

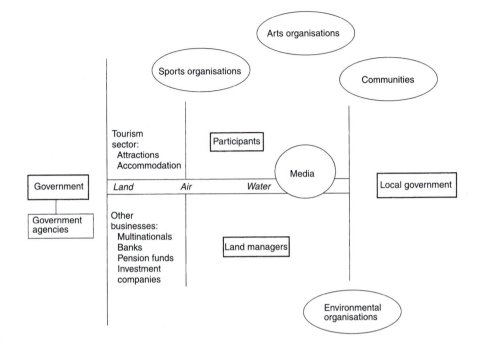

FIGURE 3.2 Who is taking part? Putting all the different actors on one sheet of paper helps to suggest where management effort is best applied

mapping, the components to be plotted will usually be assembled by interest or activity, but any complete analysis will need to take into account the mechanisms and, in some cases, other variables, such as age, background and lifestyle.

Most recreation is a social activity. For this reason, the social environment is as important as the physical environment. Mapping the group dynamics will help us understand the process and may help to reveal zones where action by the manager can be most effective, and where action must be avoided. A good deal of recreation is concerned with the freedom to associate with people of one's choice. All members of the group will not necessarily have an equal share of such freedom. A family group setting out on a drive through the country to visit a castle or a beach, in a car, will include different states of commitment. The driver may have a degree of control over the itinerary, the timing and venue of stops or the end destination. Other members of the party, the navigator, the father or mother, as well as the children will have important contributions to make at different stages of the recreation event. In determining the size of the group captured in the mapping process, transport often plays a defining role. The motor car or people carrier comprises a relevant unit of study, whether as members of a family or as a group of friends. Similarly, for more distant journeys and experiences, the coach or aircraft may define the relevant unit.

Leisure and recreation can be provided and regulated in different ways, often mediated through markets, in which goods and services are traded, most often for a convertible currency. There will also be many other important exchanges that involve things other than money (Kotler 1984: 4–14). Where the knowledge is perfect, the goods (or services) are discrete or private, markets should work well in delivering what consumers want. Often the consumer is unable or unwilling to spend time in endlessly comparing offers, and will seek recommendations through media, agents and word of mouth.

Where the goods or services are relatively costly, the traditional role of the agent is now at risk, if not usurped. We may not have perfect knowledge about leisure opportunities, but the acceleration into high-speed communications (and especially the Internet) has changed the rules. Knowledge that used to be the sole preserve of the agent can now reside networked in an expert system, or at a remote location. The web takes the place of the agent. Information and transactions can be completed almost in real time, without the purchaser leaving the computer or telephone. As a result, where markets operate, business is more competitive than ever.

The size of the markets is enormous, and some of the variations are caused by factors beyond our control. Travel and tourism alone, in 1999, accounted for 11 per cent of GDP in Europe (WTTC 2000). A recent poor summer reduced Scotland's tourism industry revenues by an estimated £30 million in one year. This loss in revenue is a function of the weather, people's preferences and changes in the exchange rate. Travel to mainland Europe became easier for Scots, and travel became harder for Americans intending to visit Scotland. The rules that govern world trade apply to leisure just as to any other sector. There are some areas where markets do not work so well. At times, intervention may be required, if the objective is to deliver societal as well as personal benefits. Where public goods are concerned different rules are needed for optimal solutions. The market may encourage shorter time preferences, and be distorting the choices open to our children and future generations, 'cheating on our grandchildren', and threatening the very basis of sustainability.

Predicting how people will behave in response to such interventions is notoriously difficult. The tools are blunt: grants, interest-free loans and tax advantages are among the financial ones. In times past, when societies were closely related, and cooperation was more important than competition, peer pressure will have far outweighed any financial levers. It may be that, in times to come, levers other than financial ones will again take hold. Markets can distort, where public goods are sold as if they were private goods. In such cases, governments step in by regulating activity, to ensure fair play for society. Information is provided to encourage or discourage activity.

The private sector

The intermediaries between the resource holder and participant are often (as has been described in Table 3.1) from the private sector. Such organisations have sometimes banded together in federations and associations to safeguard the different sectoral economic interests, in the business of leisure and recreation. In this section we explore some standard categories – accommodation, attractions, tourism and travel, and downstream businesses – and consider the issue of clubs and exclusivity.

The providers of accommodation include multinational companies, great hotel chains, and resort owners. These can exert considerable market power – so much so that it may require governments to act with considerable resolve if they are to restrain the activities of such channel captains. At the other end of the spectrum are those who provide accommodation on a part time basis within their home, on the farm, or as bed-and-breakfast providers in Britain. At this latter end of the market, operators can work together to build critical mass, and wield power. Outside the usual definition of the market, but accounting in many countries for a considerable portion of the business, friends and relatives provide many bed-nights each year. Whilst, individually, these private sector operators are quite small, together they account for a substantial portion of the market. Some of the trade organisations speak for combined interests which would make some countries' GDP look small (e.g. the British Licensed Retailers' Association).

The owners and managers of visitor attractions and facilities also range from the very large to the very small: companies owned by multinationals, many independent operators, and the self-employed guide. Associations have been formed to amplify the voice of independent operators. In the farthest south-west corner of Britain, managers of attractions formed the Cornwall Association of Tourism Attractions in 1974. There were thirty-four members of the association in 2001 (CATA 2001). When it was formed, peer reviews of the attractions led to increased quality, which in turn will have been responsible, in part, for some significant rises in profitability (of the order of 15 per cent or so). Many such associations operate on a geographical basis, like CATA or, for example, the Association of Scottish Visitor Attractions. Others operate on behalf of a particular interest, like the Farm Holiday Bureau.

Airlines, railways, bus companies, shipping lines, tour operators and travel agents of all kinds exert power in the recreation web. The audacious entry of Thomas Cook into the market, with his tours, joint ticketing, travellers' cheques and international rental agencies (in the nineteenth century) had significant impact which has been underrated in its importance (Butler 1999: 100–1). By emphasising particular routes or giving discounts on certain journeys, such players can do much to shape the market and determine what consumers demand. The frequent fare wars and development of different niches have been a constant feature. Often power has been achieved through vertical integration, either through incorporation or through franchise agreements. Governments of some countries have been working hard to wrest back some of the control which such market power has achieved (Klemm and Martin-Quiros 1996: 126–44). On the other hand, some agents are now feeling something of a draught as the Internet provides a rival mechanism.

Businesses that supply equipment, upon which leisure and recreation depend, can themselves be major players in the field. In the more affluent northern hemisphere, and in the West, a large part of the publishing business is concerned with the production of material that is read at leisure. In Britain alone, there are more than 1,300 newspaper titles. More than £2,200 million is spent each year on advertising in the regional and local papers. Advertisers and marketing experts have conspired to

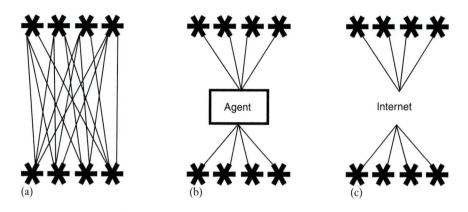

FIGURE 3.3 Development of market makers. Transmitting information through a free market (a) has been made more efficient by working through agents (b). In some cases their place has been taken by the Internet (c), which allows easy access and the servicing of thousands or even millions of people, whom a single agent could never support alone

explore markets associated with leisure. Designer labels are positioned to align with particular lifestyles. Shoes originally developed for use on the running track have taken the world by storm. Clothing originally developed for use in active sport is being used by a wide range of people, including those who have no intention of engaging in any strenuous exercise whatever. Mapping can be used to pinpoint the different downstream businesses and activities that are significantly affected by others in the web. Asking what is necessary for each link to be maintained in the web is one way of exploring their relative importance.

As the population increases and our world becomes ever more crowded, people seek to find ways of buying space, at least for a while, to secure their chosen sport. A keen fisherman can choose to buy into a syndicate, so that he will have a rod on a given beat, say on a famous river in Scotland (such as the Tay or the Tweed). For this he may pay several thousand pounds, or alternatively he could buy a day permit, costing only tens of pounds. A golfer could pay hundreds if not thousands of pounds to join a golf club (and face a waiting list lasting anything up to ten years), or just pay a few pounds for a round of golf on a municipal course. The same is true of swimming, where a swim at the local baths might cost a pound or two, but there are plenty of private clubs that cost several hundred pounds to join, for a year's membership. Someone who wants lunch could buy a pie and a bottle of beer and have change from a £5 note. Alternatively, he or she could take a meal in an exclusive club, where change would be unlikely from a sum five times that, and for which there would be a hefty annual subscription. It is not the fishing, the golf, the swimming or the meal for which people are paying. They seek a particular social environment in which they feel comfortable, and often a degree of exclusivity.

The voluntary sector

Where groups of people have common concerns they often achieve more by working together. Such people have formed charities, trusts, partnerships and associations of all kinds to further their aims. There are three principal groups of voluntary organisations: those representing owners and managers; those representing enthusiasts; and the lobbyists and activists. Their interests, motivations and actions reflect those of the parent groups, and may overlap, but using these basic groups will help in mapping and understanding the process.

The Country Land and Business Association represents the interests of landowners in England and Wales, and a similar organisation, the Scottish Landowners' Federation, in Scotland. Both organisations work across a wide range of issues. Over the last few years, the Association has been campaigning vigorously for the application of the voluntary principle in respect of access, in a campaign entitled Access 2000. However, the government (Westminster) has introduced a right of access to open ground in England and Wales through the Countryside and Rights of Way Act 2000.

BOX 3.1 Governing bodies

Almost every recreation that describes itself as a sport, within a very short while develops a governing body, which looks out for the interests of members. Emerging sports need the strength that comes from such organisation to gain a place in the queue, for succession (Sidaway *et al.* 1986: 13–18) in access to the resources it requires. One of the clearest benefits that such an association often bestows is insurance, taking advantage of the economies of scale, and with special knowledge of what the particular sport requires and entails. Governing bodies develop codes of behaviour to help gain an edge in their quest for resources, which depends on tolerance by others. Further, associations can collate sufficient information to develop strategic plans to gain a better share of any grants or revenues that might be available. The membership of associations is often made up of federations of clubs, but some (in part because of the nature of the sport, e.g. cycling and the Cyclists' Touring Club), provide for individual members. The more active of these governing bodies may also serve as lobbyists. These governing bodies pool their energies and network through organisations (such as the Central Council of Physical Recreation, in Britain, which has several divisions, and an overall membership in excess of 100 associations.)

The Scottish Landowners' Federation has taken a rather different point of view, working within the Access Forum in Scotland. There the Forum has roughly equal membership from the owners (resource holders), the user organisations (participants' representatives) and the public bodies (agencies). It has been seeking to broker an agreed way forward, to make workable the introduction of legislation (through the

---

BOX 3.2 The Ramblers' Association (in Britain)

Has something like 120,000 members, and achieves a great deal, campaigning vigorously for the right of access. It is perhaps surprising that the membership is not greater, given that walking is cited as the most popular form of recreation in countless surveys of outdoor recreation (see UK Day Visits Survey 1998). Arguably, the groups concerned with the environment have been rather more powerful, probably because their membership has been drawn from typically better paid (and connected) membership. The Royal Society for the Protection of Birds (RSPB), Europe's largest wildlife conservation charity, has more than 1 million members and an annual budget approaching £50 million. The RSPB clearly represents those who, amongst other things, watch birds as part of their recreation. The National Trust has more than 2.5 million members and a budget in excess of £166 million (1996/97). There are many such organisations but the different recreations and interests are not equally represented.

Scottish Parliament) to give a right of access, responsibly taken, to land and water in Scotland for recreation and passage (see Box 8.2).

Many recreations are not so represented and are indulged in by people who chose to avoid joining clubs. Inevitably, these are under-represented.

The public sector

Wherever recreation is seen as having a benefit for society, or is concerned with aspects of equity, then government at some level will inevitably be involved, seeking to correct perceived imperfections in the market. Rarely are its actions seen as central in the process (see Table 3.1), and often when it is most active and successful its profile is hardly visible, but the power and influence of government and its agencies should not be underestimated. The following survey considers four different levels, illustrated for Britain, but which will have analogues in many countries, particularly in Europe.

Global

The United Nations Commission on Sustainable Development has, since the Earth Summit in Rio, 1992, had increasing influence on how member states approach the environment, the use of energy and other effects, notably climate change. The United Nations Environment Programme has as its mission statement: 'To provide leadership and encourage partnership in caring for the environment by inspiring, informing, and enabling nations and peoples to improve their quality of life without compromising that of future generations.' Much of the progress has been made through implementation of Agenda 21.

The World Trade Organisation regulates agreements on trade, seeking to ensure free trade.

Regional

As part of the European Union, the United Kingdom of Great Britain (England, Scotland and Wales) and Northern Ireland is covered by European legislation. The European Community has enormous influence through its directives, such as 85/337/EEC Environmental Impact Assessment as amended by 97/11/EC, implemented by the regulations. The European Environment Agency serves as a gateway to information. There are many directives which affect leisure and recreation, for example those concerning the quality of bathing water, harmonisation of collecting statistics on tourism, and habitats and species action plans.

National

Government has many concerns related to leisure and recreation in the realm of social, economic and environmental issues. Some of these will be implemented by government departments themselves, and others by their agencies. Since 1999 some aspects of government in the United Kingdom (of Britain and Northern Ireland) have been devolved from the Houses of Parliament in London, at Westminster, to the Scottish Parliament in Edinburgh, at Holyrood, and the Welsh Assembly in Cardiff. Here, 'national' includes both these levels of government. In Northern Ireland there is the Assembly at Stormont, with a complex raft of support to encourage close working with the Republic of Ireland. At the time of writing there is considerable interest in what forms of government may develop in England, and whether there is scope for developing assemblies or similar mechanisms in each of the eight government regions in England. After all, the single county of Yorkshire has a population of more than 5 million (compared to the population of Scotland at 5.1 million).

Department of Culture, Media and Sport (Westminster) had a budget for 2000/01 of more than £1 billion. It is responsible for about fifty public bodies, including many quangos (quasi-autonomous non-governmental organisations) which carry out the work of government, at arm's length, for example in relation to heritage, the media (including film), sport and tourism. It is also responsible for the National Lottery, which raised more than £7 billion over the first five years.

Department of Environment, Transport and the Regions (Westminster) is concerned with conservation, landscape protection, the enjoyment of the countryside and the economic health and well-being of communities in rural and urban areas, as well as all transport issues and those which relate to local government in the regions. This is also the host department of the Countryside Agency and of English Nature. In Scotland devolved issues are the responsibility of the *Scottish Executive*, whose agencies include Scottish Natural Heritage and Historic Scotland. In Wales, the Welsh Assembly has comparable agencies, the Countryside Council for Wales and Cadw.

Ministry of Agriculture, Fisheries and Food (Westminster) is concerned with the countryside at large in relation to primary production in these areas. In Scotland, similar functions are brigaded within the *Scottish Executive Rural Affairs Department* (Holyrood), and in Wales within the *National Assembly for Wales Agriculture Department* (Cardiff).

The Forestry Commission (Westminster) is a government department which spans Great Britain. While certain responsibilities such as the funding of research and international issues are reserved to Westminster, forestry is devolved, and the national offices of the Forestry Commission in Scotland and Wales work through the Forestry Ministers of the Scottish Parliament and Welsh Assembly. The Forestry Commission manages the national woodlands, through Forest Enterprise, one of its two agencies (the other is Forest Research). The national woodlands represent one of the largest recreational resources out of doors, and the Forestry Commission also promotes best practice across the forestry industry as a whole.

Many other departments have an interest too. The *Treasury* seeks to contain expenditure, and some other departments, such as *Health*, will recognise that leisure and recreation (and certainly exercise) may help their objectives.

Examples of quangos in:

The arts

- *Arts Council of Great Britain*, which co-ordinates the activities of the Regional Arts Boards, in promoting all the arts in Britain.
- *Arts Council of Wales*, funding and developing the Arts in Wales, and also for the distribution of lottery funds.
- *Scottish Arts Council*, 'creating a dynamic arts environment which values the artist and enhances the quality of life for the people of Scotland'.
- *Scottish Screen* 'is responsible to the Scottish Parliament for promoting and developing all aspects of film, television and multimedia in Scotland through the support of both industrial and cultural initiatives'.

Heritage (natural, built and cultural)

- *British Waterways*, managers of most of the country's canals and waterways, for navigation and for leisure.
- *Cadw* (from the Welsh word to 'keep') looks after and welcomes visitors to a diverse collection of monuments and castles, and has a role in protecting heritage.
- *Countryside Agency*, concerned with the protection of England's landscape, its enjoyment and the social and economic well-being of those who live there.
- *Countryside Council for Wales* is the government's statutory adviser on sustaining natural beauty, wildlife and the opportunity for outdoor enjoyment in Wales and its inshore waters. It has taken the lead in developing cross-agency work in developing comprehensive whole farm grants (Tyr Cymen and Tyr Gofal) to take into account nature conservation, farming and forestry objectives.

- *English Heritage* protects the historic environment and promotes the public understanding and enjoyment of it. English Heritage funds archaeology, conservation and repair, and looks after 400 or so historic properties.
- *English Nature* is the statutory service responsible for looking after England's variety of wild plants and animals – the country's biodiversity – and its natural features.
- *Environment Agency*, working for a better environment in England and Wales for present and future generations. Aside from research, monitoring and advice, the agency has a role in recreation and access, controlling over 1,000 sites for recreational use. It also has a general duty to promote the recreational use of water and land throughout England and Wales.
- *Historic Scotland*, 'safeguarding Scotland's built heritage and promoting its understanding and enjoyment'.
- *National Museums of Scotland*, showing Scotland to the world and the world to Scotland through extensive collections built up over more than two centuries.
- *Royal Parks* manages the royal parks to offer peaceful enjoyment, recreation, entertainment and delight to those who use them; and protect them for this and future generations.
- *Scottish Environment Protection Agency* 'protects the land, the air, the water, the core elements forming the very fabric of our environment'.
- *Scottish Museums Council*, promoting a network of private, voluntary and municipal museums and galleries across Scotland.
- *Scottish Natural Heritage*, promoting the care and improvement of Scotland's Natural Heritage, its responsible enjoyment, its greater understanding and appreciation, and its sustainable use, now and for future generations.

Sport

- *Sport England* 'leads the development of sport by influencing and serving the public, private and voluntary sectors, for more people to be involved in sport, with more places to play sport, and more medals through higher standards of performance in sport'.
- **sport**scotland, 'promoting sporting opportunities for Scots at all levels, whatever their interest and ability'; 'widening opportunities, developing potential, and achieving excellence'.
- *Sports Council for Northern Ireland* and *Sports Council for Wales*.

Tourism

- *British Tourist Authority*, especially in promoting Britain overseas. The management of tourism is an important component in managing the balance of payments. Visitors come to Britain for the cultural heritage, the variety of scenery, and sometimes to seek out their roots.
- *English Tourism Council*, co-ordinating the Regional Tourism Boards in the different parts of England and the reception of visitors in England.

- ◆ *Scottish Tourist Board*, 'generating jobs and wealth for Scotland through the promotion and development of tourism'.
- ◆ *Wales Tourist Board*. The official website provides information about travelling and holidays in Wales, the United Kingdom, including Welsh accommodation, castles, attractions and events.

Other important players include the *universities*, which have a major role in providing support through research and teaching in relation to leisure, recreation and tourism, as well as all aspects of the environment, and environmental management. The various health promotion organisations also make a significant difference to what is spent on promoting healthy exercise, sport and outdoor recreations: *Health Promotion England*, established in 2000; *Health Education Board in Scotland*, promoting health in Scotland. In Wales the *Health Promotion Division* is 'involved in a social and political process that enables people to increase their control over, and to improve, their health and well-being. Health promotion involves: *advocacy* to create the essential conditions for good health; *enabling* people to achieve their full potential; and *mediating* between different interests in society in the pursuit of health'.

In England a number of government departments now share offices at subnational level. Some of the agencies are also arranged to reflect groupings at this level: the Regional Arts Boards and Regional Tourist Boards are examples. In Scotland there are two development agencies, Highlands and Islands Enterprise in the north, and Scottish Enterprise in the south, which each comprise a network of Local Enterprise Companies. These play an important role in encouraging economic activity, and therefore are major players in leisure and recreation, especially in respect of tourism.

Local

Local authorities are concerned with every aspect of the lives of people, and typically seek to provide a range of leisure facilities and services. In the past, during times of plenty, authorities were apt to build grand schemes such as theatres, swimming pools and sports centres and develop parks. Such capital projects can serve as something of a drain on declining revenues when times are hard. In keeping with current leisure management theory, more effort is now being focused on providing settings for activities and services rather than built facilities. Involving the private and voluntary sectors, working in partnership in delivering services that local people want, is increasingly important. Private Finance Initiatives, in which private capital replaces the funding previously supplied through public borrowing, has been increasingly applied, although there has as yet been little evaluation of the results (and a good deal of scepticism). Local authorities also have a major role through education.

The different roles of the private and public sectors in different countries are exemplified in the comparison of funding of orchestras in England and the United States (Figure 3.4). Similarly a comparison between England and Germany in the funding of public theatres is also revealing in pointing to the different roles of central and local government (Figure 3.5).

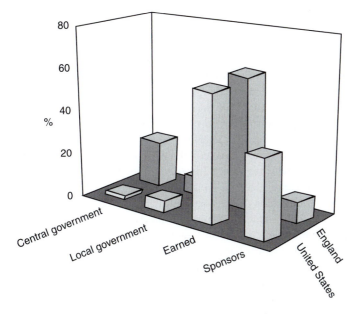

FIGURE 3.4 Funding of orchestras in England and the United States
Source: Feist (1988)

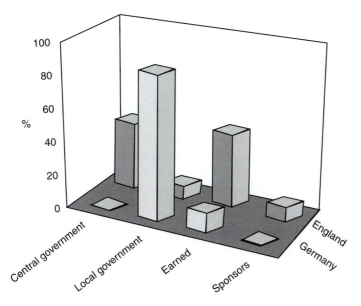

FIGURE 3.5 Funding of public theatres in Germany and of performing arts institutions in England
Source: Feist (1998)

Host community and host environment

Often, in the past, host communities have had the least power in the web to determine what should happen (Krippendorf 1987: 44–56). They have found their world changed beyond all recognition by developers and the effect of large numbers of people taking their recreation together, whether for just a few hours' or days' disturbance during an event, or for all of the time. A clash of cultures is often inevitable. Visitors seek the special characteristics of a site with natural and cultural attributes, which combine to create the character or sense of place. The balance may be easily upset. In the worst cases, the character will be destroyed by the sheer tide of visitors. People have never stood still for long. In ancient times migrations have been caused by the movements of game, a search for a suitable climate, and the need to flee from enemies. There are modern-day leisure equivalents: the search for wildlife to watch, different tastes in food, comfortable climate, and escape from everyday concerns.

Such migrations often have unforeseen consequences. Congestion may be a minor irritation for the family out for a weekend drive, but for the farmer this may be a crucial disruption to harvests that have to be brought in before any change in the weather. Other changes can be more severe, as with the pernicious spread of sexually transmitted diseases (Moorehead 1968). Depending on the themes which bind the community together and the activities which are essential for continued health and well-being, any mapping can be simplified to help focus attention on the crucial elements, or themes. The focus may be on primary production, where objectives may be in avoiding disturbance or adding to income from agro-tourism accommodation (Vaughan *et al.* 2000). Another focus might be on providing services where the income of the community is more directly dependent on the visitors but where the host culture may be under greater threat.

BOX 3.3 A tale of two resorts

St Moritz, in Switzerland, developed as a winter resort as a result of the enthusiasm for tobogganing, Alpine skiing, walking and the healthy atmosphere which British tourists in the nineteenth century found irresistible. Wave after wave of tourists arrived, and hotels inevitably followed. The setting of the resort beside the lake, which when frozen in winter provides a course for horse racing and carriage driving, is stunning. The resort is based on healthy atmosphere, spa waters and the availability of winter sports. The development of the Cresta, the toboggan run, in which people travel down a prepared run of more than a mile (and including some very tricky bends) at speeds up to 50 mph owes much to the stimulus provided by visiting tourists. The bobsleigh run followed a century later. Although resorts seem to take it in turn to be the fashionable venue for the wealthy and aristocratic, these alpine resorts rich in natural and cultural heritage seem to retain their magic against many odds.

Aviemore, in Scotland, was a very quiet village until the 1960s, when forces contrived to visit on the village a form of development that took little

---

account of the local vernacular architecture or indeed any Highland sensibilities or local people. The development was funded by public agencies working in concert with developers, to help provide a critical mass of recreation facilities and services to provide a focus to help (it was assumed) the local economy. The natural scenery, the northern corries of the Cairngorm mountains, Loch Morlich, the Spey valley and the remnants of the Caledonian Forest to be found in Rothiemurchus, Glenmore and Abernethy combine to provide a versatile outdoor playground with something for everyone. The development at Aviemore itself was constrained by budgets and the determination in the 1960s to build in a new style (which now appears brash and unsympathetic). Facilities were developed to provide alternatives for tourists, who were largely from an urban background, and attractions that could be managed to return profit. Arguments have continued to rage over the kinds of development which are appropriate, whether the skiing should be further developed with greater access to snow-holding areas. Approval was finally given in 1999 for development of the funicular railway in mountains of national significance. Indeed, the Cairngorms has been proposed as the second National Park of Scotland.

The web may extend to different environments, locations and scales. It is easy to view the web at a particular scale and remain entirely oblivious of another scale, of vital importance to communities and environments which are affected by the activities, rather than deeply involved in promoting them. The 'appropriate' scale for the map is usually set by the lead stakeholder group. The stakeholder may choose to share the map, but even so it will be drawn from the point of view of the powerful stakeholders. It is instructive to draw up alternative maps, from the points of view of other stakeholders. We also tend to view things over time scales that relate to human cycles, daily, seasonal, annual, and only occasionally periods beyond that. Rarely do we look at the effects on future generations. We naturally focus on direct or immediate effects, rather than indirect or cumulative ones.

Information: the threads in the web

Exchanges hold the web together. Our definition of leisure is shaped by our obligations, and by the revelation of the opportunities during time free from obligation. These opportunities can be seen as opportunity sets (Kent 1990: 42–62) shaped by a number of actors in the web, and certainly by those who stand to gain or lose (Stabler 1990: 23–41). But information about opportunities is conveyed by different people and through different media. Any communication has at least two components: meaning and emotion, and it is often the emotional component which attaches a value to the meaning. The human voice uses a wide waveband, between 10 Hz and 10,000 Hz (Young 1971: 537) to convey emotion as well as meaning. Music, which probably owes its origin to this emotional component of communication, can cause physiological

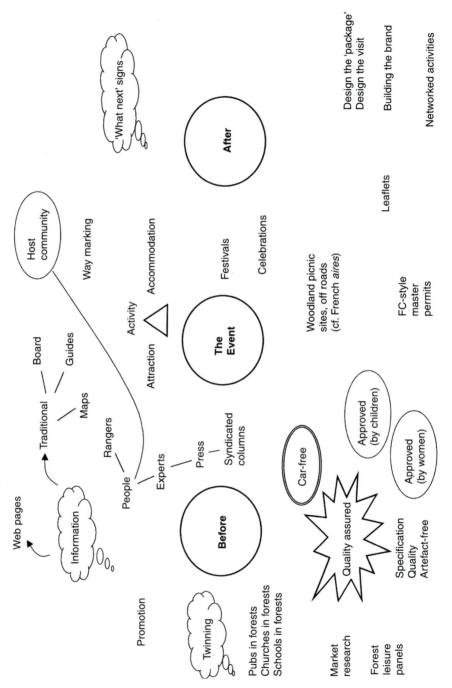

FIGURE 3.6 BEAR (Before–the Event–After Recreation) mapping, exploring interventions in visits to forests

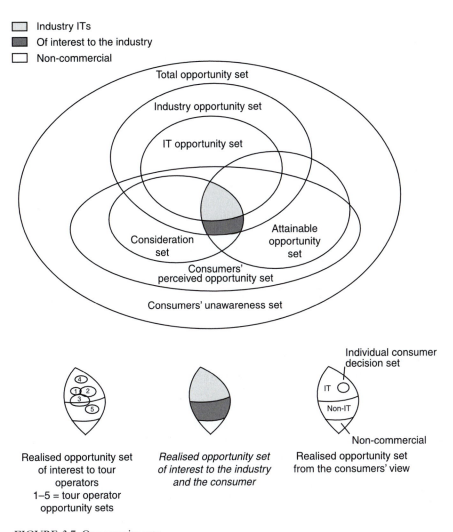

FIGURE 3.7 Opportunity sets
Source: Stabler (1990)

arousal (Storr 1997: 183). It may be in part because of this extra value of vocal communication that word of mouth is such an effective way of introducing people to new experiences or places to visit. It could also be that when communication is face to face, it is accompanied by body language, which also conveys a good deal of value-laden information. It may simply be something to do with the trust that builds up between the participant and the informer. Powerful though it is, we should be alive to the scope for misunderstandings. All the assumptions we make, the different identities we can own on different occasions (Zanger 1998), and hidden agendas all add to the potential difficulties.

Families (and extended family based social groups) are the first influence on the choices we make, which become framed by the accepted norms. Word of mouth from friends and family is a most potent source of information. Family and friends have a prominent position (Engel and Blackwell 1982: 143–86) within the web, and much recreation is taken in their company. Behaviour is a function of personality and environment. We have little choice to begin with which is not constrained by the value systems and beliefs of our parents. As we grow up, the extended family and friends begin to have a role to play in shaping our developing attitudes. These players open doors to all sorts of opportunities, some limited by circumstance, and others rich beyond compare.

As we shall see when we consider benefits, the products of recreation are not solely dependent on material wealth. The type of activity undertaken at leisure is shaped to some extent by social class, and lifestyle group. Family and friends remain, for most people, the most important influence, in suggesting or leading us into recreation activities. They may suggest taking up a particular sport or activity, or recommend a place to go walking, a book to read, a film to see, or a restaurant to eat at. Studies of consumer behaviour have produced models (Engel and Blackwell 1982: 677–90) that help to explain how we make choices based on the information before us. It is helpful to recognise the development of variety (and the complexities) in the structure of families today (Shaw 1997: 98–112), and not focus solely on traditional notions of family.

It is as though we have a library of books, many bequeathed by our close family and friends. Gifts or introductions from other sources once made are incorporated into our library. Books and recommendations are exchanged among members of the family and close friends. Our tastes develop. Most of us are creatures of habit. For people making visits to the countryside in the United Kingdom, two in three people return to the same place again and again (Social and Community Planning Research 1997). Most people take up sports or interests and focus on these over years, if not a lifetime. Older people, empty-nesters and the retired have well developed networks for sharing information. This will rely heavily on family and close friends as the major source of recommendation.

As we develop and grow, our social circle enlarges, and we gradually spend more time away from our families. This development proceeds for each of us at a different rate, and with a number of step transitions. The greatest wrench is probably when we go to school. Consumer behaviourists and others (Engel and Blackwell 1982: 167–72; Rapoport and Rapoport 1975) provide a framework that marks different stages within this development. School and the playground provide us with our first taste of peer review, as well as an opportunity to explore choices that might otherwise never occur to us. The repertoire of recreation opportunities available expands, and important patterns are established (Curtis *et al.* 1999). There will be a number of stepwise transitions, typically primary, secondary and, for an increasing proportion, tertiary education. This last step may involve leaving home to stay at college or university, at least during term time, providing a transition into adulthood. Even where mobility has helped to break down the traditional social structures, and young adults leave (often for jobs well away from home), these years in education are likely to

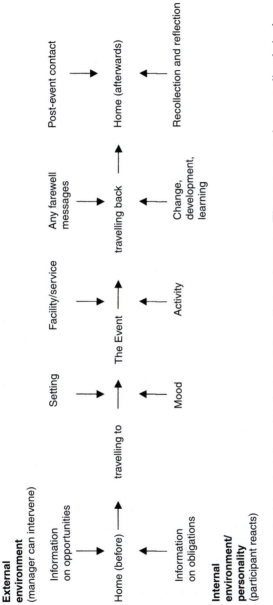

FIGURE 3.8 Mapping the information flows to explore where to intervene. The manager can intervene directly in the 'external environment' (upper) but only indirectly to bring about change in the (necessarily internal) participants' world (lower)

identify lifelong friends who will continue to play an important part in recreation behaviour.

And, in later life, work takes on a similar role, providing settings for social interaction among new colleagues. Work provides moments for leisure and recreation during breaks, and new friends to share activities outside work, giving the chance of exposure to different value systems and experiences. A new home is set up. When it comes to rearing children, the woman usually takes on most of the domestic duties, and enters what has been described as 'the tunnel of motherhood'. Opportunities for anything other than looking after the home and the family appear to be greatly constrained, partly biologically and certainly culturally (Deem 1995: 3–22). As the family develops, the new parents subsume the role of guardians of the next generation.

For certain decisions, experts will have a crucial role in the decision-making process, because they alone have the technical knowledge necessary to inform the process. Coaches in sport influence the way their charges develop their skills. This is particularly noteworthy in professional sport. Although generally not so newsworthy, the influence in amateur sports of all kinds of club officials, and of sports instructors during education is enormously important. Rangers, keepers of museums and galleries, librarians and all the other leisure professionals have a similar part to play. The influences may be given in simple conversation, or through world wide broadcast. Teachers in leisure, sport, the arts, and all other subjects will have affected their students in some way. This influence often lasts for life.

Impressionable youngsters are encouraged by rewarding experiences, and discouraged by experiences that fail to satisfy, or produce unpleasant sensations. Theories of education point to the power of experiential learning. Information gained through experience will assume relevance and importance, because it helps to build on (or be assimilated into) the model of reality that the individual has developed. This places a responsibility on the shoulders of those who introduce youngsters to any particular leisure activities. It encourages us to see the value of teachers, community leaders, parents and leisure professionals in managing the early stages of the process.

There are two distinct approaches that will have a bearing on our interest: that which focuses on the environment, and that which focuses on leisure and recreation activities. The first is usually referred to as environmental education. The second fails to fit into a neat overall box, but includes physical and sports education, music, art and outdoor education. In outdoor education, practice usually includes reference to the environment, to a greater or lesser degree. There is scope for greater integration, in the same way that education should not be restricted to the classroom, but should recognise the different contexts of learning, 'home, community, leisure and recreation, school, post-school education, the workplace' which were identified in 'learning for life' (Smyth 1998: 9).

Many different media are involved in transmitting information, some specific to each activity, participant, location and occasion. Effective publicity is all about identifying the target of communication, the message and the blend of media to be used over what period of time, at what cost, frequency and to what effect. There are many effective sources of advice on how to set about planning and achieving the delivery of

---

BOX 3.4 Edinburgh, the festival city

Edinburgh plays host to a number of festivals. It all started in 1948, with the first International Festival. Now there are festivals almost throughout the year, including:

- Hogmanay (New Year)
- International Science Festival (April)
- International Children's Festival (end of May/beginning of June)
- International Jazz and Blues Festival (end of July/beginning of August)
- International Film Festival (August)
- Fringe festival (August)
- Tattoo (August)
- International Festival of the Arts (August/September)
- International Book Festival (August/September)
- Folk Festival (November)

There are some 1,000 events in the Fringe festival alone. There are many unofficial events with brilliant mime artists, jugglers and every conceivable form of street theatre. The city swells to receive visitors to the International Festival and the Fringe, which contribute something like £9 million to the local economy, or £44 million of direct expenditure, and for Scotland as a whole, some £72 million. This creates jobs that equate with 3,000 full-time equivalents. People find out what is on offer through a wide range of media. The Festival organisers produce and circulate comprehensive programmes, which are taken up and magnified by all the media. Promoters flood the city and paste handbills and posters on any promising surface. Newspaper supplements and critics give their expert opinions, often awarding star ratings to help give a summary indication of value. Radio and television stations draw attention to the shows, performances and exhibitions they consider worthy. Individuals chat to each other, through their personal networks, as well as festival networks, swapping anecdotal evidence in restaurants, cafés, bars and public houses throughout the city. Word of mouth may be the most important element in this exercise, but almost all media are applied to make this major event entirely memorable for all concerned.

---

information (Denman 1994). In many cases, the web will only link the different players by means of the information flows, which allow decisions and actions to be taken.

In relation to countryside recreation and heritage presentation, the word 'interpretation' is used to describe the approach that conveys the significance of a particular site. Interpretation relies heavily on involving the participant in appreciation developed, as far as is practicable, from personal experience, the heuristic approach.

Media are chosen (Aldridge 1975: 16–20) to help the simulation of realism, and in relation to the interpretation of history (Binks *et al.* 1988), by bringing it alive through re-enactment or dynamic storytelling. Interpretation relies heavily on both components of communication: the factual and the emotional (Uzzell and Ballantyne 1998). Inappropriate communication can spoil experiences by conveying the wrong emotion. When people are exposed to experiences which stimulate a feeling of awe, any attempt at communication will usually appear clumsy. Under these circumstances, silence and an absence of commentary is the only solution (Aldridge 1975: 2). The expertise of psychologists has much to offer the practitioner to help match the experience to the effect sought (Uzzell 1998: 242), and tools to evaluate the success (Lee 1998).

People engaged in stimulating interest in the cultural or natural heritage are realising the great opportunities for leverage which can be gained by imaginative use of the arts in a wide range of projects (Carter and Masters 1998: 37–8). The power of the media in influencing where people go and what they do is reflected in the influence of the screening of films and subsequent changes in patterns of visits to film locations, with increases of 40–50 per cent over four or five years not uncommon for best-selling films.

The role of information

Psychologists and consumer behaviourists have worked hard to unravel how attitudes are formed and what influences us, as have philosophers (Swinburne 1974: 103–21). Family, friends, teachers, colleagues, experts, media and experience all have a part to play. In short, everyone in the recreation web (whether mapped or not) will have an effect on us. People can be thought of as dynamic information processing systems, continually adjusting and adapting to their surroundings, to the biophysical and socio-economic environment. We are voracious in our appetite for information, and often attempt to take in information from a number of sources, simultaneously (Rojek 1993: 216). Some information provides fuel for instant decisions, between different choices, taken under conditions of low levels of involvement. Other decisions require a higher level of involvement (Engel and Blackwell 1982: 21–40), and draw on memory and attitudes to provide a framework to aid choice. Our belief systems build incrementally from information coming in from a number of sources, to provide a model of the world. The information is stored in memory and contributes to the development of attitudes. An attitude is a learned predisposition to react in a consistent way towards given attitude objects. Our perceptions will be affected by moods, and these can be affected by the moods of others. They can also be affected by chemical balances, induced by hormones, or by drugs.

Our perceptions may alter according to mood, age, stage in the life cycle or weather, among other things, but attitudes are more constant. These attitudes represent a reasonably stable set of beliefs, some of which become deep-seated, and not easily overturned. Our attitudes are also crucially affected by the opinions of other people. Beliefs give rise to attitudes, and in turn attitudes give rise to intentions, and intentions are converted into behaviour. This causal relationship has been well tested (Fishbein and Azjen 1975; Tuck 1976: 92–101) and elegantly described (Tuck 1976: 74–91). Fishbein's theory states that the behavioural intention is dependent on the attitude to the act under consideration and the subjective norm. The attitude to the act is the degree to which the subject believes that the act will lead to a positive outcome. The subjective norm is the extent to which the subject believes that people important to the subject will approve of the act.

Individuals also belong to dynamic groups, to networks and to communities (of neighbourhood and interest). Under these circumstances they may behave differently, and in some cases may take on specialist roles to serve the combined interests of the group. Not everyone in the web will behave predictably, but there are enough examples to reinforce our views of what people are likely to do in any one situation. Some of us are better at putting ourselves in the position of others and thinking through or predicting what they may do. The more we know about the different actors the easier it becomes to think (or act) out their roles. Thinking our way into the way people may act or react is important in testing out ideas (de Bono 1983: 191). One way to do this is to use a disciplined approach, working through the different tendencies people have: optimistic, pessimistic, analytical, lateral and so on (de Bono 1990a). Making a map of the web helps our understanding of how the actors relate to each other and how the play may unfold.

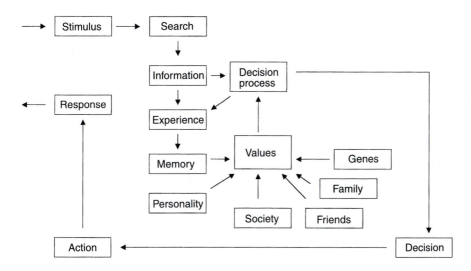

FIGURE 3.9 Influences on decisions
Source: Adapted and simplified from Engel and Blackwell (1982) to show the various forces acting on the beliefs, attitudes and intentions which help to shape our values

Mapping the web

Work through these simple steps:

- What, at first, seems to be the problem or opportunity? What is the reason for the manager to intervene?
- Who is the principal participant, audience, visitor or group of visitors?
- How do people engage in the activity, in what groups, and where are the linkages?
- Who holds or controls the resources?
- Within the recreation chain or web, who has the most power to supply or deny the leisure experience?
- Who can help the process along, speed it up or slow it down?
- What determines the speed of delivery? Who is involved?
- Who stands to gain, and by what? (There is usually a network of people whose interest is served by the principal participant's enjoyment of his or her leisure.)
- Who stands to lose, and from what? (There are sometimes people in the chain or web, who bear some of the costs – disturbance, opportunity costs, damage, and in extreme cases, injury – and who are not necessarily recompensed.)
- Now, on reflection, what is the real leisure or recreation problem or opportunity?
- Why is the intervention necessary?
- Where will the manager intervene, and in what way?

A more informative (and complex) map will result from considering the issue from the viewpoints of the different actors in the web; some complexity will need to be sacrificed to keep the model workable. Many people find it easier to work on such problems using diagrams and flow charts, but each individual will have a preferred way of working. You can either choose the (left brained) logical mapping steps described above, or choose a more (right brained) intuitive style. Set down your ideas on a single piece of paper, of a size that you feel comfortable with, using mind maps (Buzan 1974, 1993; Russell 1979). Or, if you prefer working on a larger scale, move pieces of paper about the floor or wall (using adhesives with care!) (Straker 1997). Start by setting down the components of the problem or opportunity. Add in factors that increase or decrease the likelihood of any result. Work on this until you have a measure of agreement about what factors work, in which direction, and in which way (de Bono 1994; Kosko 1994).

The principal participants will have different requirements depending on a variety of factors, including the availability of time and money, as well as motivation. The time–money matrix helps identify some of the scenarios that different combinations can give rise to. While it may be natural to think that more time and more money will lead to extra portions of recreation and therefore of happiness, experience seems to suggest that this is not necessarily so. However, when time or money is in very short supply, survival and other needs predominate, as suggested by Maslow and others (Gross 1996: 95–118). Seen from the individual's point of

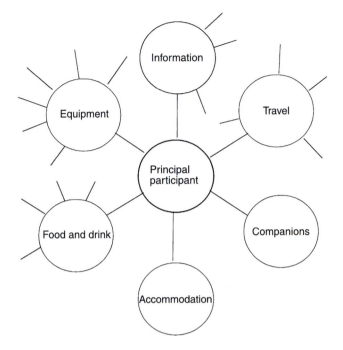

FIGURE 3.10 Participant-centred mapping. In this approach the diagram works outward from the participant and allows exploration of his or her different requirements

view, the time-poor/money-poor combination provides fewest opportunities. From society's point of view, this sector may provide the greatest opportunities, where a relatively small amount of effort will lead to comparatively greater payback. From the viewpoint of individuals, the time-rich/money-rich sector will provide many opportunities to enjoy leisure. Here, arguably, society need have least concern, as the market should deliver what is required. A further variable to consider is health. Health certainly affects the degree to which different age groups can fully engage in leisure and recreation activities. The focus of our argument is that the full range and extent of benefits and costs are very rarely taken into account. Society will have an interest in almost all cases. Matrices suggest a digital or black-and-white world, whereas experience reflects a more variable, analogue and even fractal world. Trace back what factors influence our own decisions (concerning leisure and tourism) to reveal how sensitive such decisions may be to initial conditions (Gleick 1988: 11–31).

To enjoy chosen activities participants must have access to the components that together give rise to the experience. Resource holders, who have the power to grant or deny access, are affected by agents (facilitators or catalysts) who enable recreation to take place, speed up or slow the rate of activity depending on their effort, or concentration. Some principals will be actively seeking opportunities for leisure and recreation. Others may have obligations which release blocks of time sporadically (Kay

1996: 143–59) or be unwittingly involved with leisure, as a collection of smaller events, interwoven as part of everyday life, and sometimes taken for granted. Once we have mapped the recreation process and the web of actors and interactions, we stand a better chance of identifying the really crucial or critical elements of the web. Maps can be developed to illustrate what happens over time, remembering that leisure and recreation are multi-phasic (Stewart, W. 1998: 390–400; Hammit 1980: 115–30). Mapping will help you identify some common critical points, and more importantly show you how you can develop your techniques to help unravel and understand the issues you face.

Most of life depends on transferring information, whether in the form of genes, from one generation to another, or through linguistic and non-linguistic communication. Recreation is no exception. Some information is critical. Only if an individual is aware of an opportunity does it exist. Some actors provide information, some modify it and still others withhold it. All these can be critical. Information on its own does not make the opportunity real. The resource holders – such as the owners of the natural resources on which countryside recreation depends, or the sports industries that make the equipment which participants will use – have enough power to permit or deny access (Shoard 1980, 1987, 1999; Wightman 1996, 1999). At the global level, it has been recognised that in respect of environment and development generally, more needs to be done to provide stronger linkage between the different parts of the web, to reduce the disparities between the northern and southern hemispheres, and between and within nations. Pointing out the importance of education, both formal and informal (UNEP 1999, Agenda 21, chapter 36) and collection and sharing of information (ibid. chapter 40). Whether at the global, macro or micro level, more effective communication, throughout the web, will improve our situation (if we communicate the right information).

Ideas for further study or work

1   Select any agency or organisation that has a place in the web, or influence over leisure and recreation, and identify how the influences take effect. How has this changed over time?

2   From the point of view of a participant, reflect on any recent holiday or recreation event, and construct a mind map to show the interactions between the many players. What were the critical elements, which made the difference? Try constructing maps from the points of view of others in the web.

3   Select a group (or community) affected by leisure and recreation, and work among them to generate a jointly prepared map. Ensure that the rich diversity of views is expressed, without losing the opportunity to work for consensus, if that is the desire of the group.

4   Select an event, and carry out a BEAR mapping exercise to reveal the information flows: before, at the event and afterwards. Identify the strength of flows and critical points at which influence can be exerted on particular players.

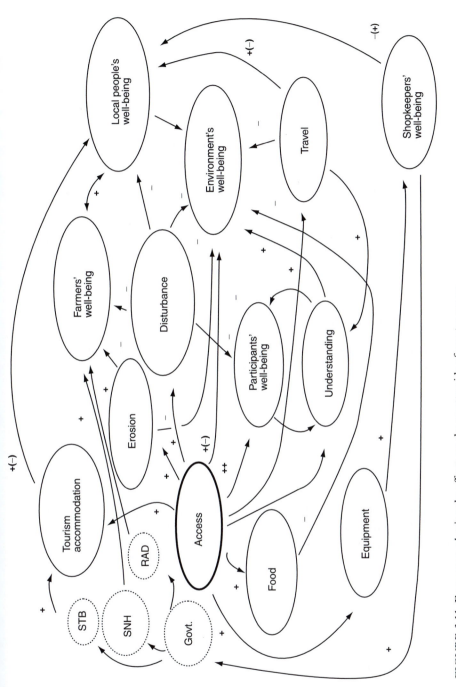

FIGURE 3.11 Fuzzy map plotting the effects on the countryside of greater access

Note: RAD = Rural Affairs Department, SNH = Scottish National Heritage, STB = Scottish Tourist Board

5   Governments have (since Roman times, at least) exerted considerable influence over how people take their leisure and recreation. What are the motives for governments to intervene? How do they intervene?

Reading

Actors and players

In respect of actors in the public sector (and the voluntary sector too), the motives are often recorded as aims, goals and objectives in policy papers, corporate plans, annual reports. In the past this was often recorded in rather dry documents, but now more often there is up-to-the-minute detail on the web sites. Occasionally you may be lucky enough to find the history of such an organisation, but with the current pace of organisational change, this is may be rather less likely in the future. For the motives of individuals, read everything you can. It is certainly difficult to tease out the 'real' reasons why people behave as they do, as will become apparent when research is considered. Biographies are also a great source.

Information and markets

Kotler provides a standard, thorough and accessible text on marketing, which among other things provides a framework for understanding the way in which different parts of the chain or web interact. Study the best writers, speakers and journalists to explore how information is transmitted. For an insight into the world of advertising, try Ogilvy.

Mapping

For basic mind mapping, Buzan and Russell are both excellent starting points. For development of flow charts and diagrams, read also de Bono, and for development of flow charts which show values and relationships, Kosko. Straker shows how to solve problems and keep the maps fluid, by using post-its which can be moved at any time.

Chapter 4

Benefits, costs,
settings and trends

Overview

Every action results in some sort of redistribution of benefits and costs. One person's loss or cost may be another's gain or benefit. Some benefits can be grossed up for the good of a community, while others are seen more in the realm of the individual. To encourage more thought about the effects of our interventions, a basic typology of benefits is presented.

Personal benefits include those relating to physical, mental and emotional health; enjoyment and human development; as well as quality of life. We touch on different aspects of happiness, from the meaningful and powerful depths of flow experiences to the apparently more frivolous superficial levels of laughter. Societal benefits comprise more than just the sum of all the personal benefits, and include reducing and avoiding dysfunctional behaviours. Social cohesion arises from individuals sharing experiences and building on each other's skills to create something altogether larger, whether at the level of family, groups of friends or community.

Another realm of benefits is concerned with human development. Play is crucial to the development of child into adult, and playfulness itself is an essential ingredient for adults. Leisure and recreation provide a vehicle for lifelong learning.

The environmental costs (or impacts) are to some extent dependent on the requirements of participants from the physical, social and emotional environments. Some activities depend on the physical characteristics of the environment, and others much more on the social characteristics. Managers need to be sensitive (and responsive) to what people require.

At times, people seek out personal spaces, and mountains have always held a special place in people's affections for their spiritual qualities. In countries without high mountains, other places are important: moorland, forest and coast among them. Closer to towns, playing fields provide special places for people to enjoy being outdoors, being active and meeting others. With increasing pressure on spaces in towns, and increasing land values, managers and planners have been focusing on outputs and making these areas give more enjoyment per hectare per minute, taking advantage of technological gain. In some towns the requirements of the motor car have tended to dominate, but as car ownership and levels of traffic have increased, so too has the desire to win back space for pedestrians, cyclists and horse riders.

A raft of social trends is changing the way we look at costs and benefits, and at settings. There are a number of contradictory trends, because we still seem to be fragmenting, with different groups within society looking for quite different outputs and outcomes from leisure and recreation. The effects of some trends, such as the ageing of the population in some Western economies, have been difficult to appraise. In the past, many groups (apart from white middle-aged men) have not been well catered for, and laws and custom are changing as we become more inclusive. The disability discrimination legislation has changed our approach, and increasingly, more effort is being expended on seeking out the views of all groups.

Leisure or recreation activity generates benefits and costs. All activity will result in changes to the environment – either to the socio-economic or to the bio-physical environment, and more often to both. These changes or impacts will result in a redistribution of costs and benefits. Each change will be perceived differently depending on the point of view of the individual concerned – whether resource-holder, participant, agent or other. This chapter deals with the social, economic and environmental costs and benefits, for individual and society.

A framework for considering benefits and costs

Work which describes needs and motivations is extensive. However, 'the language is difficult and there is no clear agreement on how many human needs there are, or what their relative importance might be' (Driver, Tinsley *et al.* 1991: 283). There are many frameworks which seek to bring order to the array of possible benefits, such as the Paragraphs about Leisure scales or the Recreation Experience Preference scales which include forty-four components (Driver, Brown *et al.* 1991). There have been formidable attempts to review benefits, particularly from the late 1980s onwards (Driver, Brown *et al.* 1991; Parks and Recreation Federation of Ontario 1992). This has developed with a further publication, the *Benefits Catalogue* (Canada Parks/ Recreation Association 1999) and the establishment in Canada of a National Benefits Network, linking the increasing number of people using the Benefits Based Approach. The *Benefits Catalogue* has sought to arrange the benefits around eight marketing messages, to provide a user-friendly catalogue accessible to professionals working in different fields. The messages focus around:

- personal health and well-being;
- balanced human development;
- quality of life;
- reducing self-destructive and antisocial behaviour;
- social cohesion;
- avoiding, or reducing, expenditure on health care, social services, policing and justice;
- contributing to economic regeneration;
- contribution to ecological functioning.

Sport England has also contributed to the advocacy of recreation, in relation to the value of sport to the nation, in social, economic and environmental terms (Sport England 1999a–d). Without quantification and evaluation, provision was likely to remain undervalued and consequently under-resourced (Smith 2000).

Different activities will be valued differently by each individual. A rock climb can be for some individuals the nearest thing to heaven, and to others the embodiment of hell on earth. Our preferences determine whether any change or activity is seen as a

TABLE 4.1 Typology of leisure benefits, related to activities and needs

Need (Maslow's hierarchy)	Personal benefit (need satisfied)	Social benefit	Activity
Level 1: Physiological			
Food and drink	Survival Taste	Survival Socialisation	Picnic Meal out Visit to the café Visit to the pub
Oxygen Clean air	Survival	Survival	Walk in the fresh air
Sunlight	Survival	Survival	Sit in the sun
Temperature regulation	Comfort	Comfort	Swim/paddle to cool down Find beach with cool breeze, on hot days Sit in the shade
Activity and rest	Survival Well-being		Walk, run, jump Sit, lie down, sleep
Sex	Satisfaction, ecstasy	Survival	Use your imagination!
Level 2: Safety			
Physical safety, avoiding risk	Survival	Survival	Climb with ropes
Exposure to risk	Learning and development	Development	Free climbing
Avoiding fear, anxiety	Survival	Survival	Learn under instruction Test skills
Exposure to fear, anxiety	Learning and development	Development	Seek thrills

Level 3: Love and affiliation			
Group affiliation: Family Colleagues Friends Neighbourhood Communities	Mental (emotional) health and therefore physical health (absence of stress) when freedoms balanced with obligations	Social cohesion, and therefore survival	The art of conversation, communication, and contact Play with the family Joke with colleagues Share experiences with and learn about other individuals and societies – sport, the arts, countryside and travel
Trust	Mental (emotional) health, avoidance of stress	Efficient functioning of groups and society	Co-operating and working together in teams
Giving and receiving affection, and love	Mental (emotional) health, avoidance of stress, and selection of mate	Efficient functioning of groups and society, and therefore survival	Through expression, play and interactions of all kinds
Level 4: Esteem			
Self esteem – a sense of competence	Development of skills Mental (emotional) health	Development	Increasing skill level in any activity, from art to zoology
Esteem of others – respect	Mental (emotional) health	Cohesion	Exhibiting skill in competition or co-operation
Level 5: Self-actualisation			
Knowledge and understanding	Mental (emotional) health Striving for meaning, purpose, religion	Development	Study, awareness, appreciation, understanding
Curiosity, leading to exploration and discovery	Stimulation through novelty and development	Development	Physical travel, meeting new people, being in new places, new cultures, new surroundings Mental and spiritual travel, searching for meaning and reason, religious beliefs

TABLE 4.1 – continued

Need (Maslow's hierarchy)	Personal benefit (need satisfied)	Social benefit	Activity
Appreciation of beauty – in art and nature Form, colour Symmetry, rhythm Balance Order, pattern Harmony	Pleasure and mental (emotional) health	Development	Look at nature whilst walking; admire a painting, whether abstract or not; enjoy the sky, the colours of the flowers; enjoy the rhythm of a poem, the sound of waves crashing on the shore; the smells of pine woods, new-mown hay . . .
Expression through creative activity and communication	Pleasure, learning and development, enhancing self-awareness and relation to the natural and cultural setting	Pleasure, development of ideas, development of culture	Write, paint, photograph, dance, act or engage in a thousand different activities
Realising potential, and therefore development to provide greater potential still – the Holy Grail	Intense pleasure, and sense of fulfilment Physical, mental, emotional and spiritual health	Sustainable development, and survival	The full range and depth of developing activity; the sum of all the leisure activities carried out (mind-centred, body-centred, and with others) in natural and man-made environments.

benefit or a cost. Generally, though, as individuals, we choose to take part in activities that give us a range of personal benefits which may improve our health and sense of well-being: our physical, mental, emotional and spiritual health. Often we take part in activities because they just make us feel good, almost at an instant, and are more concerned with stress reduction, stress avoidance and the generation of sheer pleasure. Other forms of activity may contribute to long-term development. Play is essential to allow for the complete development of child into adult. The adult strives to develop his or her potential, to achieve fulfilment through self-expression (or self-actualisation). In each and every case, we are making a long-term investment, as the experience not only gives pleasure now, but plays a part in our development over time.

Personal benefits

Physical health

Increasingly (in the developed countries of the West) people are spending more time working at desks, facing computers or in otherwise physically inactive settings, studying, at work in offices and shops, at leisure in the home or travelling by car, train or plane. The human frame is not suited to being inactive for long periods of time, or to sitting in chairs (Cranz 1998), although designers are increasingly deploying ergonomics to help us keep postures that will avoid the worst problems associated with our sedentary lifestyles. This is exacerbated in many Western countries by the increasing number of journeys made by car, rather than on foot or by cycle. Physical recreation is of enormous importance in promoting movement (Cashmore 1990: 57–78), and in maintaining health (Physical Activity Task Force, 1995) and in avoiding and reducing stress (Glyptis 1993: 4). Of course, sport and recreation are of themselves neither good nor bad. For physical health, a programme of sport activity is required, and for enjoyment the participant's skills need to be in balance with the challenge faced (Csikszentmihalyi 1992: 71; Wankel and Berger 1991: 121–44).

Children need physical activity to develop cardiovascular systems, muscles, joint function and motor skills, quite aside from social and mental attributes. In adults, activity promotes good health, prevents ill health, enhances mental health and well-being, and maintains and enhances physical capacity. Specific exercises can be applied to reduce the severity of particular disorders, and minimise the deleterious effects of ageing. Continued activity keeps systems working and improves the ability of the individual to deal with the physical demands of everyday life (Bird *et al.* 1998: 1–16). Of course, physical activity and sport include risks (Lee 1981: 5–16) and hazards (MacAuley 1999) as well as benefits, but the benefits–risk ratio is likely to be very high (Ball 1998). A few people become addicted to particular kinds of exercise, and to the endorphins produced naturally in the process, but for the most part physical exercise taken as recreation does enormous, and for the most part unmeasured, good.

TABLE 4.2 Activities alphabet, to prompt choice

Sport	Art	Outdoors	In-house
Athletics	Act	Abseiling	Arm wrestling
Badminton	Batik	Bicycling	Bottle collecting
Cricket	Composing or carving	Canoeing	Conversation
Decathlon	Dance	Drawing	Daydreaming
Exercise	Etching	Expeditions	Eating
Football (Rugby and Association)	Figurative drawing	Fishing	Forfeit
Golf or gymnastics	Gilding	Gardening	Games (cards, board games and gambling)
Hockey	Handicrafts	Horse riding	Humming
Ice-skating	Instrument playing	Ice climbing	Imagine
Judo	Juggling	Jogging	Joking
Kick-boxing	Knotting	Kite flying	Kissing
Long jump	Lithography	Langlauf skiing	Laughing
Marathon	Mime	Mountaineering	Model making
Netball	Naïve art	Nature study	Nothing much
Orienteering	Opera	Open-air picnic	Ornery things
Polo	Painting	Photography	Play
Quoits	Quoting Shakespeare	Quad-biking	Quizzes
Running	Recording	Rafting	Reading (and Radio)
Shinty	Singing: choral, opera, lieder, jazz, popular	Swimming	Storytelling
Tennis	Theatre	Tai-chi	Television
Underwater polo?	Understanding hidden meanings	Underwater swimming – subaqua	Unwinding
Volleyball	Video filming	Very quiet wildlife watching	Video watching
Wrestling	Writing: letters, prose, poetry, plays	Walking	Whistling
X-country running	Xylophone playing	X-country skiing	X-word
Yachting	Yodelling	Yachting again!	Yoga
Zooming	Zither playing	Zoological photography	Zzzzz!

Mental health

People in developed countries now live longer and enjoy much better physical health than ever before. Life expectancy has dramatically changed; in 1900 it was just forty-seven years. The proportion of people over 65 years of age is steadily increasing. In 1900 only 4 per cent of the population was over 65 but by 1990 12 per cent were over 65; and this is expected to rise to more than 20 per cent in 2020. But longer life is not necessarily accompanied by better mental health (Greenfield 1997: 3). Mental health is not just a concern among older people. Urbanisation, together with the speeding up of communications and transport that drives development, also brings pressures, especially among those in work, and those commuting to work. Not everyone is so well equipped to deal with the resulting time famine. The endless striving towards greater or more intense activity, the continual change, the time pressure, the greater mobility and consequent collapse of the natural order of society have served to put unbearable pressure on some people. No longer do people live with the seasons, relying on the natural length of the day to govern what is done. Our ability to generate light and heat has made us physically independent of the natural clock (Future Foundation 1998). Surveys in the United States, Canada and Europe have all indicated the need to find ways of reducing the pressure of time (Goodale and Cooper 1991: 34), and some (more than half in a European survey) have also indicated a preferred choice for more time rather than more money. Taking part in sport and recreation can help reduce stress and increase self-esteem (Bird *et al.* 1998: 49–61), although the portrayal of increasing professionalism may make this a less attractive proposition for people who merely want to enjoy a sport, without the pressure to excel. Active participation in recreation activities of all kinds promotes good health, whether exercise for physical health, or participation in the arts or other activities for mental health. The call has gone out for a closer relationship between the leisure and health professionals. Walks are being prescribed in some areas of Britain, and more than 300 general practitioner referral schemes operate in leisure centres in England (Emmet 1999).

Leisure can be an important means of establishing self-identity, especially since the decline of many institutions (Clark *et al.* 1994: 30–1), and recreation provides symbols for the development of an identity image (Haggard and Williams 1991: 110). Physical health and mental heath are intertwined, and many authors point to the holistic nature of the benefits (Chubb and Chubb 1981: 5). The evidence for such health benefits is growing (Fentem *et al.* 1990; Driver *et al.* 1991a; Canada Parks/ Recreation Association 1999).

Emotional health

Emotions are complex reactions to stimuli, involving a synthesis of activity in the brain with the release of neurochemical transmitters, and of activity in the body with the release of hormones and physiological responses. There are some 400 words for emotions (Stewart S. 1998: 10). Included in this list are fear and anger, and such emotions will be accompanied by physiological responses, like increased heart rate and blood pressure. Smiling and laughter can be potent in defusing fear and stress, and so help the immune system to resist erosion by stress (Hodgkinson 1987: 14–17). The escape from stress, through the search for peace and tranquillity, can take the form of tourism, and visiting the countryside and peaceful settings. One survey in England revealed that most of the benefits of visiting the countryside were perceived as being psychological benefits (HPI Research Group 1997). Settings which include the coast, running water, hills and wind are favoured. Listening to natural sounds can be helpful. An alternative is to search for inner peace through listening to music, or losing oneself in self-expression whether through painting, making music or other artistic activity. It seems that certain patterns and rhythms of music work better than others, and these effects are well known to musicians (Robertson 1994; Storr 1997: 25–31, 67–76). Recreation provides relief from stress, which can otherwise become unbearable (literally, where stress greatly contributes to the risk of heart disease). Self-expression and creativity are important means of expressing emotions. This is well understood in therapy, and it is an area where we could focus more attention to ensure that people have the opportunities they need.

Leisure must allow for fun and frivolity, but its underlying purpose is anything but frivolous. It is vital. Even where the Protestant work ethic is still alive and well, and where people are fully employed, lives are not necessarily centred on work. Individuals often now have less allegiance to work, the Church, the family and other institutions. In these circumstances, people may centre their lives on leisure, and begin to seek an identity and a lifestyle described by what they do in their free time (Clark *et al.* 1994), and perhaps because in this area they have not ceded control to someone (or something) else.

Happiness

We all aspire to be in a state of happiness. The US constitution goes so far as recording the pursuit of happiness as an inalienable right (p. 38). While we all may wish to pursue it, comprehensive definitions of what it amounts to are rather harder to come by. Contentment, being at ease, satisfaction, well-being, and joy are some of the words that people use to describe it. Argyle (1996) suggests three aspects worth focusing on:

- satisfaction with different domains of life;
- joy;
- absence of distress.

and goes on to reveal what most of us intuitively know, that looking at things in a positive light generally seems to make people happier. However, continually trying to see things in a positive way when all the other signals reaching the brain suggest that the outlook is rather bleak may actually be unhelpful. The dissonance so produced, unsurprisingly, leads to stress. There is some evidence that richer people are margin-ally happier than poorer people, presumably because of the reduction in levels of distress, but what seems to matter more is the status or self-esteem which comes about from relative differences in wealth within peer groups (Tietenberg 1994: 383). Happiness includes elements that relate to intentions being carried through to reach achievements, and expectations met to yield satisfaction; avoiding stress and distress; and intellectual and social motivations (Beard and Ragheb 1983: 19–28; Murray and Nakajima 1999: 57–65). This happiness will be mediated by chemical and electrical activity in our brains. We may not easily be able to map the processes, but we have developed drugs that can restore a balance where it has been lost, to treat mental health. The procedures are still rather crude. We can also manage settings (physical, social and psychological) and produce environments that do not induce stress, and which enhance the likelihood of a happy outcome.

Csikszentmihalyi, in the preface to his book exploring the psychology of happiness, writes of the 'positive aspects of human experience – joy, creativity, the process of total involvement with life I call *flow*' (1992). His book is concerned with the more achievement-satisfaction oriented happiness, and less with happiness associated with relaxation, frivolity or fun. One of the most telling images in his book is a graph (1992: 74) which summarises the hypothesis he propounds, and explains why he believes that the complexity of consciousness increases as a result of the flow experiences. His ideas follow on from interviews with people engaged in strenuous and demanding physical recreation, but this approach would seem to fit other activities so long as they are totally absorbing, whether painting and similar creative work, or for that matter paid work (Rojek 2000: 1–15) for many people. In an earlier work he pointed out that there were mini or micro-flow occasions, often occurring in the interstices of busy days (Csikszentmihalyi 1975: 201), perhaps in a coffee break, or other conversation. The characteristics of the peak experience have been a constant source of fascination (Maslow 1967: 20–35; Lipscombe 1999: 267–88).

Some happiness is rooted in simple achievements and the resulting contentment,

	Low challenge	High challenge
Low skill	**Flow (or peak experiences) possible**	Anxiety, fear, tension
High skill	Boredom	**Flow (or peak experiences) possible**

FIGURE 4.1 Skills and challenge matrix: conditions for peak experiences or 'flow'. As participants become more skilled they generally seek greater challenges until they come close to, or are enveloped by, 'flow'
Source: Based on Csikszentmihalyi 1992

giving the relaxation after the previous tension. More fundamental still are the pleasures associated with the senses, those concerning sight, smell, taste, hearing, and of course touch. Many of these seem tuned to our physiological needs, the first (rather than lowest) level of Maslow's hierarchy, and these are of enormous importance to our survival. The taste of food and drink and the pleasures associated with touch, as well as sex (Oppermann 1999; Valentine 1999), are included in this group. Whilst this book does not directly seek to show how we can manage environments to enhance such pleasures, there is no intention to diminish the perception of their worth.

Smiles, laughter and humour

One of the most common manifestations of happiness is the smile. Apart from the familiar curve of the mouth, real smiles are accompanied by movement in the muscles around the eyes (Hodgkinson 1987: 34–5), hence 'smiling with your eyes'. Such muscle action arises involuntarily, while facial smiles (of the mouth) can be put on, or engineered. Even though a false smile can easily be distinguished from a natural smile, the effects of a false smile are still beneficial, albeit not to the same degree as a natural one. Smiling tightens muscles (notably the zygomorphic muscle) and this causes endorphins to be released. These alter moods, making events seem more pleasant. Laughter and humour are natural accompaniments to smiling, and their beneficial effects have long been charted. Smiling makes you feel good. And when you feel good, you smile, so reinforcing the upward spiral. Smiling makes us calm. When we feel good and smile at someone, we make them feel good, they smile at others and the ripples of goodwill spread. Smiles often give way to infectious laughter.

Laughter makes us happy, and a happy work force has been estimated to be as much as 30–40 per cent more productive (Borland 1999). Making people laugh can be a serious business (Chapman and Foot 1996) and is certain to become an important weapon in medicine. Laughter clinics have been set up to combat some of the most horrific diseases such as cancer, and for some people, laughter can lead to recovery from the most serious condition (Cousins 1979; Hodkinson 1987: 106–9). Tears of laughter are different, in chemical composition, from tears caused by onion peeling or exposure to the wind. Tears of laughter include enkephalins, natural painkillers (Greenfield 1997: 57–8).

Stress is often implicated in facilitating various diseases and conditions. Laughter reduces stress. And as a remedy, being altogether inexpensive, laughter has a good deal going for it. Laughter is as infectious as any disease (Goleman 1999: 1). Humour has an important role to play in many social situations, often serving as a way to break down barriers, and put people at their ease. There are many kinds of humour. Most forms seem to rely on bringing apparently unrelated or novel ideas together or in juxtaposition. There are many formulas (Fry and Allen 1996: 246) devised to explain the variables involved, e.g.:

$$\text{Humour} = \text{Salience (State} + \text{Trait)} \times \text{Incongruity} + \text{Resolution}$$

(Derks 1996: xiv)

A joke is often a very brief story that leads the listener to an apparently logical conclusion, only at the last moment to reveal an alternative and surprising ending. Humour allows otherwise taboo subjects to be raised in a non-threatening environment, and can also reduce stress by attempting to make light of otherwise horrific circumstances, as in the black humour of medics. Because humour relies on bringing novel ideas together, it is often extremely valuable in stimulating and accelerating creative thinking (de Bono 1990b: 80–3), in which one is also looking for novel arrangements and solutions. Such creative thinking may be indulged in at leisure or it may be the subject of sustained effort during work. Shared humour can also bond people together, and the complementary effect is that it may set some people apart, discriminating typically against people of neighbouring towns, counties or countries.

Addressing dysfunctional leisure?

Many leisure activities seem only to give rise to benefits, and no perceptible costs to society. To be even-handed though, a few people engage in activities that we would classify as dysfunctional, and which appear to give rise only to costs to society. Such activities must be perceived, by the participant at least, as giving rise to personal benefits. Distinctions are sometimes made between (so-called) recreational drugs and others (hard drugs) which are perceived as carrying greater risk. The motivations are complex (Plant and Plant 1992: 113–21), and such people may well assess risks differently (Palmer *et al.* 1998: 132–47). Even though there will be costs to society, there are clearly monetary gains to others in the chain; the growers and suppliers, who may well be operating in different countries and cultures. Worldwide, an estimated US $400 billion is spent on illegal drugs (Booth 1998: 8–11). Social norms (which will differ from one society to another) determine what activities are socially acceptable.

Taking alcohol has been encouraged in many societies for thousands of years, and while many enjoy their alcohol and gain benefits, a few bear the horrific costs associated with alcoholism; great personal and societal cost. For some people drinking becomes too easy a response to stress, and ultimately they become dependent. This may be because of irreconcilable stress, because the individual has some genetic predisposition to addiction, or among northern nations due to the long dark winters. Defensive spending on the consequent medical treatment, the loss of productivity and emotional hurt concerned is substantial. Until recently, many countries allowed companies concerned with alcohol, tobacco and gambling to sponsor sports and recreation activities, sending very mixed messages especially to the younger participants and spectators. In one imaginative programme in Australia, a programme has been initiated which uncouples this link, and ensures that some of the money spent goes toward sports and health programmes, but without the advertising (Galbally 1996: 115–20).

Vandalism is another clear example where the benefits must be personal, and the costs are societal. The cost can be substantial (estimated at $500 million per year in the United States alone, in 1981; Christiansen 1983), and the solution seems to be to act quickly to put things right (Welch 1991: 99ff.). The motivations for vandalism are complex. A helpful starting point may be to recognise different kinds of vandalism: acquisitive, tactical, ideological, vindictive, play and malicious (Cohen, in Iso-Ahola 1980: 158). The predisposing factors to vandalism and drug abuse appear to be similar; a feeling of helplessness and of being powerless. The dysfunctional behaviour in question may well be more a question of the way in which parents and society are behaving during the development of youngsters who later exhibit such self-destructive behaviour (Iso-Ahola 1980: 159). Great costs are borne by the vandals and drug abusers too. There are, unfortunately, all too many different forms of aberrant and dysfunctional behaviour in the realms of drugs, sex, violence and gambling. There are also antisocial aspects of sport, which have been reviewed (Coalter 1988). This is not the place to describe them in any detail, or to analyse the costs of these antisocial behaviours, many of which are criminal. What is clear, and pertinent to this book, is that providing recreational opportunities for young people otherwise disadvantaged can be important in providing, 'positive lifestyle choices and alternatives to self-destructive behaviour' (Parks and Recreation Federation of Ontario 1992: 38–9).

Spiritual health

Many people associate spiritual experiences with formal religion. Given the declining membership of Churches in some Western countries, it is indeed fortunate that such experiences appear to be within reach in many different leisure settings (McDonald and Schreyer 1991: 179–94). Recent research indicates that many people seek to visit places, especially in the countryside, which can act as a focus for emotional energy (Tresidder 1999: 136–48). Exposure to the elements – particularly sun, wind, sea and snow – can be incredibly invigorating, reminding us just how small we are, how we fit into the world, and just how wonderful life can be. For many, then, these experiences of being close to natural and cultural heritage have an almost spiritual significance (HPI Research Group 1997), and provide a setting within which we can explore this interdependence, which comes close to religious experience. There are arguments that some formal religions have been at the root of all our ecological problems, because they so obviously place mankind above all other creatures (Tietenberg 1994: 45), rather than alongside them. In many cultures, such experiences are more explicitly linked with sacred places out of doors (Wondolleck 1997: 257–62), and have had such a link for thousands of years.

Appreciation of the arts also comes close to a spiritual process for some. Perhaps this is so because of the prospect of intense and personal experience, which appreciating (and perhaps more so, making a work of) art can give. Visiting an ancient building, and stepping in the footprints of people of the past, can make extraordinarily strong bonds with times gone by, bringing history to life. Looking at a painting created by a figure who may now be almost legendary can have a similar effect. It is not easy to explain the pleasure, the sense of elation or of well-being that such appreciation can bring about. With the techniques now available to the neuroscientist, it is possible

BOX 4.1 Clearly of national significance, but whose nation?

Clashes of culture happen everywhere, but Wondolleck (1997) has written up a particularly poignant case in the United States. In essence, the story is simple. Devil's Tower National Monument, Wyoming, is the United States' first National Monument, established in 1906. In this case, of course, 'national' refers to a site chosen for the United States rather than because of any association with the first nations. It is in breathtaking scenery. It also presents some very interesting climbs, and has become something of a magnet for climbers, with well established routes. Over the years it has accumulated many pitons and pieces of hardware to help climbers, and it is popular amongst young and fit climbers in particular.

    Long before the mountain was selected by the US federal authorities for its scenic value, this same area had accumulated hundreds, if not thousands of years of significance for the native Americans, representatives of the first nations, as a religious site. There are special characteristics here that draw people of the first nations. At particular times of the year they come and offer sage and tobacco prayer bundles, and burn them so that the smoke rises and takes their prayers away. The US National Park Service have been working with all parties to devise a solution which gives the best practicable outcome for all parties (http://www.nps.gov/deto/).

to explore further how the brain reacts to visual, auditory or other stimuli. Some very interesting work has shown how music can excite different areas of the brain (Roberston 1994). This falls rather short of demonstrating how it lifts the spirit, but empirical work comes to the rescue (Bygren *et al.* 1996).

    On the ground, at least, we can choose to operate any system we wish. With this in mind it seems surprising that people in northern countries during winter should set off for work in the dark and return to their families after darkness has fallen again. How could we be surprised that some people have developed conditions such as Seasonal Affective Disorder, where lack of sun makes them depressed in midwinter? The mental health bill is rising in Europe at an alarming rate. People who take part in physical recreation enjoy better mental health, and there is every indication that other recreational activities such as singing, and more spiritual activities (walking in the countryside or worshipping) not only safeguard people's mental health, but prolong life (Bygren *et al.* 1996). The Stress Foundation recommends laughing your way out of stress, in this way:

◆   Listen or watch comedy programmes on the radio or television.
◆   Draw a smiling face on a card and look at it when feeling down.
◆   Use humour to counteract stress at work.
◆   Lift spirits by listening to music.

(*Scotland on Sunday*, 11 April 1999)

BOX 4.2 Changing patterns of thought about leisure and happiness

As each wave of philosophical thought has swept across our cultures, each has been reflected in our developing approach to leisure. In Western countries a handful of perspectives have been particularly influential (Goodale and Cooper 1991: 25–35). In Aristotle's day leisure was considered the highest state of being. With the advance of Calvinism (after Jean Cauvin, France) in the sixteenth century and the birth of the Protestant work ethic, an alternative emerged which encouraged the view that leisure was equivalent to idleness. In the late eighteenth century Jeremy Bentham (whose mummified body is still on view at University College London from time to time, under the rather strange terms of his generous bequest) and in the nineteenth century John Stuart Mill helped to change the map once more with the development of Utilitarianism in England. Adherents of this school of thought judged any action right if it promoted happiness, and this has been a very important influence ever since. A more collective view prevailed in France, and at the end of the nineteenth century Marxism began to spread a different set of values across some countries, and stimulate thinking elsewhere. Existentialism encouraged the view that people could determine their own destiny, as lives were not in any way mapped out in advance. Of course what happens on the ground takes something from each of these waves of thought, as with the expression of Utilitarianism through preferences expressed through markets, tempered by Kantian morals and democracy (Sagoff 1996: 897–911).

Utilitarianism enveloped the notion of utility, as the pleasure gained from the fulfilment of some need, by means of the consumption of some good or service. The theory of demand built on this foundation to develop the idea, expressing the concept of marginal utility, where (for example) the first taste of beer is worth rather more than the second, and the second rather more than the third, and so on. Neat though the theory was, an alternative that rather more closely resembled experience developed. This was the concept of ordinal utility, in which it was recognised that the consumer could at least rank preferences in order, even if they could not always be measured. This marches more comfortably with the notion of choice as described by consumer behaviourists (Engel and Blackwell 1982) and psychologists (Tuck 1976; Fishbein 1975). Although the theory of Utilitarianism was initially based on individual behaviour, the aim of government was interpreted as the delivery of the greatest happiness of the greatest number. But what of minorities? With the development of global markets, we can see uncomfortable decisions looming, in relation to equity, not just in respect of future generations, but also here and now.

Recent work has reinforced the notion that the requirements for exercise to maintain the health and well-being of society are not onerous. Regular aerobic exercise of a moderate nature (Physical Activity Task Force 1995), even as little as thirty minutes of walking every other day, has significant beneficial effects for those whose work does not provide the necessary physical component. In middle age, physical recreation and exercise maintain the muscle tone required. This helps avoid lower back pain and similar ills. People who indulge in exercise feel healthier and less stressed. With the benefits of modern medicines and wealth, people in developed countries are living longer. One of the most obvious benefits of physical recreation is the way in which it enables people to maintain their health and quality of life. Whilst health may not be easy to value, nevertheless in order to ration the expenditure of public funds on health, economists have devised techniques to support decisions in this area, relating quality of life to the cost of treatment or care. While this helps to focus on relative costs, it does little to help ascribe the value placed on health benefits, by the individual or family concerned.

The power of the media is very great. Advertising is believed to play a major part in shaping fashions and setting up role models. An increasing problem among the better-off Western countries has been the development of eating disorders, aided by the portrayal of super-slim models in fashion magazines, and the British Medical Association has called for a more responsible attitude among the media (Morant 2000). These disorders are complex in origin, but they are believed to be related to low self-esteem, an area where leisure and recreation could play a role in helping to reverse the trend. Similarly, fashion seems to be playing a role in increasing smoking among adolescent females (Wearing and Wearing 2000: 45–58), which is now being accompanied by the inevitable rise in cancers of the lung.

Social benefits: cohesion

We are not alone. While we often tend to think of ourselves as wholly individual, we truly are attached to everything and to everyone else. Depending on where you live, and your background, you may perceive yourself to be strongly tied to a number of different groups of people. Where there is low car ownership, and public transport is scarce, communities of neighbourhood remain of great importance. Increasing personal mobility, and the ability for remote communication without the need for conversation, have enabled new configurations of society. With this increasing personal mobility, much of it due to the invention of the internal combustion engine, communities no longer have to be founded on notions of family, extended family and neighbourhood. Developments in technology, such as the invention of the telephone, radio, television, video, computer networks and the Internet have led to an explosion in communication. People communicate, live, work and play among communities of interest, a virtual neighbourhood. Families play an important part in leisure and recreation (Kelly 1993: 127). Nevertheless, greater numbers of people are living on their own. Children are often being reared, or cared for, by a single parent.

Clearly, leisure and recreation give rise to benefits for the community at large (society), in many different ways. Leisure provides settings within which people can meet, make friends and be intimate (Kelly 1993: 23). Taking leisure together increases social cohesion. Leisure and recreation become the social glue, with almost universal application, between spouses, in families, among teams, groups of friends, and also at work (Lieber and Fesenmaier 1983). As patterns of work and living do not require (arguably) as much proximity of individuals, the role of leisure (in binding people together) seems set to increase. Taking part in joint activity leads to the development and sharing of values, and shared values have a role in ensuring the survival of any group as an entity. Health benefits resulting from leisure (whether physical, mental, emotional or spiritual) reduce the call on funds often administered centrally on behalf of the community or society.

Many pooled resources are subject to unhealthy competition because individuals see the possibility of reaping more than their just rewards, even though the community as a whole suffers a loss. This will also lead to a long-term loss for all individuals, and the problem has been eloquently described by Hardin in his essay on the tragedy of the commons (Hardin 1968; Russel 1996: 269). Using iterations on computer, possible outcomes have also been simulated to explore what might happen under different scenarios (Deadman 1999: 159–72). On the other hand, and more refreshingly, there are also situations in which individuals have been ready to sacrifice personal gain for the sake of the community as a whole: the altruistic surplus context (Faulkner and Tideswell 1997: 3–28).

Freedom is what people strive for in their leisure; liberty and choice. People are subjected to all kinds of authority (or a sense of obligation) at work, and for that matter in the home within the family. Moments of freedom are naturally sought by all, as refreshment. Although often painted as all or nothing, freedom surely comes in degrees. We can be free within the rules that govern our society, within whatever degrees of freedom our society formulates. Managing environments for recreation and leisure, for freedom, may seem a paradox, but the aim is precisely that. It is to manage settings, where people feel free; as free from restraint as is possible. In this state, individuals can feel that they have almost total control over what they choose to do.

People may use leisure pursuits and companions to build, or rebuild, links with society, and provide some meaning or an identity. They may explore at leisure links with people in the past, or the culture that has been bequeathed by those people: stories, legends, literature, heritage, historic buildings, paintings or other works of art. Others (or the same people at other times) may choose to use visits to the countryside to become closer to the natural world of which we are a part.

Human development

Babies learn a great deal from play. All learning comes from experience within or close to the family, initially. As babies grow into children, the degree of freedom is gradually extended and play becomes an important instrument in the developing individual, inextricably linked with learning (Parker and Stimpson 1999: 275 ff.). The home range of play and understanding expands as the child develops (Moore and Young 1978: 83–130). There are different kinds of play, during which different skills are learnt. Through movement children learn motor skills and hand-to-eye co-ordination; through playing with others they learn social skills; and through other games they learn cognitive skills (Timothy Cochrane Associates 1984: 9). All these forms of play contribute to development, and play is essential for mental health. In adulthood, play still contributes benefits, not least in the development of youngsters who may be within the family or other social unit. Play lets children, and adults, escape from the ordinary world (Huizinga 1949: 10). Make-believe can create a world peopled by anything you want. It allows people to de-stress. Many of the skills learnt can be read across into other areas of life. There is evidence that play provides youngsters with skills for use in problem solving (Parks and Recreation Federation of Ontario 1992: 36–7). Skills learnt at play are applied to study, or later at work, and may pave the way to new interests, and (in adult life) the way into a new career. There is a risk that the anxiety of parents coupled with dependence on the motor car will continue to reduce the opportunities for children to play normally.

Leisure activities can contribute to personal development, increasing skill levels, as well as self-esteem and feelings of well-being; and also to community development, by forging new links between members of the community. Leisure can help develop shared values, by increasing the effectiveness of the community, and developing the overall skills base of the community, whether the skills are technical or social. Recreation provision can often be a basic social service. It is not always easy to make the distinction between recreation and work, or between leisure and obligation, as all these activities become swept up into the life of the community. Some projects have an immediate payback to the community, as with conservation projects or the putting on of community arts events. At the other extreme, in some countries, lifesaving is carried out on a voluntary basis. It may be pursued by those who feel a sense of obligation, or because it offers excitement and the chance to improve skills, as in the Mountain Rescue Services, Lifeguards and Royal National Lifeboat Institution.

Around the shores of the British Isles is a network of 224 lifeboat stations, with crews who regularly put to sea and rescue people who have underestimated the conditions, were blissfully unaware of what the sea could be like, or who simply got into difficulties. That the Royal National Lifeboat Institution (founded in 1825), which runs these maritime rescue services, depends entirely on voluntary contributions is remarkable. Anyone inclined to help should contact them at RNLI, West Quay Road, Poole, Dorset BH15 1XF. They need all the help they can get. According to a recent leaflet, 'One in every four people who use the sea for recreation has been involved in a life-threatening situation.'

Businesses have been quick to realise the value of recreation in developing management skills. Putting people in an unfamiliar environment, with a challenge, provides an atmosphere in which all members can develop their skills and contribute in a special role for the team. The social cohesion which follows shared recreational activity is important. Everybody's contribution and skills become visible. It is not always the leader who excels in every department. Team members may reveal social or technical skills that once recognised can be deployed at work. Self-esteem is raised,

TABLE 4.3 Dimensions of leisure. The multidimensional nature of leisure could be described by the following sets of dimensions, which when combined give a wide range of experiences

Arousal – stimulation *Freedom (individual)*	*Relaxation* *Affiliation (group)*

Activity:
    physical, mental and emotional
	Rest:
	    physical, mental and emotional

Excitement, tingle	Security, comfort
Expression	Appreciation
Discovery, novelty	Understanding, meaning
Joy (intense)	Humour (light)
Watching	Taking part

often, among those whose worth has been overlooked. Such extramural activities allow all team members to show a different side of their natures. The range of skills displayed is usually formidable. This should be no surprise. For, after all, an individual trusted at work with a small budget of several thousand pounds may be entrusted at home with several lives, and the equivalent of a small enterprise, and major assets.

Environmental costs and settings

Identifying and measuring the benefits and costs of the impacts of leisure and recreation in economic, environmental and social terms is important, if we are to make best use of our resources. It will also be important to identify the costs of keeping the natural and cultural resources that we treasure secure in the face of the pressure from so many tourists, visitors and residents.

Recreation has direct impacts on the environment. There is a perverse general law that suggests that the more precious a place, and the more scarce the recreation opportunity, the greater the risk of pressure of popularity. We are in danger of loving our special places to death, whether they are natural or cultural treasures (Department of Employment 1991; Shipp 1993). The direct impacts that we have to be aware of, in respect of recreation and conservation in the countryside, are: disturbance, trampling and erosion. These all bring about unwanted change. We will return to impacts and conflicts in greater depth in later chapters when we focus on how to resolve them.

What is required and valued by the participant, will clearly depend on a number of factors:

◆   *Physical environment sought*: scale, slope, aspect, complexity, rock, soil, water, degree of compaction; exposure to elements: sun, wind, rain, snow; space, temperature, humidity, water and air quality, lighting; ability to change environment: availability of energy, power, water; and systems to supply materials and extract waste.

◆   *Social environment sought*: purpose: activity or mix of activities; people taking part: number, ages, socio-demographic characteristics; preferences of different groups and individuals; owners and managers: public, voluntary, private; degree of control; motivation (e.g. financial return required).

◆   *Emotional environment sought*: ambience, warm/cool, stimulating/peaceful, intimacy/solitude.

For many sports, the physical characteristics and setting are crucial. We can change our environment to provide the settings we require, e.g. in respect of snow skiing: downhill, cross-country, on artificial snow, on artificial slopes, on tracks, on roller skis, on simulators, outdoors and even indoors. The physical setting will be important for other forms of leisure activity too. A room of appropriate size with blackout and noise insulation might be considered essential for showing films. Certainly, a room with a poor blackout would be ineffective, but lateral thinking produces the drive-in, with earphones to provide sound on a one-to-one basis. Some activities need a controlled environment. Swimming pools require careful maintenance to ensure that the water stays in the right place, that the changing rooms are a comfortable temperature, and are clean. The water quality in the pool will be maintained, the temperature within a comfortable range, and chemical additives or physical treatment put in place to ensure that swimmers are not put at risk of catching communicable diseases.

Risk-averse tourists may feel more comfortable taking their own culture and values with them when they travel, for example staying in Western-style hotels (often part of large chains) when visiting countries in the Far East, and eating Western food rather than trying the local delicacies. Such impacts will be revisited in later chapters. Examples abound of attempts at establishing resorts where temperatures can be kept at comfortable levels, such as in the Himalayan hill resorts. More than this, we have produced enclosed complexes to keep the rain out as in the Center Parcs resorts in Europe (Box 7.3, p. 210), or to provide air-conditioned oases away from the sun in the Middle East. The cost in energy terms to provide these sought-after environments is great today, and of course contributes to the costs we will bequeath to future generations in terms of the outputs contributing to global climate change. On the other hand there are resorts that are being promoted on the basis of being eco-friendly, such as Couran Cove (on South Stradbroke Island in Queensland) which promotes itself as the South Pacific's premier eco-tourism destination, (http://www.couran-cove.com.au/).

People tend to travel and enjoy their leisure in company, even in trips to experience wilderness, and certainly closer to home. Settings need to be selected or provided to reflect the likely composition of the social group, and provide alternatives for groups of different sizes, and different preferences. Often it is possible to provide modular units, the simplest example being the provision of dining tables in restaurants, where it is usually possible to rearrange tables to suit the customers. Sometimes it is helpful to provide adjacent areas where, for example, parents can enjoy reading or chatting with friends on a bench reasonably close to, for example, the favoured play area. In this way they can watch over their children playing without appearing to be policing them (Timothy Cochrane Associates 1984). In some cases, the setting will be dynamic, and the activity requires a sequence of settings. Then it will be important to run checks as the group flows through these settings. These are the skills which architects, landscape architects and designers employ. To reach an understanding of what needs to be specified, we can use virtual reality (computer-aided design). The best computer package stills falls far short of the advantages of running images through the mind's eye, or 'skull cinema'. Put yourself first in the mind of a would-be participant, and simulate a visit by viewing each and every stage of the process; then stop and have a brief break before re-running the film and viewing from another perspective, of the manager and so on.

Where relaxation is desired, borrow from nature. Incorporate a fountain or other moving water feature. This provides a moving image and natural sounds, and can be especially effective in bringing a slice of the natural world into man-made environments. Introducing plants can also do much to soften environments (as well as increase the air quality). Because freedom is at the root of leisure, we need to emphasise freedom in the way settings are presented. The garden or park has for a long time provided an extension of the house or home, and provides a sense of refuge (Appleton 1975) from the elements, and (increasingly important as we become more crowded) from being overlooked. By careful design of the landscape (at whatever scale), by managing plants and thoughtful sighting of artefacts, individual settings can be provided which can offer framed views, and some privacy (or enhanced sense of refuge). People generally stick to paths, without any need of fences (save to protect from hazards). The degree of apparent freedom is enormously enhanced through the absence of fencing. Similarly, people experience great freedom walking through a wood which is reasonably open, but may be apprehensive in situations where the views are cut short by dense planting, and the sense of enclosure is too strong (Burgess 1995). Sinuous paths that lead into the distance, and out of the picture, literally lead us on, and encourage us to explore, to see what is around the next corner. Keeping man-made artefacts to a minimum and ensuring that they are in keeping with the environment will do much to keep alive the illusion or sense of place. At the micro-scale, indoors, colour treatments as well as mirrors can be used to give the illusion of space. Within our towns and cities, churches and other places of worship provide a sacred refuge from the hustle and bustle of everyday urban life. Art galleries, museums and some gardens or parks can also provide little oases of calm.

Personal space and personal places

We are often searching for our own personal space. For many, being in the elements, up in the hills (Ferber 1974) or on the coast brings a deep sense of relaxation, and well-being (HPI Research Group 1997). Wind blows away cobwebs, and exposure to the elements helps to put things into perspective. Contact with nature, with the rest of creation, and with our natural and cultural heritage reminds us of our place in the world, and gives a sense of being at one with the world, and ourselves. The relative peace and tranquillity which can be found in the remote places, up in the mountains and on our coasts, is much valued, and with the landscape, creates sacred places. Some, like the Himalayas, Ayers Rock (Munt Uluru) and the Devil's Tower National Monument (Wondolleck 1997), Wyoming, have been sacred for thousands of years. Others, including the national parks in Britain, serve a similar purpose, as a 'depository for the emotions' (Tresidder 1999).

Increasingly, our communities are urban. Originally, towns developed from farmsteads and buildings clustered together on enclosed land, growing organically over time. The lanes became streets, and the traffic increased (originally horse-drawn). Space was allocated for public functions, for meetings, discussions, markets, parades and festivals. Some old towns were extremely pressed for space, being squeezed between features such as rivers and hills, as well as the developing network of roads. (The most pressed spaces lend themselves to pedestrianisation, as in Siena.) Edges, and nodes, attract pedestrians, such as promenades along coastal defences, piers (including the great pleasure piers of Victorian times) and the humble bridge – all close to water, a universal attraction.

As lessons in the new style of town planning spread across Europe, spaces were designed in, distributed across the town, so that everyone should have some space near by. Some of this requirement arose out of concerns for health. The early sewage systems (save for those based on Roman structures) often amounted to no more than an open drain in the street. The residents of the upper flats emptied their buckets out of the window into the street below. As horse traffic increased, the ammonia must have been powerful on nose and eye. Now in many towns, cars have replaced the horses. The pollutants left behind may not be so immediately obvious, but the noise and fumes together have quite changed the nature of towns and communication between residents. There were a few attempts to plan new towns or villages in England, with improved conditions for working people, in the nineteenth century, such as Bournville, but they were exceptional. By the close of the century, ideas were taking hold on the development of garden cities, the first of which was Letchworth, begun in 1903, and this approach spread, but other concerns were soon to overtake this thinking. With the enormous increase in car ownership in the 1950s and more especially the 1960s, cars were in danger of taking over the public spaces. In Britain an influential report (Buchanan 1964) drew attention to a way of giving towns back to their inhabitants. Many cities are following the leaders to pedestrianise at least some of the older and more central parts. Regrettably, some of these schemes were later to be neglected as local authorities' budgets became constrained.

Bonn, in Germany, as long ago as 1953 denied the motor car access to its charming town centre, which is now once again a truly public space. There are still areas where use of the motor car is squeezing pedestrians out. Public space out of doors is at a premium, and the price of land is so high that infilling is common. At whatever station or terminus people change their mode of transport there is usually a substantial public space, often covered, or indoors, and now often associated with commerce too. Shops are provided in airports and railway stations. The humble bus stop is a public space, perhaps worthy of more attention, given the time people collectively spend there. In among the thoroughfares, and in the small spaces such as at street corners, public art thrives. Some sculpture remains *in situ* for hundreds if not thousands of years. At the other extreme, the needs of street theatre are rather more transient. In most towns, on special occasions and festivals, the town square and streets may be temporarily closed to traffic.

In times of plenty, Britain (like other countries) has erected magnificent buildings in which to house treasures accumulated during campaigns abroad, as well as archaeological and historical treasures from home. The end of the nineteenth century in Britain was particularly rich (for Britain), and most towns and cities have substantial buildings dating from this period to house their treasures. Entry to most has until recently been free of charge, but recently, argument has raged over pricing and admissions policies. These periods of relative wealth have be accompanied by a boom in capital spending, as in the 1960s and early 1970s, which left a legacy of leisure centres and other facilities, which now provide the manager with a challenge to raise the necessary revenue funding. Some agencies have used computer modelling to help in planning where new developments should take place, to avoid these problems. Technological developments and the increasing (perceived) value of time have combined to stimulate the development of large stadia, which can be open to the elements one moment and totally shielded from them just twenty minutes later (Weston 1996: 106–8). In some cases the task has been simplified by making the stadium entirely indoor.

If we used cash turnover as an indication of the value and importance of any one category of leisure facility, and used Britain in calibrating the pages of this book, then perhaps half, or more, of the book would be given over to the public house. Something like £17 billion is spent annually in some 60,000 public houses across Britain (British Licensed Retailers' Association 1999). Arguably, one of the best ways to study public houses is at first hand, because of their rich diversity (Smith 1995). Despite many pubs being linked (vertically integrated) with brewers, there are still some 18,600 free houses that have no such allegiance. Some public houses have a wealth of history, stretching back hundreds of years. Others have atmosphere constructed by the square metre, overnight. Themed public houses (notably 'Irish' pubs, complete with live music) have been very successful financially. Most countries have a network of informal public spaces that revolve around refreshment, whether drink or food. The tavernas of Greece, the cafés of France, and of other countries in mainland Europe perform a broader social function. They tend to include all ages, whereas the English pub was once seen (and still is by many) as a haven for men to escape domestic responsibilities, and is still rather dominated by males (Hey 1986).

Before industrialisation there was little pressure on space, and people were engaged in physical activity in everyday work and life. The need for playing fields emerged as industrialisation concentrated people, and people needed designated spaces in which to play. In Britain, parks and playing fields have been provided in waves reflecting societal concerns. Jubilees and other royal celebrations provided the impetus for a series of moves to acquire recreation grounds for public use. King George V recognised the importance of play, that children 'may be helped in body, mind and character to become useful citizens', and had (as Duke of York in 1925) set up the National Playing Fields Association. As a memorial to King George V, in 1936, King George's Fields were set up in towns and villages across England, Scotland and Wales, many of them protected in perpetuity. The National Playing Fields Association fights hard to protect some 1,015 playing fields, including King George's Fields, and campaigns for the provision of safe but challenging play for children, and the provision of well-maintained playing fields for all (National Playing Fields Association 2000).

Schools in many areas shared playing fields with communities, but some have their own playing fields. Land of this kind in urban areas (and space for development) is at a premium, and there may be considerable pressure to sell. Budgetary pressures on local authorities, changing political ideas, and new ideas about asset management, have all combined to encourage the selling of spaces previously attached to schools, often to be lost to development. Planners, in the past, have provided pitches according to standards usually described in terms of so many pitches per thousand head of population. Such an approach is input driven. The National Playing Fields Association still campaigns for six acres (roughly 2.5 ha) playing fields for each 1,000 residents. Grounds maintenance is a labour-intensive process, and no matter how brilliant the maintenance team, weather conditions will render pitches unplayable for many days in the year. Technology has provided a solution in the form of all-weather pitches, using synthetic materials and a good deal of engineering. Less space can now provide for more matches, and render recreation programming easier. The target is now about output: number of games played (and enjoyed).

Urban parks vary greatly in kind and size. They provide different settings within the one area for different age groups to enjoy the wide open spaces. Over time, the treatments have changed to keep pace with changing desires, and with changing practice in management. They reflect the different cultures from which they spring. There have in some cultures and at some times been attempts to formalise the amount of park provided for each inhabitant (in Britain, for example, in the early to mid 1970s) but the effort is now directed to making the parks work harder. The drive to contain costs has encouraged waves of reduction in staffing, with consequent waves of concern over the lack of surveillance and management, (Greenhalgh and Worpole 1995). This has encouraged the development of schemes such as the Green Flag (ILAM 1997). This award recognises good practice, and seeks to assure quality. The provision of parks varies enormously. There are real concerns about their future management and the erosion of skills arising from decades of neglect. These skills are necessary to restore the kinds of park people most seem to like.

Value is added, for wildlife and for those seeking space for leisure and recreation, if these parks, greenways or corridors are integrated as a network. This can be developed linking existing green spaces together, and can lead to enhanced ecological functioning of such spaces. Keeping development back from the edge of water provides for enhanced buffering, so increasing the quality of the water. Providing more space for water to be filtered back into the river, rather than running straight off the hard surface, will slow down water run-off and so aid flood control. Green spaces also include all the planted areas, alongside our houses and roads.

Judicious planting provides a healthier setting, producing oxygen, filtering noxious chemicals and contributing to a richer biological environment. The plants can also provide insulation against noise, but they do need managing, and such work is labour-intensive.

The distinction between rural and urban areas is not recognised in the same way by all other species. Foxes are now quite common in urban areas in Britain, and elk use the spaces along railway lines to migrate through Moscow, to follow routes otherwise cut off by roads and development (Chebakova 1997). Occasionally roads and railways serve to isolate spaces, which may become valuable reserves. In the heart of London the London Wildlife Trust has a number of reserves which provide oases of (relative) calm. A piece of ground cut off by railway lines, the Gunnersbury Triangle ironically ensures that (other) disturbance is minimal (London Wildlife Trust 2001). Between King's Cross and the Regent's canal lies the site of an urban reserve called Camley Green Natural Park, which provides all-year-round access for those who are fascinated by pond life.

The architecture and decoration of buildings, churches (Randall 1980) and other places of worship, markets, eating places and meeting places, all have links with the particular area. The vernacular architecture arising from this collection of characteristics enables historians to trace the social history of the area (Hoskins 1967). The small towns and villages of rural areas retain their diversity of character. Local distinctiveness is reflected in speech (language and dialect) food, clothing, sports and ways of life. The structure and appearance of the buildings themselves reflect all the character and distinctiveness that existed at the time of construction. Local distinctiveness is in danger of being eroded with the rapid increase in travel and communication. Culture accumulates as succeeding waves of migrants bring something of their heritage into the area in which they settle. The speed at which ideas can now bounce around the globe will surely confuse future field historians.

Visitors to historic buildings present the manager with particular difficulties in balancing conservation and recreation interests (Binks *et al.* 1988). The amount (and distribution) of activity is obviously dependent on the distribution of surviving riches, but also on the distribution of circumstances and motivations to conserve such sites. The recreational value can be greatly affected by interpretation, either enhanced or destroyed. In promoting certain sites it is also easy to destroy the images that people have in their minds.

The innumerable associated artefacts become valued in their own right as contributing to the character of the area. Boundaries are often marked out by characteristic styles of fencing, hedgerows, turf banks, stone walls and ditches the styles being blended as a result of the kinds of material available, underlying rock, the fertility of the soil, local species, as well as the cultural characteristics of the local people or work force. Artefacts also include such items as telephone boxes and signposts. In the interests of economy, there is often a move toward a standardised approach. Standardised roads, and standard roadside furniture, can insulate the traveller on motorways from the real character of the locality through which the road passes. On the other hand, for the traveller in a hurry, some standardisation and the development of branding, reduces risk and gives reassurance.

BOX 4.3 Sculpture in the forest and sense of place

In 1977 the first three-month sculpture residencies took place in Grizedale, a forest set within the Lake District National Park in England. Who would have guessed that, twenty years later, this forest would boast, beside the residencies, a sculpture trail, a Theatre-in-the-forest, a gallery, craft workshop and a play-ground where a fanatical sculptor had been let loose to create a playground which was carefully crafted and sculpted. Behind this has been a partnership forged amongst the Grizedale Society, the Forestry Commission and the Northern Arts Board, with many other organisations and individuals playing a part from time to time. Many of the foresters who had a spell of duty at Grizedale found that the forest got a hold of them. This was certainly the case with Bill Grant, who was the Chief Forester at Grizedale. A Churchill Travelling Scholarship to the United States and Canada encouraged Bill to develop his ideas on recreation. Later, on retiring, he found a new role as Champion (Director, anyway) of the Grizedale Society, which has provided opportunities for 'actors, musicians, sculptors, painters, crafts people, poets, writers, teachers, and natural historians' amongst others (Grant and Harris 1991: 7–10). The forest provides the physical and dynamic settings within which many of the works can be placed, or built. Sculptures grow, flourish and are allowed to decay gracefully. From small acorns do great oaks grow. The forest has been instrumental in the development of many sculptors and environmental artists, from Kemp and Goldsworthy to Matthews and Frost. Frost subsequently created a forest sculpture play area in Moors Valley, which has proved immensely popular. The forest continues to inspire people of all kinds to appreciate and express themselves through art (Davies and Knipe 1984; Etchell 1996b; Goldsworthy 1996; Matthews 1994).

Special places with special values

High places have always been sought out, for spiritual experiences. They are special places valued for their wilderness qualities, and they are very vulnerable, epitomised by the plight of the Himalayas. Mount Everest, the highest mountain in the world, has been described as the highest junkyard or rubbish dump in the world, with the highest litter trail. In the 1996/97 season, Sagarmatha National Park (Everest) received more than 30,000 tourists, Sherpas, and porters, and one in five of the major trails was adjudged to be either moderately or severely degraded. Because of the difficulty of climbing at that altitude, equipment litters the area, and what is bio-degradable at lower altitudes remains literally frozen in time and space. Between 1979 and 1988 some 840 expedition teams resulted in more than 400 tonnes of disposable rubbish, 140 tonnes of non-biodegradable rubbish and 200 tonnes of oxygen cylinders. These problems are part of a much bigger picture, and unless there is monitoring and management, problems will multiply through the ecological, socio-logical and economic realms. Tourism is an important part of the economy of Nepal, worth 3.8 per cent of GDP and 18 per cent of foreign earnings, set against a background of population growth, poverty and environmental degradation. Tourism has been growing quickly (from 4,000 tourists to Kathmandu in 1961 to more than 450,000 in 1998) and it has induced problems, not just of waste disposal but of deforestation, changes in land use and agriculture. The effects have even impinged on the monks of Thyangboche. There has been more success in other areas, like Annapurna, where participatory techniques have been deployed to good effect to ensure that local people are more involved in the decisions that will affect their communities, now and into the future (Nepal 2000; http://www.thinkquest.org/tqfans.html; http://www.thenepaltrust.demon.co.uk; Nepal Trust 2000).

By definition, therefore, mountains need great protection to ensure that they do not suffer degradation:

◆   *spiritual* degradation that may be caused by a cheapening of the image, which will affect all those who hold these values (whether use, option, existence or bequest);

◆   *cultural* degradation, for example by the insensitive introduction of new cultural influences and values;

◆   *social* degradation, with the introduction of new diseases to which host communities have little or no resistance;

◆   *biological* degradation, as soils are usually very thin, and the ecology of such areas is finely balanced;

◆   *chemical* degradation, or pollution, from accumulations of debris and human waste (Tribe 1995).

In Britain areas of open hill and moorland (which may in other countries be considered low ground) take the place of mountains in some respects. Mountains are defined by the relative topography of the surrounding land, although in some countries, people have devised classification systems (Poucher 1965).

Forests and woodlands, for many areas of the world, represent the dominant natural habitat, characterised by the predominance of a treed landscape. In many parts of the world (e.g. Amazonia and British Columbia) people are concerned about how to contain harvesting so that the old growth (or a proportion) is preserved. Voluntary organisations in the developed world have started all kinds of charities to help address the problems relating to forests in other parts of the world (WWF, FoE, Greenpeace) to mitigate the effects of deforestation. People seek to develop or maintain the species mixture that delivers the stream of benefits we seek, balancing the needs of local, national and global communities while taking account of objectives for conservation and recreation. The goal is the sustainable management of forests. Countries have been seeking to interpret the principles agreed at Rio for their own setting (political, economic, sociological and technological). Forestry Ministers in Europe have agreed principles (Helsinki 1993), and the United Kingdom has sought to integrate this thinking within a single framework to describe the sustainable management of forests, the UK Forestry Standard (Forestry Commission 1998). In the more populated parts of Europe, forests have a very special role in providing for recreation. In Canada the Forest Service has been working through the Canadian Model Forests Programme to develop an approach to sustainable forest management. Many of the greatest difficulties being resolved hinge on the different values held by different groups, particularly by the urban communities and the communities that depend on the forests for their livelihood. On balance the ecological problems seem simpler. Generally, woodland can absorb the sight and sound of people like no other habitat, and can be managed to provide settings for many different activities. Within woodlands and forests, there is a framework of open spaces, and of water features that are essential to the proper ecological (and recreational) functioning of these spaces.

Water, and more especially the edges alongside water, attract people and wildlife in great abundance. The banks of rivers and lakes are susceptible to erosion and the rich resident wildlife prone is to disturbance. Increasingly water planning and management take account of the needs of wildlife and of recreation. There can be significant conflicts associated with water management projects, for example where flooding exchanges one habitat for another, but often with greater recreational potential. The maintenance of the quality of water and rates of flow remains a major concern. Nevertheless, the impacts of leisure and recreation are usually very slight indeed, compared with the effect of adjacent land management (or water management itself). Farming practices, run-off (which can harm fishing and spawning areas), pesticides, herbicides, nitrates and other chemicals can upset natural balances. There are places, though, where the potential level of recreational activity requires very careful management (Peter Scott Planning Services 1997) to avoid damaging the river systems. In other areas as yet undeveloped, cruising opportunities on inland waterways and rivers provide an opportunity for a welcome growth in tourism, such as on the Shannon–Erne waterway in Ireland (Guyer and Pollard 1996). Waterfront developments are much sought after, but the effects can destroy relatively susceptible habitats.

In much of Europe, productive land is very intensively farmed, particularly in countries like Denmark, northern Germany, northern and eastern France. Arable land is often farmed right up to the margin of the field. Fields have been increasing in size and woods and copses are scarce. In the west of Europe it is wetter and livestock are more common. Here the boundaries, such as stone walls, copses, shelter-belts and hedgerows give shelter from the predominant south-westerly winds. These features have, in turn, created a network of attractive spaces of great potential value for countryside recreation. An issue here is who benefits, and who pays the costs. In Europe the farmers might say that they bear the costs of maintaining this framework whilst the users and perhaps some local hoteliers and shopkeepers gain the benefits. The users, hoteliers, shopkeepers and other taxpayers might say that they bear the costs through the enormous subsidies (for production) which farmers receive. Changes to reflect the outputs (social and environmental) required could help. There are some issues here, in terms of environmental economics, which need to be solved, to ensure that the values we hold dear remain secure.

Social trends and changing values

What benefits and settings are required, and where, will clearly be affected greatly by social trends. The following is a selection of trends that may be more relevant to urbanised countries (Broadhurst 1997; Driver *et al.* 1999; Roberts 1992):

◆ The population is ageing, with people living longer, giving rise to an expanding group of older people who have more time, greater health and greater wealth than in the past. It is worth remembering just what a varied group this is. The needs of an 80 year old are not the same as those of 60 year old. Pensions are required to cover much longer periods than in the past. The ageing of the population with fewer youngsters and people of working age, are factors that work together to produce greater pressure on public funds.

◆ Households are increasing in their variety, with fewer 'typical' families but more people living as child-free couples, lone parents or simply on their own.

◆ Leisure time is being redistributed. Generally, people have more unobligated leisure time, with younger people delaying having families and older people living longer, although there is a sector (chiefly among professional and managerial staff) for whom the hours worked seem to be increasing. These effects, together with increasing uncertainty of employment and a trend towards contract working, exacerbate short-termism. The short break market is increasing in importance in tourism. Another group has a good deal of time, even unemployment, but little money.

◆ Whilst older people live longer and live in better health than in the past, there is some evidence that younger people are more home based and less active than in the past (Physical Activity Task Force 1995). There has also been an increase in home entertainment (electronic, video and television) and a continuing rise in alcohol consumption. Doctors, in some areas, are now prescribing walks. Another group is increasingly concerned with health, for instance with what is eaten, with diseases, animal welfare and use of antibiotics.

◆ Perhaps as many as half the diseases of today, in the industrialised West, are related to nervous stress (Krippendorf 1987).

◆ Commercialisation continues to influence the public and voluntary sectors, as well as the profit-seeking private sector.

◆ Green issues continue to be of importance among some sectors. Conflicts in personal (as well as societal) policies on the use of energy, as against the exercise of personal mobility, present real problems in setting environmental priorities, policies and programmes.

◆ The declining influence of institutions – the Church, the Crown, the state, the family and (for some) the workplace – has increased the importance of other activities and referent groups in defining identity and meaning. Leisure is now central to many people's lives.

◆ People are developing specialised lifestyles, and tastes, with each activity fragmenting into niche markets. At the same time people are expecting ever increasing standards and quality of experience.

- People are living in a world of increasingly high-speed communication, looser and more dynamic connections, putting (in some cases) a greater distance between the real and perceived world. At the same time, information technology is fuelling a migration to more rural locations for those who have climbed aboard.
- The overall demand for countryside recreation remains relatively steady, but masks underlying changes, such as the increasing demand for very active recreation. For woodlands, cycling (away from the public highway) and the demand for access for all will continue to be important trends.

Leisure and recreation interests now help to define lifestyles and therefore have a part to play in defining market segments. Each segment, defined by lifestyle, will have a different set of expectations.

Paid holidays (in Britain) have increased from one or two weeks in 1951 to four weeks or more in 1991 (Whitby and Falconer 1998). Legislation is one of the instruments used to determine how recreation opportunities are distributed, in time (statutory holidays) and space. For example, many of the laws that govern people's ability to move around the countryside arise from interpretations of property law. The incoming labour government in Britain in 1997 made it part of its programme to give people greater freedom to enjoy access to the countryside, whilst respecting the needs of managers and agencies concerned. Access to resources for the arts, and for sport, is

---

BOX 4.4 *Allemansrätt*, **everyone's right**

In parts of Scandinavia the old laws have been passed down, so that access is allowed across the countryside so long as privacy is not invaded nor damage done. There are a number of caveats, and each of the countries does things slightly differently. The principle is similar to the belief enshrined in the old Gaelic law tracts (Hunter 1995: 121–47), and for that matter the beliefs of the native Americans. Chief Seattle was alleged to have said, 'The Earth does not belong to man; man belongs to the Earth' (van Matre and Weiler 1983, although academics seem to agree that Ted Perry, an enthusiastic scholar and playwright, was the first to 'reveal' this in a script written for an environmental documentary). Whoever wrote the line, it does succinctly put a point of view. In similar vein, the Scandinavian laws bestow rights and responsibilities, which have been interpreted as providing the opportunity for everyone to enjoy the countryside for their own recreation so long as it does not disrupt other activities unduly. It grows out of the Germanic tradition where the right to use land was superior to the right to own land (Vistad 2000). The model has proved to be of considerable value for other countries such as England, Scotland, Wales and Ireland that have been reviewing their laws on access, although each has taken something different from it (Mortazari 1997; Peter Scott Planning Services 1991, 1998).

also often circumscribed by laws. Laws govern provision for education, a requirement for a syllabus, the establishment (and entitlement to a proportionate share of the proceeds from) the national lottery, and the setting up of agencies. Some of the agencies will have grant-giving powers to support particular activities. Laws also provide duties and powers for local authorities to provide for their communities.

Nobody has access to recreation (or any other) resources if they are unaware of the existence of such resources. They need information, tailored to the target audience (Countryside Commission 1994). Experience shows that in urbanised and highly mobile societies this is more difficult than it first appears. Being aware of the opportunity is not much good unless it is physically accessible to you. Legislation such as the Disability Discrimination Acts, in the United States and in Britain, have done much to encourage a more thorough approach which requires audit followed by appropriate interventions. In respect of countryside, the principles behind the Recreation Opportunity Spectrum (p. 165) have been used to suggest what standards of accessibility people might find (Fieldfare Trust 1997). Communicating what may be found is more difficult in relation to the countryside, but the approaches favoured are twofold:

◆ describing the physical properties of the site, perhaps as standards, to allow individuals to determine, given their individual abilities, preferences and sources of help available to determine for themselves what might be feasible;
◆ describing the level of difficulty, giving some sort of shorthand symbol or grading in which the expert (who may be advised by an access group including representatives with different disabilities) makes the judgement of what is accessible.

The first method allows the individual the freedom to make the choice, but describing the physical properties in the detail required to facilitate that is difficult. There is always a temptation for the manager to make generous allowances for just how accessible the facilities are. The individual seeking access needs to have a very good understanding of his or her capabilities, but who is better placed? Some may be dissuaded to attempt access under such conditions. The second method is considered by some as giving too great a steer as to what is accessible, and risks taking away some freedoms. Apart from physical accessibility, there is the question of programme access: access to the activity and to any essential information. On balance the first is preferred as meeting the spirit of encouraging personal responsibility and more access. Considerable progress has been made in relation to transport, access and programme access for all activities, but leisure managers need to be at the forefront of such effort to ensure that all people have access to all the benefits of leisure.

Participation in recreation is not distributed evenly across the population. Many studies (Cushman *et al.* 1996; Glyptis 1991) have tended to reveal that groups not represented so well, among those taking part, include:

◆ children;
◆ disabled people;

◆ ethnic minorities;
◆ older people;
◆ women.

This list is similar to those who generally do not do so well and have often been excluded in the past. In Britain, white Anglo-Saxon (even if not necessarily Protestant) middle-aged men are well provided for and take part in recreational activities, perhaps because they are the ones planning it, deciding what should be provided, and managing it. Clearly, if we are to provide for the whole population, we need to find out why these disparities occur and take all reasonable steps to correct the anomalies. It may be inappropriate to even out the provision across the entire spectrum, but there should be a complete range of opportunities available to all who want them, if we wish to make the most of the societal benefits of leisure.

We like to be in control, and there is always the risk that by gaining control over a simple part of the overall process we ignore the things that really matter. On the socio-economic side, Bullock and Mahon (1997) quote the Howe-Murphy and Charboneau approach, which emphasises the different aspects of people – the physical, psychological, social, cognitive and spiritual facets which make up the total person. This model also relates the total person to other systems – the biological, family, agency, neighbourhood and community. The divisions are angled towards better understanding the needs of disabled people. In the past there would have been a tendency only to look after one aspect, the easiest or most obvious one, probably the physical, without recognising that any real solution demands more. This total approach, which some may think of as holistic, is something rather more. It is not black-box management. The result depends on dealing with all the different aspects, and may be holistic, giving outcomes that add up to more than the sum of the parts. The management is still focused, on the different areas. The same approach can be taken to the biophysical environment too. In practice, we need to continually shift our gaze across different scales of time and space, so that we are alternately integrating and differentiating our knowledge and actions. The holistic view can be very useful, but there needs to be a balance (Barrow 1999: 130–1).

Ideas for further study or work

1   Call to mind a recent leisure or recreation event of personal significance to you, and identify all the personal benefits and costs involved, describe their nature and the time over which they would accrue.

2   For the same event, identify all the benefits and costs (their nature and the time over which they would apply) to others. Who are the others?

3   Are there the equivalent of transaction inefficiencies, where (at the present) costs are greater than they should be, or where benefits are sub-optimised? Characterise the settings where these are likely to arise and suggest remedies.

4   Survey the area in which you live, work or study, and note the public spaces (inside or outside but accessible without charge) where people gather, to meet

and to talk. What groups are attracted, and what constraints seem to apply? What interventions could managers make to increase benefits (or reduce costs)? Who are the managers concerned?

5　In a selected area, close to where you live, identify which of the listed trends are likely to have greatest effect. Describe what effects these might be, and list other trends or issues which will be of particular relevance in relation to leisure and recreation.

Reading

Benefits and costs: from the psychological and sociological viewpoint

Writings extolling the virtues and benefits of leisure and recreation stretch back a very long way. Plato, the Bible and Cicero are among these sources. Most writers now tend to specialise and see the benefits from a particular perspective. The fundamental psychology has been touched on in Chapter 1: Gross, Csikzentmihalyi, Argyle, etc. Greenfield gives some understanding of the physiology and biochemistry involved. Beyond that, Kelly sets out a framework, and with Godbey looks at the sociology of leisure. Driver has encouraged academics to bring together examples to cover the complete span of benefits, including those that relate to health. This is an approach adopted and developed through a collaborative process in Ontario, by the Parks and Recreation Federation, compiling the *Benefits Catalogue*. At the more personal end, there are powerful examples such as Cousins, which encourage us to take a broader view.

Settings and trends

Better to enjoy the many different settings than read about them, but there is no shortage of material. A popular starting point has to be the federal or government agencies, which look after the wide open spaces, and often manage large areas. For many of the indoor spaces, too, there are still arms of government, whether national or local, which are natural starting points. Travel writers have produced any number of guides, and most areas have been well researched. The local library and university will have details. For trends, read the news, and study the statistics sites. In Britain the Office for National Statistics (http://www.ons.gov.uk) produces regular reports, and many departments publish details of their statistics on their web sites, so look through gateway sites (http://www.open.ov.uk). For tourism, the joint site (http://www.star.org.uk) is a good place to start.

Attaching values

Overview

There are many different kinds of value. When seeking to measure benefits and costs, economists have a range of methods under differing conditions. But measuring is difficult. By looking at some of the methods economists and others use, we reveal different facets of value. Where goods and services are priced, the answers should be simple (or simpler) to understand. When dealing with goods that have no proper market, the methods are usually anything but straightforward. Usually the method seeks to evaluate the benefit or the cost in terms that are closer to the market. In this way the value of a visit may be related to the cost of travel. Other studies rely on asking what people prefer. The trick is to ask the right question, and to be able to interpret the answer. Economists vary in their view of which methods work best, under which conditions, and some of the issues are discussed.

Selecting a clutch of indicators can sometimes help measure different attributes at the same time, and so give a more multidimensional view. The work of the New Economics Foundation (in revealing hidden variables in charting the rise in gross national product) encourages us to think laterally, and not be bound by convention, as has the emergence of the Genuine Progress Indicator, developed by Redefining Progress. Life Cycle Assessment techniques can be used to explore the full ecology of recreation, and we also look at benchmarking as a valuable tool to focus on relative measures.

Asking people will not always reveal what they really want, but it is a good place to start. There are several general surveys that have been carried out for a number of years, and these give a general snapshot of what is going on. In Britain, the Office for National Statistics carries out surveys to help meet the needs of government departments. One of these surveys, the General Household Survey, includes questions on what people do with their leisure. Some of the government agencies commission their own surveys to reveal more information about their own area of responsibility. A consortium of the Countryside Recreation Network agencies has co-sponsored the United Kingdom Day Visits Survey. Co-operation of this kind reduces some of the problems otherwise introduced by using different definitions and data sets. Despite the existence of a wide range of surveys, there is no one comprehensive British leisure survey, although an ambitious project is being piloted.

Attaching labels in the form of designations is another way of recognising value. This can pose difficulties as our priorities change, and yet (for the same reason) designations can provide a framework to encourage continued management when resources become scarce, by ensuring that certain priorities are maintained. More attention has been placed on our natural than on our cultural environment. Professionals have had the leading role in setting values in the past, albeit modified by politicians. There have been attempts to involve some of the groups usually excluded, and an example is reviewed which involved children. We have also tended to mark out certain times as special – festivals, and celebrations – but these have been under threat from commercialisation.

The experience in Europe is that laws, alone, fail to adequately safeguard the quality of our environment. The development of Environmental Impact Assessment as a means of gaining planning approval through the Community's Eco-management Audit Scheme should work well for the larger projects, but the scheme does not include smaller projects and will not take account of the cumulative or compound effects. Curiously, the forces which led to much of the damage to our environment can be deployed to good effect though Environmental Management, which is being adopted readily where large savings can be made, and profits increased. This can be used to help develop the Environmental Management Ethic, which once in place should be the most reliable safeguard of what will become shared values.

Using economics: benefits and costs

Economics can be used to ascribe values to activities of all kinds in order to help us understand the effects of these activities, or choose those with the best outcome. Inevitably, this means we have to decide from whose point of view we are looking. Positive economics can be invoked to ensure that we are merely attempting to describe the process rather than suggest what the desired outcome should be. Put to use by politicians, whether at local or national level, value systems are applied which will tend to favour particular groups or systems. Describing leisure and recreation solely in terms of economics and benefits may be deeply offensive to some, particularly those who consider 'economic' to be synonymous with 'financial'. It may suggest a reductionist approach where everything can be measured and put in a box.

An alternative view is that many benefits defy simple measurement, and the qualities on offer will often be viewed differently across the many philosophies and religions. From this viewpoint, economics is just one tool, and an imperfect one at that. Where measurement is too difficult or costly, description may be a better way forward, and one that allows us to share different understandings, and values. When talking to policy makers, economics is useful. When dealing with individuals an approach that takes account of the emotional significance of choices may be appropriate. Individual decisions about leisure are often made intuitively.

BOX 5.1 Man's best friend

By 1990 it was estimated that in the countries which were then in the European Community there were 170 million pets, and on average, just under half of all households had a pet (Burger 1990: 44).

In the industrialised countries of the West, dogs are still used for working

---

with sheep, for hunting (although values here seem to be changing) and in crucial work guiding blind people, helping deaf people, and in more specialised applications. Apart from work, dogs are now increasingly valued as pets.

The behaviour of dogs is markedly different from that of wolves. There is increased representation of infantile behaviours, including play and submission, and disruption of the natural stalking and predatory routines (Bradshaw and Brown 1990). The process of domestication of animals as pets and selective breeding continues, and the amount of time and money devoted to the process is substantial.

Beyond the obvious pleasure they give their owners, controlled studies have demonstrated that pets are beneficial to the health of humans. Studies report repeated beneficial effects in the short term (Serpell 1990: 1) and in the longer term, notably among coronary care patients (Friedmann 1990: 8–17). The benefits noted include:

♦   encouraging physical fitness through exercise;
♦   decreasing anxiety and stress, by providing an external focus for
     attention, promoting feelings of safety, and providing contact comfort;
♦   decreasing loneliness and depression, by promoting companionship and
     an interest, and by encouraging a desire to nurture.

There have also been many successful programmes of keeping pets among patients in long-term care, to beneficial effect. Elderly people are less likely starve themselves or become hypothermic in winter because, in their effort to care for their pets, they themselves benefit.

On the other hand there is no doubt that the keeping of pets has an effect on the natural environment, and in some parts of the world pressure has been rising to contain the impact of domestic cats (in Australia). Dogs too can cause concern. They can spread diseases, chiefly through human contact with their excrement, and this is obviously a concern, particularly for youngsters. *Toxicara canis*, a worm that dogs can carry, can cause blindness. The bacterium *Escherichia coli 0157* can cause real problems, and indeed until recently, rabies (a prime reason for keeping quarantine regulations for pets entering Britain) has given rise to concern. Meanwhile, people who unintentionally share their bathing water with rats (in canals and sometimes rivers) may risk catching Weil's disease. People walking through bracken and heather, where deer have been, may risk the attachment of a small tick, and if even less fortunate may risk catching Lyme's disease, which if not treated early can be fatal. The risks are very low, but exist. There has been rising concern too about diseases the infectivity of which crosses over from one species to another (zoonoses) following the outbreak and spread of HIV (which it is thought was previously carried by chimpanzees) and CJD (from cattle), and this may help shape attitudes to animals in the future. The benefits and costs associated with pets may be rather more difficult to identify than many of us assume.

Einstein is credited with saying that 'The things that can be measured don't count, and the things that can't be measured *do* count' (Black 1996). 'You can measure everything.' Of course, but not necessarily to reveal or add any meaning. Some joker has come up with a unit: one millihelen = the amount of beauty required to launch a single ship (http://funnies.paco.to, 21 August 2000), but more seriously:

a major problem in cost–benefit analysis is the evaluation of certain types of cost or benefit. The first problem is that of measurement in physical units. We may measure savings in travelling time in minutes and hours, and noise nuisance in decibels, but how do we measure the 'amount of pleasure' derived from a particular piece of scenery? The second problem is that of reducing all costs and benefits to a 'common unit of account' so that they are comparable with each other.

(Bannock *et al.* 1984: 97)

The temptation is to measure those things that can be measured easily in units that are available, rather than to seek to measure those things in which we are really interested, but which seem to present difficulties. The two conflicting viewpoints – that everything worthwhile can be measured, if we just work at it, and the opposing view that no experiences of very great value can be properly measured – give an idea of the scope of the problem. Of course this is a gross oversimplification: the best studies develop from a sound understanding of values, through the application of sociology and psychology, before economics is set to work. The temptation is to cut corners.

In the examples quoted above, the physical units can be only proxy measures. Most such measures will be on some scale that if not absolute will have some reference point against which the scale is set. These measures may be on a continuous scale or may be a series of steps, i.e. analogue or digital in nature. They may be nominal, ordinal, interval or ratio. At the very simplest level we must choose the appropriate scale, and this will vary according to exactly what is being measured (Tull and Hawkins 1984).

Assumptions also need to be made about the scale of measurement. After all,

TABLE 5.1 Measurement scales

Scale	Use	Examples	Manipulations
Nominal	Classes	Adults and youngsters	Percentage Mode
Ordinal	Ranking	Preferences	Median
Interval	Index	Distance	Mean Range
Ratio	Proportions	Fractions	Proportions

Note: Depending on the different scales used in measuring and gathering data, different manipulations can be employed in interpreting the results depending on the scale of measurement the coast of the United Kingdom can boast a length of anything between thousands of miles and infinity. This is because the length depends on the rule you use. If measuring the distance by running a ruler round each indentation of each rock around the coast, the length is great indeed. If measuring around each atom which makes up each molecule which makes up each crystal which makes up each rock, the answer will be infinite. None of the answers is wrong, and each will have a different value and relevance. We should also make sure that we do not spend more money, time or energy on gathering the information than it is worth. Finally, we must make sure that we do not destroy, or even just change, what it is we are measuring by the process of measurement. Other questions are suggested when ascribing values. Are all minutes saved of the same value, for each individual or between individuals? The nuisance value of noise (Bond 1996) is not merely a matter of the decibels, but depends on whether those decibels induce pain or discomfort. Thresholds vary, and the nature of the decibels varies as to whether they are wanted decibels or not. People who enjoy flying will treat the noise of an aircraft rather differently from those who do not.

Economists employ a number of analytical techniques in evaluating the benefits and costs, working in a linear sequence of rational thought to solve the problem. These are modes of working that are typically dominated by the left hemisphere of the brain. Increasingly breakthroughs are being made where the intuitive mode of thinking, typically dominated by the right hemisphere of the brain (Russell 1979) is also applied. The emphasis is still in recognising a pattern, and a logical explanation.

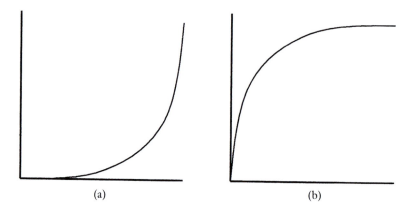

(a)   (b)

FIGURE 5.1 Total utility and marginal utility. (a) Each increment of time brings more benefit, so that total utility builds up over time. (b) The marginal utility (the utility added by each extra unit of time devoted to the activity) gradually declines

The scope for measurement

Measuring the benefits and costs of leisure and recreation can help in the decision making of managers, resource holders, politicians, communities and occasionally participants. Usually participants do not go through the formal steps we shall be considering, but take their decision intuitively (measuring intuitively), or through negotiation with others. Where resources are limited, we need the extra information (and formal steps) to help us determine where best to invest, or intervene. Where activities are governed entirely by markets, preferences are revealed through the prices people pay for the goods and services concerned, which are governed by supply and demand. It is this that partly explains the paradox of value, that people seem willing to pay relatively higher prices for what seem to be luxuries and yet much lower prices for necessities. The fundamental assumption in utility theory is that people will seek to maximise their total utility, the total satisfaction, from the consumption of some good. Each successive unit is usually of less value than the last (the concept of

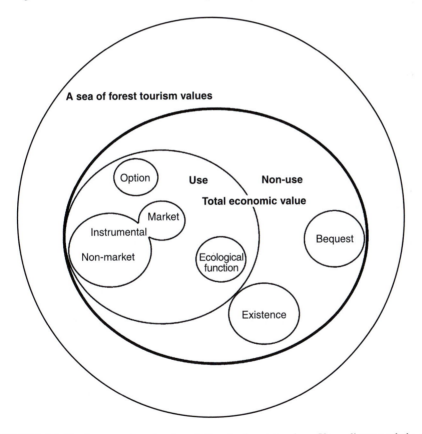

FIGURE 5.2 Total economic value in relation to forest tourism. Venn diagrams help to explain that economic value is not all, and that there are different kinds of economic value
Source: Broadhurst and Harrop (2000)

marginal utility), and will add rather less to the total utility. This gives rise to demand curves which show how prices vary with demand. Consumers will pay successively less for each extra unit (and still gain consumer surplus) so long as the price paid remains below the value held. The relationship between demand and price can exhibit elasticity where the extra price paid is related to marginal utility. The relationship will be inelastic if the item is considered basic (where, for example, the consumer considers the good essential) and increasingly elastic as the package is seen to be a luxury.

Where there are public goods involved, i.e. goods and services or benefits that cannot be withheld from all the consumers, or quasi-public goods, then no market

TABLE 5.2 What to measure, when and how? Some easy ways to measure performance, in this example relating to an evening out at the theatre

Before, during, and after the event	What to measure? (But think, 'Why?' before measuring at all)	When, and how?
Read review	Awareness (of play, of cast, of theatre), expectations	Household, telephone survey, focus group
Book tickets	Number, rate, value and distribution of bookings	Automatic, from ticketing software
Arrive at theatre	Expectations	Personal and very brief interview
While waiting for friends: drink and snack in the theatre bar, before the play	Takings, and products sold	Automatic from tills with PLU set up
In the auditorium	Proportion of seats taken	A quick glance from the front, just before the play starts, backed up by the automatic details collected through ticketing
During the play	Level of satisfaction, clapping	Personal judgement
After the play	Level of satisfaction, clapping	Length of ovation, volume of clapping, buzz among audience on the way out
Word-of-mouth advertising	Referral	Survey of those buying tickets, of how they found out about the play
Well after the play	Sales of scripts	Through booksellers and publishers

exists, and a different approach is required. Economists and others have developed an array of techniques (Clawson and Knetsch 1971; Glass 1997; Walsh 1986). There are at least six areas where we should be asking questions, when thinking about what to measure:

◆ *What benefits and costs should we measure?* Things are usually more complex than we think. 'We think of causation as being a single line, when it is in fact a multidimensional net spreading far and wide' (Rowe 1991: 40). Are we concerned only with the first order costs and benefits, or second and third order ones too?

◆ *Why are we concerned?* It will help inform the process if we think what we are going to do (differently than we would otherwise do) once we have the information. What kind of decisions will be taken, and for what purpose: advocacy, rationing or allocation of resources? Then we can decide which aspect of the process to measure: perceptions, attitudes, preferences, intentions, behaviours or outcomes.

◆ *When should we measure?* Any event can be investigated in at least three crucial parts: beginning, middle and end. Making a map of the event (see BEAR mapping, Figure 3.6) over time will help to confirm what time envelope is most appropriate, and what stages are most important. The classic anticipation–journey–event–journey–recall model (Clawson and Knetsch 1971) can be very useful. Because events are multi-phasic, decisions need to be taken about when the data are best collected, without destroying the experience (McIntyre and Roggenbuck 1998).

◆ *How should we measure?* Measurement techniques, methodologies and approaches abound. Decisions have to be made about what is best in any one situation. How should substitute and complementary products, services or sites be taken into account?

◆ *Where, or over what area, should we measure?* Questions concerning the space, area and scale of resolution for any measures need to be specified.

◆ *Who?* Whose costs and benefits are we measuring? Is it at the level of the individual, families, community or society?

Most actions have unexpected as well as expected results, and each of these may take effect over different time periods. Consequently, it may be an advantage to use a basket of measures. In addition, we often consider that an action will result in either a cost or a benefit. Most actions result in a multitude of benefits and costs (Rowe 1991: 50), and leisure and recreation are no exceptions. Individuals will take part in their chosen leisure and recreation activities according to their own preferences, which can differ greatly, as in the example of rock climbing (Box 4.1). When public money is being spent to provide leisure and recreation, we want to be sure that we are using the money to best effect. Where a public asset (such as the national forest) is managed for a number of benefits, or multiple benefits, there is considerable interest in identifying the value of non-market benefits (like nature conservation, biodiversity, cultural heritage, non-market recreation). This will help ensure that decisions are not skewed toward the requirements of markets (in forestry, timber and market recreation), at the neglect of what can be very considerable non-market benefits for the nation. This is a notoriously difficult area, but it has been exercising economists and policy makers for a number of years. However, managers make implicit judgements about non-market benefits in their normal day-to-day work (Millar *et al.* 1992).

In the 1970s it was common for local authorities in Britain to demonstrate the amount of leisure and recreation that they were providing merely by announcing how much money was allocated to this function. Such an input-based measure may sometimes be the best available, but it is the most extreme proxy measure possible to gauge the outcome from the input. The value may be more, equal to, or less than the cost. All other things being equal, the two will be related, and the manager's task is to work to give the greatest value for the least cost. A similar misreading is possible at national level, where for some time gross domestic product has been used as a measure of our standard of living. The more we are able to spend, the better off we are. That was the logic. The work of the New Economics Foundation (Mayo and MacGillivray *c.* 1992) has demonstrated that such a measure can be misleading, as it takes no account of any defensive expenditure, and alternatives can be developed. Nevertheless, there has been increasing interest in the use of cost–benefit analysis, and the development of techniques to put a monetary value on non-market goods (Pearce *et al.* 1989). Of course there are also those who consider that this approach is part of the problem, rather than the solution (Adams 1994).

In the past people have tended to focus on *use values*, for example an evening at the theatre (market good), or a cycle ride through the countryside (non-market good).

There are also values associated with the proper functioning of systems. They will include the ecological function values, and also the socio-economic function values. This allows us to recognise some of the difficulties of reconciling the different values placed by individuals operating out of self-interest, and by individuals when

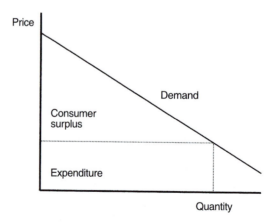

FIGURE 5.3 Consumer surplus and willingness to pay. The traditional representation of consumer surplus: the difference between what an individual would be prepared to pay and the price actually paid

pursuing the interests of the group. This echoes the problem rehearsed in the prisoners' dilemma (Kuhn 2000), where individuals would have to behave counter-intuitively to escape sentencing. Garett Hardin in his essay 'The tragedy of the commons', written in 1968, deals with (what he sees as) the inevitable destruction of pooled resources, and this has been simulated in many iterations and studies (Deadman 1999). As individuals we have many roles, in which we may act out different parts. As consumers we may express different preferences than as citizens (Owens 1992: 48).

Non-use values, on the other hand, have been found to make up a large proportion of some valuations, as much as 40 per cent for wilderness, and more than 60 per cent in relation to water (Hanley and Spash 1994: 66). These include:

- ◆ *option* value, such as that attached to a favourite book which is on the shelf and available to read, should I so wish;
- ◆ *existence* value, such as for the humpback whales, which we may never have a realistic chance of seeing, but to which we ascribe a value for their continued existence;
- ◆ *bequest* value, for the opportunity to pass to our children and grandchildren opportunities, perhaps to see Venice, and enjoy it as much as we did.

Preferences and values

By observing the behaviour of consumers under different circumstances with different prices, preferences and values can be explored. Where there are market transactions, it is easy. Where there are no markets, the preferences for such non-market goods can sometimes be revealed by relating them to goods for which there is a market (e.g. time/labour, travel, housing), using techniques such as the Travel Cost Method, Hedonic Pricing or the Dose Response Method. Where this is not possible, it becomes necessary to ask people what they might do, to seek their expressed or stated preferences, through the Contingent Valuation Method or Choice Experiments. Each of the methods has advantages and disadvantages, and is better suited to identifying particular values.

Travel Cost Method – revealed preference

Popularised in the United States in the 1960s (Clawson and Knetsch 1971), this method relies on looking at the change in demand across zones or distances from, say, a visitor attraction and looking at the costs involved in making the trip. The underlying principle is simple. Generally (all other things being equal), a visitor's preferences and valuation of a visit can be related to the costs incurred in travelling to the recreation site, in paying any entrance fees and the opportunity cost of the time spent in travelling. The propensity to travel is related to the disposable income available to the individual, and adjustments can be made for the socio-economic characteristics of

TABLE 5.3 Framework for summarising non-timber benefits of trees and woods. Examples are given to illustrate different categories

Benefit category	Type of value				
	Actual use values		Non-use values		
	Direct	Indirect	Bequest	Option	Existence
Ecological functions					
Carbon sequestration		Carbon sink Alternative to CH_4 and N_2O emitting agriculture	Knowledge of provision of woodland to reduce global warming	Increased tolerance to temperature change?	
Air filtration		'Green lungs'			
Hydrological effects		Regulating of drain and storm water run-off			
Control of micro-climates	Windbreaks			Local temperature control	
Soil protection/land restoration	Sewage sludge disposal Treatment of contaminated land	Soil stability Nutrient fixing			
Non-timber products and services					
Products	Revenue from: mushrooms, herbs, countryside products	Net employment generation Economic diversification		Genetic prospecting for unknown products	

Formal and informal recreation

Formal recreation	Walking Cycling Riding Orienteering Visitor nights, etc.	Breeding grounds for hunting, birdwatching		New types of recreation	
Informal recreation	Rights of ways for walking, dogs, jogging, etc.				Knowledge of right to roam

Nature conservation and biodiversity

Conservation and biodiversity	New habitat creation Increased diversity	Buffers between other habitats Changes in aquatic ecosystems (increased organic matter)	Ecological value of mature trees, decomposed wood, etc., in future		Knowing ecosystems exist

Landscape

Landscape character	Diversity, form, shape, point and linear features	Increased attractiveness to inward investment			
Local distinctiveness		Increased property values			

Social and community benefits

Education	Visits by schoolchildren	Inclusion in environmental studies curriculum			
Quality of life/well-being	Forest walks considered stress relieving	Soothing visual and sound impact			
Local economic effects	Local employment				

Source: Environmental Resources Management (1968: 6)

visitors. Surveys are conducted on site to reveal the number of trips made to the site each year, and the actual costs involved in the journey.

The following example is taken from a study by Benson and Willis (1992), commissioned to update a previous (1972) valuation given by the UK government's Treasury, to establish the monetary value of a visit to public forest managed for the nation by the Forestry Commission. The annual value of informal (non-market) recreation on the Forestry Commission estate could then be derived, by multiplying the value of a visit by the estimated number of visits made each year. Having such a value to hand would help in decisions on the amount of support which the government should give the Forestry Commission to manage its estate for informal recreation, and help in the development of policy. The Travel Cost Method was selected for its utility in determining direct use values.

The equation which describes the basic model is:

$$V_{ij}/N_i = f(TC_{ij}, T_{ij}, S_{ij}, A_{jk})$$

where V_{ij} = number of visitors in sample from area i to forest j, N_i = population of area i, TC_{ij} = travel costs from area i to forest j, T_{ij} = time costs from area i to forest j, S_{ij} = socio-economic characteristics of residents at area i who visit forest j, A_{jk} = attributes of forest j in relation to other substitute recreation sites (k).

The study produced an average net benefit per person of £2.00 (1988 values, equivalent to £3.16 in 2000), ranging across the fifteen clusters of forests from £1.34 to £3.31. This, multiplied by a conservative estimate of the number of visits gave a value for the estate of some £53 million, (equivalent to £84 million in 2000) which was more than five times that allowed just two years earlier by the National Audit Office (Benson and Willis 1992: 38–9).

Economists seek simplicity, and would be among the first to recognise the crude simplifications that have to be made in reaching such monetary evaluations. A marketing specialist or psychologist would take a different view, seeking to explore the rich variety in human behaviour of the different segments, or groups of individuals.

The pattern of trip-making in Britain, involving the car, reveals a bimodal preference. One interpretation is that this indicates two kinds of trip: one where the experience sought is at the end of the journey, and another where the journey is itself the experience. This fits with the notion of the car providing a setting within which fellow passengers enjoy a day out, and the social interaction is as important as any other benefit. For the first group, the travelling time represents a cost, and there are standard formulas to allow a fixed proportion of the worth of time (allocated according to the participant's salary). For the second group, time spent in this way may be a benefit and not a cost. Nevertheless, in countryside recreation the Travel Cost Method has been widely used to obtain economic evaluations, and been found to be reasonably accurate (Walsh 1986: 217–30). Up-to-date reference lists can be found on the web (http://www.sscnet.ucla.edu/ssc/labs/cameron/nrs98/tcostinv.htm).

The development of geographical information systems has enabled much greater sophistication in the use of these techniques (Bateman *et al.* 1996, 1999; Brainard *et al.* 1999; Lovett *et al.* 1997).

Hedonic Pricing

Although the value of leisure may be very difficult to evaluate convincingly, there is another way of capturing the value. In a sense it is an extension of the Travel Cost Method: a special case, in which the costs of moving to an area to gain the leisure or amenity benefits are identified through the relative cost of housing in that area. People choose to live in places where they are close to facilities that they can enjoy, whether these are pleasant outdoor wooded environments, areas close to the sea, within culturally rich cities, or places close to a range of leisure facilities. Comparing the difference in house prices close to popular places or leisure facilities, or tracking changes in house prices in the same area with developing facilities, will (other things being equal) reveal the value which residents ascribe to those facilities. Clearly, we would need to make corrections to allow for changes caused by other variables and general economic effects.

One of the problems the economist always faces is that 'all other things being equal' is an extremely rare event. Nevertheless, Hedonic Pricing is one way of tracking changes in market values, and the market value captures a basket of otherwise apparently non-market benefits.

Contingent Valuation – stated preference

Where no real markets exist, or the benefit being explored is related to non-use values, Contingent Valuation is a helpful technique. Researchers simply ask people how much they would be willing to pay to take part in a certain activity (or to be compensated for not taking part) in a particular scenario. There are variations in the way in which the questions can be framed or asked:

◆ *Open-ended*. How much would you pay for . . . ?
◆ *Framed*. Would you pay $£x$, $£x + 1$, $£x + 2$ . . . $£x + n$?
◆ *Shared*. If you were given $£x$, how would you distribute these across the following activities . . . ?

Intuitively this seems a very reasonable approach, but there are problems here too. The first is that the questions are asked without the respondent having any real reference point. Who should be asked the question is another problem (Owens 1992: 47). If the person does not already take part in the activity, they may have little idea what to bid (like a novice in a card game). Reality is temporarily suspended, and the respondents have to put themselves in a hypothetical situation. They may behave rather differently with imaginary money than they would with the real stuff. What people are willing to pay will also depend greatly on their own personal disposable income, as well as on their valuations. The perceived wisdom is that results using Contingent Valuation have often yielded much greater values than the Travel Cost Method, although a comparison (Carson *et al.* 1996) of a large number of studies failed to support this contention. One study (Bennett and Tranter 1997: 220) using

Contingent Valuation has produced a cost–benefit ratio for the provision of access for recreation in woodland as high as 1:17. Such a result must surely help encourage managers to continue to maintain such access, but such techniques should only ever be a part of the support for decision making (Spash 1992: 73). The use of Contingent Valuation in the *Exxon Valdez* oil spill contributed to its respectability, but also demonstrated that its use according to the guidelines drawn up was prohibitively expensive, except in the biggest cases.

Choice Experiments (or conjoint analysis)

This is a technique which has been much favoured in marketing, where often preferences are being elicited for products which have yet to be produced, and is now frequently used in environmental economics. Respondents are asked to rank a series of combinations of attributes, to reveal overall preferences. There is a limit to the number of attributes that can reasonably be explored in this way without fatigue setting in. There are also other techniques using sophisticated mathematical techniques such as multidimensional scaling which can reveal relationships between choices, and display them in a visual way, thereby helping to identify relative strengths and ways to position products in the market (Tull and Hawkins 1984: 314–20) and this has been applied to nature tourism values (Holmes *et al.* 1996).

Capital: natural and man-made

For a long time accountants and economists have distinguished between capital and revenue expenditure. In recent times, innovators have sought to apply these ideas in human resource accounting. The basic idea is that some assets will give rise to a stream of benefits, and should be regarded as capital. In economic terms resources can be accrued and built up as capital and then invested to yield revenue. In human resource accounting the individual or team may be the initial capital, that can be built up as knowledge and skills are added. Such human resource accounting has been used by some organisations compiling annual accounts, and most notably by professional sports clubs, perhaps unsurprisingly where transfer fees can run into several million pounds for a single individual. The same kind of thinking can be applied to the natural environment, and has successfully raised the political profile of the impacts on our natural and, on occasion, cultural heritage.

Some environmental capital cannot be easily replaced, and for some, there are few alternatives. Our environment is in continual change, although the periods of time involved can be rather humbling. Fossil fuels are being created, albeit at a very much slower rate than we are consuming them. Similarly mountains are still growing and other parts of the earth are being eroded. Nevertheless, viewing the earth in our own time-scale, erosion of this capital is serious (at least for our survival). *Critical Natural Capital* has been defined by English Nature as:

consisting of assets, stock levels or quality levels that are:

- highly valued, and
- essential to human health; or
- essential to the functioning of life support systems, or
- irreplaceable or unsubstitutable for all practical purposes (e.g. because of antiquity, complexity, specialisation, location).

(Gillespie and Shepherd 1995)

Constant Natural Assets, on the other hand, represent the tradable or replaceable elements, which can be offered up in land use change, but for which there must be full compensation.

Removal of forests in one area may be replaced by planting forests in another. A forest is more than a collection of trees, but the principle is clear. There are some natural resources which exist as patches or matrix and of which we need a certain amount, but it need not be located in exactly the same place for all time. There is a need for these resources to be conserved, although they may be distributed in different ways, so that together, they add up to the Constant Natural Assets required. This kind of thinking provides a bridge between the natural and social sciences, and economics and landscape ecology, allowing a comparison of benefits and costs. The approach is not without its critics (Adams 1994).

Another popular concept in deriving values is discounted cash flows, or other benefit streams. We all prefer to have benefits now rather than benefits in the future; 'jam today is better than jam tomorrow'. This gives rise to the artifice of interest rates, whereby through markets we invest money now for increased benefits in the future, to make up for the perceived (and therefore real?) difference in value. Within a generation this presents few problems ethically. Projected forward, across generations, it becomes literally (and in a different sense) a discounting of the future, and economics has not yet addressed this major problem, although there are strenuous efforts (Pearce *et al.* 1999). It suggests that the options available to our children are less important, or valuable, than options available to ourselves. This is not morally defensible, and economics has yet to confront the ethical issues involved (Hanley and Spash 1994: 145).

Capturing externalities

Much effort in environmental economics has been directed at attempting to ensure that costs as well as benefits are distributed equitably. One way of doing this is to identify the externalities (Tietenberg 1994: 36–7), and to set about ensuring that they are captured and brought back into the system. Often these reflect gains by individuals and consequent costs to other individuals, or to society. A popular articulation of this is the derivation of a 'polluter pays' principle, where previously the polluter may well not have paid. Britain has been steadily increasing taxes on motor vehicles to attempt to reduce the emission of greenhouse gases by curbing the amount of traffic, and in order to recapture externalities by charging the polluter for the damage being wrought. Benefits can be externalities also, as Figure 5.4, shows for a woodland project (Bateman 1992). Making decisions that do not take account of these externalities may well lead to problems of distribution.

A problem, loss or cost for one individual may be an opportunity, gain or benefit for another. Many of us seek out holidays in exotic places, in a quite different natural and cultural environment. For those from developed countries, the excitement is often in visiting a less developed country, to experience something quite different. Under these circumstances, we are often surprised by the low standard of living in the host country, and the people's approach to life. Sometimes there may be some guilt attached to our holiday, a concern that it might be exploitative. We may recognise that our visit will have a lasting effect on the host communities, and that their future choices may be constrained by the changes wrought by tourism (Goodall 1992: 60–74). At other times, we may be happy that at least some of the proceeds of our holiday spending will be entering the local economy. There will be mixed emotions. Even when we remain at home, lifestyles in the West may appear to be rather overheated compared with those in other parts of the world. What are the opportunity costs of our consumption? We may be spending more on a holiday than our hosts spend in a decade on keeping a family. This problem of distribution is not confined to leisure and recreation. In developing projects and programmes the intention is to maximise the net benefit, the additionality, usually striving for the greatest benefit for the greatest possible number of people.

When considering the economic impact of leisure, recreation and tourism on local communities, multiplier studies (Mathieson and Wall 1982) take account of

- *first order* direct expenditures (e.g. what someone spends on the holiday at their hotel);
- *second order* indirect expenditures (e.g. what the hotel does with that money, paying salaries and for goods, as cost of sales);
- *third order* induced effects (e.g. how hotel staff spend their salaries in the local economy or elsewhere).

By relating the expenditures between the different levels, a multiplier can be determined. There is though the problem of leakage, which can be of many kinds. Profits of the hotel may be remitted back to the home base, which may be a large chain, and components of some expenditure made locally will include taxes, which will head back to the appropriate level – municipal or national government. For each pound spent by a tourist in the area, we can know the indirect and induced effects, and how such spending adds further value to the local economy. Studies of this kind were becoming rather disreputable, as the figures tended to become larger and larger. There are also difficulties of defining the boundaries of the economy under study, although there have been studies which make the distinction. A study by Archer, quoted by Stabler (2000), revealed a muliplier for the United Kingdom of 1.78, and a lesser figure of 0.24 for Keswick, a holiday area in the Lake District, for example. Such studies can be resource-intensive, although there have been attempts to use Proportional Multiplier Studies, using smaller (and more carefully selected) samples

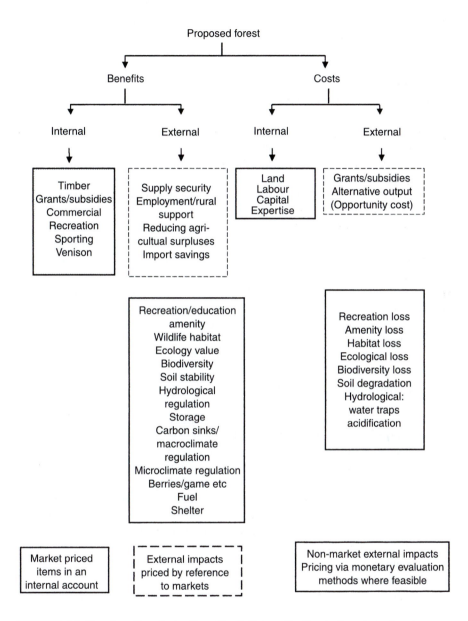

FIGURE 5.4 Cost–benefit analysis of woodland. 'Internal' refers to the private forestry company's costs and benefits while 'external' refers to costs and benefits accruing directly to society. Shadow pricing must be applied to all values
Source: Bateman (1992)

to reduce the cost (Vaughan *et al.* 2000). There remains the acknowledged problem that these techniques almost entirely ignore costs (Stabler 2000), and, consequently, it is perhaps not surprising that such studies are often commissioned by advocates of tourism.

At a more local scale the benefits and costs generated by recreation may not be evenly (or equally) distributed. Is the expenditure at one site truly additional, or merely displacing expenditure at another site altogether? In Britain, we have to employ great skill if we are to ensure that the farmer, forester or other land manager will continue to manage the landscape to produce the scenery that the visitor seeks and values. There has to be some linkage in the environmental economics to make the systems work, and we require to understand the motivations of the different actors involved. In our example, it is usually the land manager who bears the immediate costs (and some less immediate costs such as the cost of public liability insurance, and associated safety checks, and the visitor who gains; although this may be offset by grants to the owner (or cross-compliance) and taxes on the visitor. This complex issue is usually resolved through a mixture of market forces, aided by government intervention through a complex web of taxes and grants. In encouraging activity in one part of the country, by giving grant aid to, say, the development of an art gallery or museum in one area, we need to consider the effect on galleries and museums in neighbouring areas. The problem of displacement is a vexed one, although in relation to some activities there are now computer models that can be invoked to help decision making of this sort.

Alternative approaches

Quality management has been a rapidly expanding field of enquiry, and has spawned many different theories and approaches; Buttle (1993) gives a very useful summary. Among the standards institutions of the world, several quality management systems have developed: in Britain BS 5750, in Europe CEN 29000 and the international programme ISO 9000. These approaches do not of themselves guarantee quality, so much as a quality. Other approaches such as the Servqual instrument developed by Berry, Parasuraman and Zeithaml provide measurement scales to measure service quality against five factors commonly associated with quality:

- reliability;
- responsiveness;
- assurance;
- empathy;
- tangibles.

The BPZ team developed the Gaps model to identify individual discrepancies which when aggregated explained the difference between customers' expectations and their perception of performance. For an approach to using these techniques in facilities management see Alexander (1996), and for a review of the reasons why quality management is still gathering momentum see Robinson (1999). Sport England and a number of partners have championed a quality system called Quest for use in the sport and leisure industry. From the early quality systems, environmental quality systems developed, first as BS 7570 and thereafter as the ISO 14000 series, and in respect of Europe as the Community's Eco-management Audit Scheme (see p. 155).

Another approach is to identify a number of key indicators for the range of goods or services with which we are concerned. Keeping these indicators firmly in view, we can strive towards ever more efficient, effective and economic approaches when we manage. However, precisely because performance indicators do encourage sharper focus, they can be extremely dangerous. There is a temptation only to set down indicators that are more easily measured, and neglect some of the qualitative aspects that we need to bear in mind. A number of studies exist that point to quite different priorities as between managers and users. In some cases it is merely that the two groups score qualities differently, and in others it may be that different qualities are included. In one study of providers of activity holidays, managers identified comfort as the number one factor, and the customers' concern with safety was over-looked (KPMG Peat Marwick/Tourism Co. 1994).

Indicators are powerful. We become accustomed to their use to the extent that they sometimes act as a screen. For many years, one of the most powerful indicators of economic well-being has been gross national product, a measure of the total flow of goods and services (gross domestic product) combined with the total value of invest-ments. As can be seen in Figure 5.5, for Britain, it has been rising ever since the 1950s. However, as with any indicators, the tool is rather crude. It is based on what people spend their money on. It is therefore concerned with inputs. The implication is that

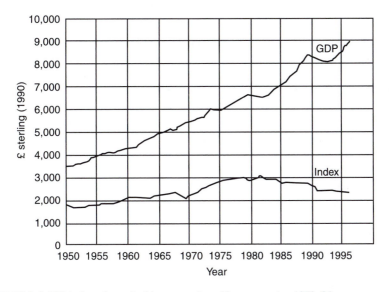

FIGURE 5.5 UK index of sustainable economic welfare *per capita*, 1950–96
Source: Mayo and MacGillivray (*c.* 1992)

the index measures wealth and by inference well-being. The measure is concerned with prices, and not values. The New Economics Foundation (Jackson and Marks 1992) has carried out some imaginative work, unpicking the way in which the gross national product is aggregated and restructuring an index to take account of defensive spending, to create an Index of Sustainable Economic Welfare. This shows a different picture. There has indeed been a very modest rise in this index since the 1950s, but more important there has been a decline since the 1970s. The two different curves, extrapolated, will lead to very different results, and this is instructive. Adjustments include taking account of any long-term environmental damage, income inequality and the value of household labour.

Often it is easier and more relevant to make comparisons between different options or ways of managing environments than it is to seek to measure absolutes. Finding truly comparable examples against which to benchmark may not be easy, but the process of benchmarking is straightforward. Putting together a matrix of the most important customer requirements against the performance measures of the organisation and a handful of competitors, in respect of each requirement, presents a powerful map of what needs to be done (Department of Trade and Industry 1992: 21).

To the manager of a National Park or Monument, designated because of unique features, or of one of the IMAX cinemas, it might be hard to find a comparable example for such purposes close at hand. However, if you identify the components of the recreation activity, or in this instance day out, you will find many elements for which comparable examples are at hand. Focus only on the really crucial customer requirements that make the difference between the good and the best, say the top three, to make the process feasible.

Either identify competitors who you think are leaders, and chart your perform-

---

BOX 5.2 A visit to an IMAX cinema: customer requirements to consider

- Advertising
- Signposting
- Threshold
- Car parking
- Terminus of public transport
- Greeting
- Ticketing
- Waiting, queueing
- Ambience
- Contact with staff
- Availability of refreshments, associated goods and services
- Toilets
- Cinema auditorium

- More waiting
- Choice of film
- Quality of projection
- Quality of sound
- Comfort of seats
- Cleanliness of auditorium
- Temperature of auditorium
- Inconspicuous audience
- Lighting after performance
- Egress from the auditorium
- Farewell
- Signposting to exit, or next event

ance against their performance for the requirements of the customer and the components of the visit, or identify leaders for each component, with whom you wish to benchmark. Focus on the most crucial stages, and identify examples close to you of leaders in that sphere. Then visit those places, and glean every scrap of information you can about how they are managed, the inputs and the outputs. Keep a check on how your management of these elements matches up.

Life Cycle Assessment is a technique devised to explore the whole life benefits and costs associated with the production and consumption of (usually) a good, from first to last. The same technique can also be amended to explore the costs and benefits associated with the provision of a service, and can recapture what have often been regarded as externalities. As an audit the process demands considerable effort in data collection, but without importing the values, the approach can still be very helpful for identifying critical issues.

Figure 5.6 describes a possible approach, building on a case study that considered a holiday product (Cope and Sisman 1994), but the technique is very flexible, and could be combined with the BEAR mapping (Figure 3.6) to identify appropriate stages. This technique will help to ensure that the full development inputs and impacts are included as well as those of the leisure or recreation event itself – including waste, energy and transport considerations.

Surveys and what they tell us

To know what the population wants, and how they attach values, we need to ask a sample, in their own homes rather than in the place they choose to visit (Huff 1973). We can also ask people to complete leisure diaries to record what they do. Asking only people who visit the establishment will give partial information. Also, managers tend to ask about their facilities, whereas the product of interest to participants is the experience. Many of the experiences of greatest value to the participant will be beyond the reach of the manager. People sometimes find it difficult to answer questions fully. The difficulty may be in articulating a reply, because the process breaks the almost subconscious flow of experience. Respondents may also give you the reply they think you want. For these reasons, observing how people actually behave can be a useful auxiliary mechanism or substitute.

In the United Kingdom questions were asked in the 1980s about how many day visits were made to the countryside; Ministers of government were interested in the economic impact of leisure and tourism. The answer depended upon who you asked, because different people interpreted the question differently. The tourism agency defined a day visit as anything longer than three hours' duration, whereas the Countryside Agency defined the visit by purpose. The answers were wildly different. As a result, representatives from the agencies concerned met round a table, as a sub-group of the Countryside Recreation Research Advisory Group (the precursor of the Countryside Recreation Network), to discuss the nature of the problem. In these meetings the representatives began to shape a survey (UK Day Visits Survey), piloted first in 1994, and now running in alternate years. The survey reveals the volume and value of such visits. The results give an idea of what takes place over the United Kingdom as a whole, but provides little information at a regional level. As the data build, so regional information accumulates and, for example, in Scotland it has been possible to provide analysis at this regional level (Scottish Tourist Board 1998); see http://www.staruk.org.uk./.

The General Household Survey, an annual survey of something like 10,000

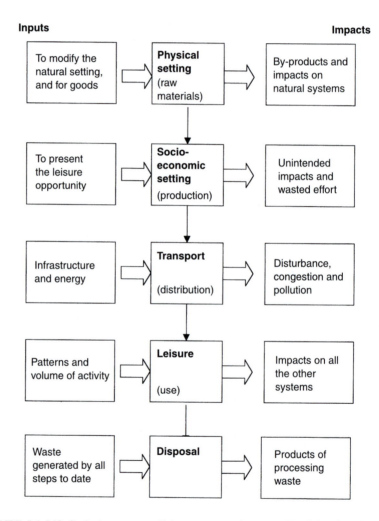

FIGURE 5.6 Life Cycle Assessment of leisure or recreation event. Traditionally this approach encourages inclusive analysis of all the flows (costs and benefits in terms of resources) over the lifetime of the product and it is very data-hungry. Applied to leisure, a period of time needs to be defined, and it is of more value in checking that all the impacts are included (for instance, travel to the site)

Source: Adapted from Cope and Sisman (1994)

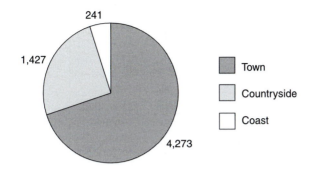

FIGURE 5.7 Day visits in the United Kingdom, 1998. (a) Number of day visits in Great Britain (million)
Source: Social and Community Planning Research (1999)

FIGURE 5.7 Day visits in the United Kingdom, 1998. (b) Spend per visit, by destination
Source: Social and Community Planning Research (1999)

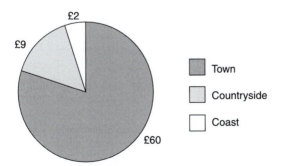

FIGURE 5.7 Day visits in the United Kingdom, 1998. (c) Total spend on day visits, by destination (£ billion)
Source: Social and Community Planning Research (1999)

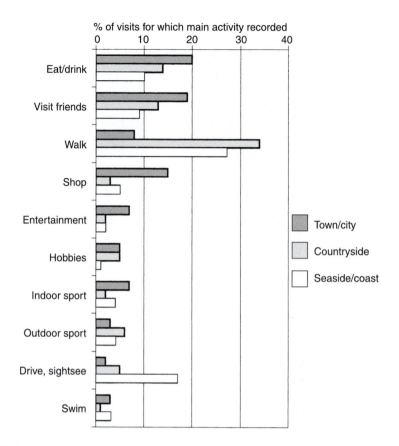

% of visits for which main activity recorded

FIGURE 5.7 Day visits in the United Kingdom, 1998. (d) Main activity on the visit, by destination

Source: Social and Community Planning Research (1999)

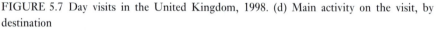
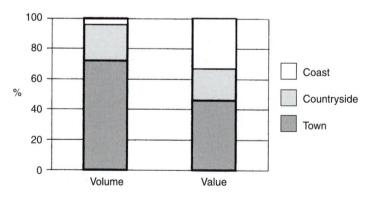

FIGURE 5.7 Day visits in the United Kingdom, 1998. (e) Volume of visits, and spend, by destination

Source: Social and Community Planning Research (1999)

households in the United Kingdom, has been conducted since 1971. Many countries have something similar. What marks this particular survey out is the size of the sample, and the fact that the data set has been continually added to since 1971. Typically, some 18,000 people (all over the age of 16) are interviewed, on a range of topics. There are questions on certain core areas each year such as income, economic activity, leisure activities, health and family composition. It remains invaluable for policy makers. As well as these core questions, there is also space to include topical questions each year. The results are reported in *Living in Britain: The Results from the General Household Survey* (Thomas *et al.* 1998) and form the basis of many articles in a range of journals, including *Social Trends*. The survey is conducted by the Social Survey Division of the Office for National Statistics (http://www.ons.gov.uk/).

What people do with their time continually changes, as do the definitions of leisure and recreation. Several studies have periodically been carried out to reveal the way we use time (Gershuny and Fisher 1999). The information is of great value to anyone considering how to manage environments for leisure and recreation. There are attempts to harmonise survey techniques and questions in order to develop a pan-European data set relating to time use. In the past our activities have been inextricably linked with natural rhythms. Thanks to Thomas Edison and others, we can now turn night into day at the flick of a switch, and rearrange our patterns of work accordingly. A telecommunications company and a bank jointly commissioned a major study into our changing habits (Future Foundation 1998). The study explored the degree to which we are developing a twenty-four-hour day, and what services and leisure opportunities consumers are expecting to be available at any time of day or night.

Commercial organisations are often in a position where surveys of this kind are

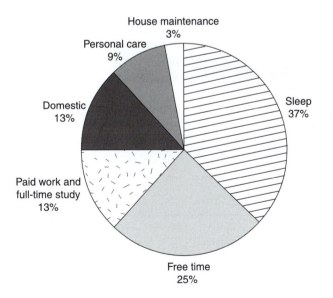

FIGURE 5.8 Time use, averaged (from 15 years to over 60 years)
Source: Office of National Statistics, *Social Trends* (1997)

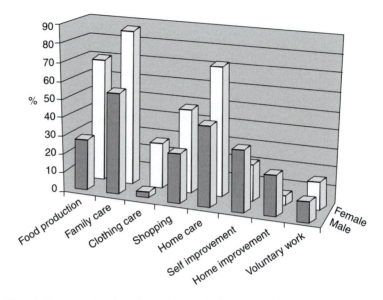

FIGURE 5.9 Time use: division of unpaid work in the household between males and females
Source: Office of National Statistics, *Social Trends* (1997)

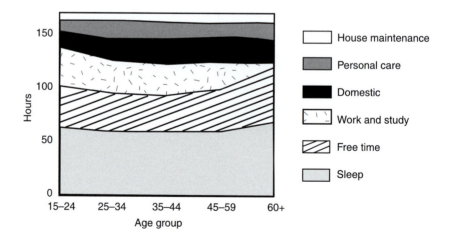

FIGURE 5.10 Time use at different stages of life, showing the distribution of hours in the week (out of a total of 168 available)
Source: Office of National Statistics, *Social Trends* (1997)

superfluous. Preferences are revealed through prices paid for tickets, services or goods. This market information may be supplemented by survey information, but this is not generally published. The information is commercially sensitive. We can, though, gain a great deal of information from on-site surveys, where profit is not necessarily the driving motive. Many organisations that manage forests (especially state forest services concerned with the delivery of non-market benefits) carry out regular recreation surveys. These usually collect data to provide a profile of the visit, and are often combined with the monitoring of visitor numbers. The typical survey in Britain gives a snapshot of what people do in the forest. A different selection of forests is covered each year to build up a reasonable picture of what is going on across Britain. In addition, certain indicator sites are monitored, to build up an idea of how the pattern of visits changes over time. Reports of the surveys are prepared principally to help managers, but the reports are made available to those who want copies, and are also published on web pages (http://www.forestry.gov.uk/).

Most countries have regular surveys conducted within their national parks, e.g. the UK All Park Visitor Survey. Many are available on the web, e.g. for Canada (http://parkscanada.pch.gc.ca/). The development of leisure, recreation, and tourism as specialist areas of study has spawned focused surveys, inevitably leaving gaps. A group of academic institutions across the United Kingdom have obtained funding for a pilot survey, which seeks to cover leisure in a much more holistic manner than before.

Of course, there are many arguments over what represents good value. Some believe we risk spending too much on monitoring. And the more anyone spends the more valuable the investment, the more attractive the idea of building up a trend series, and the more research money becomes locked up. Longitudinal studies are relatively expensive and not quite so common now, at least in Britain. Trend series will be helpful only if the relationships under study are linear in nature, or extremely simple and stable. Under those circumstances, it may be that a survey is rather an expensive way of collecting management information anyway. We may also be lulled into studying only those things that are easy to collect data for. Many interesting characteristics do not lend themselves to simple monitoring.

Special, with and without designation

If we designate a place as special, does it mean that the other places are not special? Perhaps not, but, if the word has any meaning at all, such areas will have to be treated differently from others. Planners have often used designations as a way of signifying value and directing development. Developments that would have a negative effect on conservation values are discouraged or rejected. They would otherwise erode the environmental capital in areas where it is most highly valued. An alternative might be the application of another designation, one for development, and it might become an enterprise zone. Here development would be preferred, as this would deliver socio-economic benefits where they are most needed. Designations such as this tend to be of shorter duration. Once the benefits have been delivered, the equilibrium shifts, conditions change, and action is required elsewhere. Sometimes the process is encapsulated in a formal system of planning (in Britain by means of the Town and Country Planning Act 1971).

Parks and reserves have different meanings in different cultures, the former suggesting some return of benefit to the visitor, the latter being concerned more with conserving the potential to provide a return to future generations. Those seeking a systematic approach to nature conservation in Britain developed a number of criteria to help in judging whether an area qualified as a potential candidate or not. A consensus developed among the scientific community over the appropriate criteria for 'assessing comparative site quality'. These criteria were described in a major work sponsored by the Nature Conservancy Council and the Natural Environment Research Council in 1977, *A Nature Conservation Review* (Ratcliffe 1977). The author of the introduction (also the editor of the entire work) was the chief scientist of the agency at the time, and he recognised that 'this is a difficult field in which value judgements figure prominently'. The process of review consisted of three steps:

◆ recording the intrinsic site features;
◆ assessing comparative site quality;
◆ choosing the national series of key sites.

In assessing comparative site quality a number of criteria were applied, the major ones:

◆ size (extent);
◆ diversity;
◆ naturalness (appears unmodified by human influence);
◆ rarity;
◆ fragility;
◆ typicalness;
◆ recorded history;
◆ position in an ecological/geographic unit;
◆ potential value;
◆ intrinsic appeal.

The last recognises the human bias that ensures that we rate different species according to our own particular value systems. In *Nature Conservation in Britain* (Nature Conservancy Council 1984) the underlying principles were refined to safeguard what might otherwise be lost, and provide for conservation in the wider environment. The document recognised opportunities for creative conservation, and also drew attention to the need to focus on some previously overlooked areas, such as the marine environment. Interestingly, there was also mention of the need to 'Develop a rationale for nature conservation as a socially necessary activity, including its relevance to spiritual, leisure, and recreational needs and to employment'.

Each country does things differently. In Canada the developing park system strives to include a representative selection of eco-systems across the country. In the Canadian Model Forests programme (http://mf.ncr.forestry.ca/) the Forest Service has been working in partnerships to develop local indicators of sustainable forest

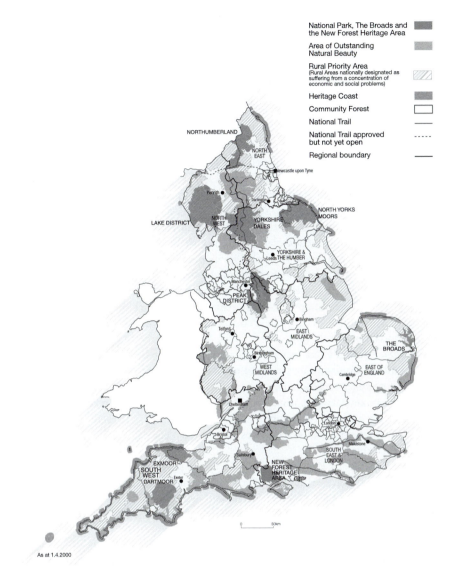

National Park, The Broads and the New Forest Heritage Area

Area of Outstanding Natural Beauty

Rural Priority Area
(Rural Areas nationally designated as suffering from a concentration of economic and social problems)

Heritage Coast

Community Forest

National Trail

National Trail approved but not yet open

Regional boundary

FIGURE 5.11 The Countryside Agency's designated and defined interests in England, demonstrating the many overlapping designations
Source: Countryside Agency, based on Ordnance Survey map

management. Under the Aboriginal Involvement Strategic Initiative, first nations involved with the programme, such as in the Waswanipi Cree Model Forest, have been seeking to develop approaches based on aboriginal values, beliefs and attitudes.

In the US National Parks the classification extends to include sites of natural and cultural significance. At the time of writing, one of the challenges the USDA Forest Service has set is that of the proposed rule to prohibit new road construction or reconstruction in the unroaded portions of the inventoried roadless areas of National Forest System lands (http://roadless.fs.fed.us/).

In Scotland some time ago a park system was proposed by the government agency then concerned with such matters (Countryside Commission for Scotland 1974). This envisaged:

- *Special parks*. The term 'special' was proposed as an attempt to get round some of the perceived problems which had been experienced in the National Parks created in England and Wales (from 1954 onwards), but the proposal to set up a park system failed. John Muir, from Dunbar in Scotland, may have been the founding father of the American National Park system, but in 1974, despite a good deal of discussion, Scotland still had no National Parks. By now the first National Park in Scotland should be in place in Loch Lomond and the Trossachs, with a second planned in the Cairngorms for 2002.

- *Regional parks*. These might serve the populations of different local authorities, because the boundary of the park might straddle the borders or be at such a position that it was well used by a neighbouring authority's population. Such parks have been set up: Clyde Muirshiel Regional Park (just south of Glasgow) and, after much debate, the Pentland Hills Regional Park (on the southern boundary of Edinburgh).

- *Country parks*. These were usually operated by the local authority. The Countryside Commission for Scotland provided support: grant aid, as well as training for ranger services and managers. The thinking was that these areas would act as honeypots that would be well managed, and would provide a stepping stone towards the real countryside. A number of private estates (and some in the ownership of voluntary bodies such as the National Trust for Scotland) also fulfilled this function.

- *Urban parks*. In Britain, as a whole, urban parks have been undervalued for some time. Largely funded by local government, urban parks lack the support of an agency focusing on their needs, and managers envy the support that parks in the countryside receive from the countryside agencies. Designation is seen as a valuable tool.

The park system for Scotland had been proposed as a mechanism to influence funding and to help integrate and encourage strategic planning. It failed. National Parks were to reappear on the political agenda some twenty-five years later, by which time the map of agencies had changed and the heightened interest in devolution and a new Scottish Parliament (the first for 300 years) gave added weight to the proposal, and the necessary legislation was approved in 2000.

A question which has been asked from time to time, is 'Should areas be designated for a single purpose for all time, ever?' We live in a dynamic world and relative values keep changing. The problems posed can be best appreciated by overlaying designations within a single country. The designations tend to build up over time, and few are removed. The map (Figure 5.11) of the Countryside Agency's designated and defined interests serves to highlight the problem.

For Britain, membership of the European Union brings with it another layer of designations, based on European legislation: Special Protection Areas and Ramsar Sites among them. Under the Habitats Directive there are Habitat Action Plans and Local Biodiversity Action Plans as well as plans for particular species. When resources are scarce, focusing efforts on particular objectives may bring more success. This is the principle underlying the Biodiversity Action Plans, which cascade down from international efforts to member states. Plans for endangered habitats and species co-ordinate and allocate necessary action. Problems arise when local authorities and government agencies are considering how to deploy their resources, because all statutory duties have to be discharged before attending to the other needs of society. This should work well when budgets are static or are growing. It does not work so well, when budgets are diminishing. Certain designations have a review system built in and require renotification to the parties concerned on a periodic basis. Most designations are concerned with conservation either of the natural environment or of the cultural environment. An exception is the system of World Heritage Sites, which covers a range of different forms of heritage, natural and cultural.

On a world scale, the World Heritage Committee had included 582 World Heri-tage Sites on its list by December 1998, and the number of sites, and the representa-tion of certain categories is to be extended. The Committee has signalled that it intends to broaden the representation of sites, particularly of industrial archaeology and cultural sites. The United Kingdom has published a new tentative list (Depart-ment of Culture, Media and Sport, 1999) of some twenty-five sites. Inevitably, listing is rather haphazard, in terms of distribution across the world, but includes many of the world's gems. The United Kingdom's recent tentative list includes the Cairngorm mountains, the Forth railway bridge, the New Forest, selected parts of the Great Western Railway, Shakespeare's Stratford and the Cornish mining industry. These join such gems as the island of St Kilda, the Old and New Towns of Edinburgh, the city of Bath, the Palace of Westminster, the castles and towns of King Edward in Gwynedd. The UK government has carefully described the process of evaluation, which delivered the tentative list (DCMS 1999b). Listing definitely has caché but is also likely to bring continued added pressure from tourism. The label will be used by businesses as a marketing device. Member states are also bound, by agreement, to take good care of the monuments. In respect of such obligations, actions are not always louder than words.

But in Queensland, Australia, World Heritage Sites are governed by the federal rather than state government, and this has served to ensure a measure of protection for two World Heritage Areas which form the outer part of a sandwich which contains a proposed resort development at Port Hinchinbrook, and the story has a political twist to it (http://www.democrats.org.au/issue/enhinchinbrook.html).

Visitors will not be familiar with all the different designations – how could they be? – but they will have expectations. These will be linked with the history of the land, tenure, access rights, property laws and the welcome they may expect. In this system, there are at least three important categories: open countryside, working countryside and wilderness.

A good deal of upland Britain is covered by heather moors, a form of land cover that has arisen from a combination of different land management practices. Clearance of woods and overgrazing from the medieval introduction of sheep by the monks has ensured the absence of trees except where protected from grazing, as in ravines, on top of rocks or on islands. The resulting heather is managed by burning to provide a mosaic of different ages, with areas good for cover and areas good for supporting the food system required by the red grouse. (The red grouse is a game bird native to Britain, but is here treated as an example.) The mosaic enhances the carrying capacity for this game bird, which has been a favourite since Victorian times. This form of management has given rise to large open areas. They are very popular for walking, especially those close to where people live. The Peak District National Park in England is within one hour's journey time (well, sixty miles) of 17 million people (Peak District National Park 1999).

Such moorland, along with heathland and downland provides the large open areas which lead to the definition of this first category – 'open countryside'. In Britain this definition was formally encapsulated in the National Parks and Access to the Countryside Act 1949, to include 'mountain, moor, heath and down'. In later legislation the definition was extended to include coast, rivers and woodland. The Countryside and Rights of Way Act 2000 now makes provision for access to open ground in England and Wales (using the older definition), once the appropriate areas have been mapped.

In the working countryside, visitors understand that there is a productive use being made of the countryside, with quarrying for materials, farming for food and forestry for (among many things) wood. Any welcome has to fit in with the needs of managers. Much of this countryside is enclosed by hedges, stone walls or turf banks or fences, characteristics which are integral to the character of the area. Britain is a very heavily populated country, and there is little of the land that has not seen extensive intervention by man. Nevertheless, there are parts of the countryside in which people have an experience that simulates wilderness; mountains and coast are among the land types which come closest. These three categories are not mutually exclusive types of countryside and different people will put different sites into different categories, but people may well expect different degrees of freedom and constraint in each.

Countryside in and around towns is treated and appreciated differently. Here there will be much greater visitor pressure and the need for a degree of enclosure and management. Many people in towns rarely visit the countryside, and the collapse in linkages with the countryside (milk identifiably from cows, or wood from the local forest) has eroded understanding of the rural way of life, and how it all works. It is close to where people live that the greatest need for countryside recreation will exist. Such demand is not easy to assess or quantify in scientific terms, as much of the demand will be supply-led. Many towns in Europe have their own woods on the outskirts, often owned by the community, and this will provide for recreation needs without putting pressure on farmed land. In Britain moves have been taking place to plant a number of community woodlands and forests around the country, which will in time be a major recreation resource. The recreation value was an important part of the cost–benefit analysis (Whiteman and Sinclair 1994), and already the added access such land gives to urban communities is greatly appreciated. Of course, the idea is not new. In Britain the New Forest and Epping Forest fulfil this role. In Germany and France there are few communities without their town wood. In one case in Italy the forest has served the needs of the local community since medieval days (Sims and Hislop 1998).

The growth in property rights was accompanied by a series of safeguards to ensure that communities could continue to develop and go about their business. Part of the requirement was for a continued right of passage on land and water. From this, in Britain, emerged a network of rights of way (Anton 1991). These could be rights of way on foot or on horseback. As our ability to travel increased, so some of these rights of way inevitably were engineered to take a greater volume and weight of traffic. Some were developed for motor cars (first as horseless carriages), and over time, some of these have developed as motorways. Other networks – canals and railways – were also developed. Inevitably some of these routes crossed other rights of way. As the system of roads developed, they tended to cut across pedestrian and equestrian rights of way. What we are left with in many areas is a network of rights of way, many of which have been severed by road and rail, like chopped up spaghetti. These rights have a long and interesting history (e.g. Haldane 1968) and represent a recreational resource of considerable value. Quiet roads and the developing greenways could provide a matrix or network of linked recreational spaces, within which there might be patches that were quiet and much valued (Countryside Agency 1998).

Some of our built heritage is listed in terms of architectural or other heritage value, assessed by architectural historians and other similar professionals. It is not usually listed in terms of recreation and leisure potential, or demand. This does not mean that such heritage is not special for recreation, but merely that it has been measured against a different yardstick – usually one devised by enthusiasts or professionals who see things with a different eye. Does it matter whether popular buildings are listed or not? If such a listing assures that the necessary resources are applied, then it may.

Conserving cultural heritage is something of a paradox. Culture grows by accretion, and outright preservation would stop such growth in its tracks. We usually wish to do just that to certain examples of our culture. Some of the more complex heritage sites are sufficiently extensive to allow focus on a certain period at least somewhere on site, whilst enabling the monument to continue to play a modern-day role. Canterbury Cathedral continues to welcome visitors while also meeting the basic needs of worshippers. Some of our cultural heritage is at even greater risk than our natural heritage. King's College, Cambridge, was a hundred years in the making. There is only one King's College. On the other hand there are many examples of columnar basalt of the kind which gave rise to the magnificent Giant's Causeway at the north-east tip of Ireland on the Antrim coast, Fingal's Cave on Staffa, and along much of the west coast of Scotland.

Is King's College more valuable to us than Fingal's cave? We rarely need to ask questions of that kind. When we do, it is usually the experts who determine what value systems or frameworks (see 'World Heritage Sites' above) are used in reaching decisions about when and where conservation should take precedence over recreation

BOX 5.3 Death in Venice?

For hundreds of years this staggeringly beautiful city has attracted people from all over the world. Its position on the trading route between East and West ensured the wealth of the city, especially in the golden age of the fifteenth century. We are left with a legacy of a very rich architecture and style of decoration among the churches, villas and palazzos of a city made up of hundreds of islands joined by a series of bridges, and the famous canals. The city now depends greatly on tourism, and the city fathers have to work hard to ensure that the image of Venice can withstand the millions of visitors it receives. They also have a battle on their hands against the water. High tides (*aqua alta*) swamp the square at St Mark's every year and steal the headlines, as did the burning down of La Fenice, the opera house. Raising the large sums required has often had the support of the international community and organisations such as UNESCO. Meanwhile the thousands of pieces of public art – often coats of arms, reliefs and sculpture – which adorn the city are at risk from the natural processes of weathering, as well as pollution, and even theft.

With such a wealth of priceless heritage to protect, where would you start? Priceless it may be, but the costs of conservation are great. There are organisations in many countries set up to raise funds to support the conservation work, and Venetians and Italians are themselves lending support. Fabio Carrera, a Venetian, who studied computer science in Worcester, Massachusetts, has devised a system to help in setting priorities. As Tyson (1995) writes, by describing each work, recording its location and any associated bibliography in one database, Fabio has been able to combine it with information, held in another database, about the condition of the monument. He has written a series of programs to work on the information in the databases. Items are weighted according to the nature of the problem. Priorities are assigned, taking into account factors such as age, rarity and the risk of further deterioration. Lastly, an estimate of the cost of restoration is produced. Most of these works of art could be restored for less than US$5,500 or so, and the scheme has encouraged great public support. The scheme does more than assure the continued survival of some 3,000 works, it is reawakening and rebuilding the cultural links between local residents and their local public art.

and for which aspect, habitat or species. When did you last hear of a flatworm watchers' society raising funds for a new reserve? In Britain it is the Royal Society for the Protection of Birds which is the pre-eminent voluntary body working in the field of nature conservation (with an annual budget approaching £50 million, rivalling some of the government agencies). In North America it is the Audubon Society. In much of mainland Europe it may be hunting clubs. But, in each setting, it would still be the experts who led the decision making.

When the designation 'of importance or significance' has been assigned by experts, a reasonable question to ask is: but important or significant to whom? In England, some years ago, there was a research project let (with the working title of Oasis), to explore the potential for sites of significance to sport. It was instigated by the agency charged with promoting sport, perhaps partly tongue in cheek. The concern was that the conservationists would soon have the whole country sewn up with designations that would effectively prohibit some sports activities. Sites are designated by specialists, with all their biases, and also at a particular scale. What may be important nationally may be of no interest to people living in a particular area. Similarly, what may be superabundant locally may actually be very scarce nationally. We need to encourage sustainable leisure and recreation in our planning, taking into account people living now (*intra*-generational equity) as well as future generations (*inter*-generational equity). We need to take fully into account what all people think, not just what experts think.

Sites of special importance to children

Recent years have seen a growth in the provision of extremely boring and sterile-looking facilities for children's play, spurred on no doubt by a combination of commercial interest and concern about potential lawsuits. There is almost endless advice on designing for play. Psychologists have helped to make professionals (Timothy Cochrane Associates 1984) aware of the different modes of play children indulge in:

◆ *motor play:* learning about co-ordinated movement, hand–eye co-ordination, and the pure joy and release given by physical activity;
◆ *social play*: learning to play with others, the beginnings of experiments in interpersonal skills, team work skills, and interacting with others;
◆ *cognitive play*: learning about the environment, and taking time to reflect on things.

Refreshing in their simplicity are studies (Moore and Young 1978) that provide a more informed approach to providing for children's play, by working with children, encouraging them to map their favourite play spaces. In a number of these studies minimal guidance was given, and very different maps, at different scales, resulted. The back garden, the doorstep, the street and favourite puddles feature rather more often than play equipment designed by professionals. This study eloquently demonstrates the value of using drawings to communicate, within participative appraisal. Community involvement has been revisited as a means of ensuring that the development of play is matched to the community's needs, and takes account of customers' values (Tibbat 2000).

Times of special importance

People have been just as busy designating particular times of the year as special as they have in designating places. In fact, special times came first by a very long way – unsurprisingly, as we were originally nomadic hunter-gatherers, relatively unattached to place. In times gone by, as we have seen, the passing of the seasons was of vital importance. In Britain the coming of spring and fertility were as eagerly awaited as were autumn and the time of harvest. Add midsummer and midwinter and you have the beginnings of a calendar linked to survival, that would allow for plenty of recreational activity to help celebrate these crucial times. Dependence on successful fertility and harvests was no doubt an important driver in ensuring that these events would be treated as special and imbued with religious significance.

As each successive wave of migrants replaced the previous one, festivals were adopted or adapted by the next, to fit into the next set of beliefs. So it is that in Christian countries midwinter festivals were replaced by Christmas. Most early cultures provided for time off, on these special days. In Christian cultures each saint had a special day allocated for celebration. Not every saint's day led to all people having a day off originally to be spent in celebrating that day. Each area would have different 'holy days'. Holy days became 'holidays' and, for most of us, the religious connotations have largely been lost except, perhaps, at Christmas and Easter. The recreation activities that accompanied these holy days often included attending a service, followed by a mixture of singing, dancing, feasting and drinking. Occasionally this would be added to by competitions of one kind or another. Many rules that normally governed how the community operated would be relaxed. There are some special times of the year associated with sporting events and arts festivals.

BOX 5.4 Oral heritage

There are special cultural places around the world where people have gathered for years, sometimes hundreds of years, to exchange views, thoughts and ideas – storytellers, musicians, jugglers, acrobats, and snakecharmers among them. In 1997 UNESCO launched an initiative to preserve these spaces. The idea was born out of a group of concerned people in Marrakesh, Morocco, who were worried about the long-term future of the Jamaa-el-Fna square, an extraordinarily busy and historic market square. Jugglers, dancers, healers and *hlaiqui* (reciters of holy tales and legends) vie for attention in the constant bustle. (http://firewall.unesco.org/opi/eng/unescopress/upanglo.htm).

There is also a collection of local events of old passed down by folklore associated with particular times and places. Special one-day events can be manufactured. If successfully established in the calendar, as has been achieved by Common Ground, with Apple Day (created to focus attention on the variety of apples) or Tree Dressing Day (focusing on the special qualities of individual trees), parts of our heritage, otherwise in danger of disappearing, can be preserved. As individuals, we celebrate the passing of each stage of life. Births, birthdays, reaching adulthood, marriages and wedding anniversaries are all subjects of common celebration. Significance can be attached (or reattached) by application of the arts (Gill and Fox 1996). This can help to retain ownership by the celebrants, rather than the intermediaries, retaining the personal and individual values that might otherwise be lost. Many celebrations are accompanied by the drinking of alcohol, and some by singing and dancing. All this activity helps the escape of pent-up emotions, and allows people to come into close contact with one another. By doing so, they offer support to each other, strengthening group cohesion.

Environmental management and impacts

There are more than 450 laws and regulations governing the environment across member states of the European Union, yet despite this the environment has continued to deteriorate (EMAS 2001). Enforcement was clearly no longer effective. As a response, it was proposed to adopt a voluntary system, which would be supported by the EU Eco-management Audit Scheme. The purpose was to gain compliance with environmental legislation, to prevent or reduce pollution, and to seek continuous improvement in environmental performance. Essential to the scheme is the voluntary preparation of documents on behalf of the developers. These are open to inspection by third parties, including accredited verifiers and the public. The scheme was agreed in 1993 and open to participation by companies in 1995. The system has been reviewed, and a revised system adopted in 2001. Participation will be expanded to include any organisation that could have a significant adverse impact on the environment, notably local authorities, and will provide for the application of ISO 14001, the international standard in relation to environmental management systems. It will also strengthen the role of the Environmental Statement as a device to demonstrate transparency and communicate clearly with stakeholders about what is proposed and what the expected associated impacts are. The statement should include an objective assessment of all the significant environmental issues, with a summary of the consumption of resources and materials, pollution emissions and waste; and details of policies, programmes and management system.

Statutory Instrument 1999 No. 293, the Town and Country Planning (Environmental Impact Assessment) (England and Wales) Regulations, and its equivalent in Scotland and Northern Ireland, govern what recreational developments require assessments in the United Kingdom (aside from special cases under other regulations: see p. 306 for useful web addresses). In the second schedule, leisure centres, sports stadia, shopping centres and car parks are included as infrastructure. Under tourism and leisure, ski-field sites, marinas, holiday villages, theme parks, permanent camp-sites and golf courses are included, with various caveats, as well as changes (to any existing such projects) if they should lead to significant impacts. In determining or assessing the significance of impacts, we need to have information about the nature of the impact, whether short or long-term. Distinctions need to be drawn as to whether the processes monitored are reversible; how the effects act cumulatively (additive, straight line, step, 'chaotic' or other); and how they interact with other outputs and outcomes. The value of the process must outweigh the cost of collecting the information.

Project-level Environmental Impact Assessments will not necessarily include some of the indirect or induced effects which follow on from the initial project (Harrop and Nixon 1999: 151). The impacts of projects that fall beneath the assessment threshold may interact with the impacts of other similar projects, to produce significant impacts, and these would also not be covered. Strategic Environment Assessments can be used to collect information about the impacts, which are significant at that level, in just the same way as the project-based Environmental Assessments will collect (most of the) information about what is significant at the project level.

The ideal stages in Environmental Impact Assessment have been identified by Barrow (1999: 97) and Harrop and Nixon (1999: 8). They amount to the following:

1   Proposal put forward.
2   Proposal screened, to determine whether an EIA is required:
     ◆   yes, 3;
     ◆   no, 6.

3   Scoping takes place:
     ◆   pilot EIA alternatives are reviewed;
     ◆   issues are identified;
     ◆   material is open for public inspection and comment.

4   Impact assessment proceeds:
     ◆   baseline data;
     ◆   identify and predict impacts;
     ◆   evaluate likelihood and severity of impacts;
     ◆   identify measures to avoid, or mitigate, negative impacts.

5   Environmental Statement is prepared, and after checking is made public for consultation, and views taken into account.
6   Decision to proceed or not, and if yes 7; if not, end.
7   Implement.
8   Monitor and audit, and feed back, for continuous improvement.

Unfortunately, once approval is gained for the project, not always is the value of continuous improvement included (EMAS 2001), although substantial cost savings (or saving of reputations) may provide added incentive.

In the United States the National Environmental Policy Act (NEPA) of 1969 requires that before any government agency is involved in 'actions significantly affecting the quality of the human environment', an Environmental Impact Statement is required. At that time, much of the work had been focused on the biophysical environment. However, since then, issues of social equity have increasingly been reflected in changing attitudes and legislation. In 1994, through the Executive Order 12898 issued by President Clinton, government agencies were charged to aim for environmental justice, on matters concerning health or the environment, for minority communities and low-income communities.

In the Guidelines and Principles for Social Impact Assessment (SIA), the Interorganisational Committee (Burdge *et al.* 1993) point to differences between the approach to biophysical and social environments. Managers with some knowledge of natural sciences, but not necessarily any knowledge at all of the social sciences, may find it difficult to acknowledge the perceptions of others as anything other than:

> imaginings, emotions, or perceptions rather than social constructions of population characteristics, community and institutional structures, political and social resources, individual and family changes, and community resources. The steps to take are:

1   *Involve the diverse public.* Identify and involve all potentially affected groups and individuals.
2   *Analyse impact equity.* Clearly identify who will win and who will lose and emphasise the vulnerability of under-represented groups.
3   *Focus the assessment.* Deal with issues and public concerns that really count, not those that are just easy to count.
4   *Identify methods and assumptions and define significance.* Describe how the SIA is conducted, what assumptions are used and how significance is determined.
5   *Provide feedback on social impacts to project planners.* Identify problems that could be solved with changes to the proposed action or alternatives.
6   *Use SIA practitioners.*
7   Trained social scientists employing social science methods will produce the best results. *Establish monitoring and mitigation programmes.*
8   *Manage uncertainty* by monitoring and mitigating adverse impacts.
9   *Identify data sources.* Use published scientific literature, secondary data and primary data from the affected area.
10 *Plan for gaps in data.* Evaluate the missing information, and develop a strategy for proceeding.

(Burdge *et al.* 1993)

Environmental Management is a term commonly applied to managing activities in a demonstrably environmentally sensitive way. Usually such systems are audited to clear rules, often by a third party, in order to gain recognition (such as a label or charter mark). This may give competitive advantage or entry to a market. Barrow (1999: 6) has identified the typical steps:

1   Identify need(s) or goals(s); define problems.
2   Determine appropriate action (likely to involve impact, hazard and risk assessment).
3   Draw up plan.

4   Implement and evaluate success (also feeding into 7).
5   Develop on-going management (also feeding into 7).
6   Evaluate and adjust management (also feeding back to 5).
7   Future Environmental Management.

Much of the attention to date has been focused on assessments prior to development, but there is ample scope for employing the methodology to existing management of leisure and recreation, leading to substantial savings for the operator, for example, in these four areas:

◆   *Waste*, through minimisation at procurement stages, re-use, or recycling.
◆   *Water*, through wise use, application of water-saving devices, using 'grey' water where appropriate, and ensuring that management operations do not pollute or interfere with drainage.
◆   *Energy*, through good design, reducing unnecessary heat loss in cool countries, and heat gain in warmer countries; use of fewer energy consumptive processes or materials; use of renewable forms of energy, locally produced.
◆   *Transport*, using the most effective and least disruptive forms of transport, with a presumption in favour of walking, cycling and riding (or swimming, rowing and sailing) before resorting to powered transport, and always choosing the most environmentally friendly option, within your constraints. Thinking of transport prompts an uncomfortable reminder that, currently, much leisure, recreation and certainly tourism is dependent on travel. Developing an Environmental Management Ethic that can be applied by citizens, consumers and companies should help address this, at least in part.

Most projects in leisure and recreation, and certainly most leisure and recreation management, will not be subject to the requirements of the regulation, because they will not (of themselves) reach the threshold. This is a problem, as the combined effect of the projects may well be significant. Also, the project-based Environmental Assessments often take inadequate, if any, account of impacts which are distant in space and time, or which filter through a number of other processes before they take effect. On the other hand, many small businesses (as well as the environment) would benefit from Environmental Management. The business would save on costs associated with more effective approaches to waste, water, energy and transport. Early studies of the Green Tourism Business Scheme demonstrated a 30 per cent saving in energy costs (http://www.greentourism.org.uk). Such schemes have developed at all levels, from the International Hotels Environment Initiative (1996) to a burgeoning scheme in the Highlands of Scotland for environmentally accredited destinations (Tourism and Environment Forum 2000). The thrust of the argument in this book stresses that the smallest organisations (and especially individuals) can adopt the Environmental Management Ethic, whether or not they adopt formal Environmental Management Systems, or are required to undergo Environmental Impact Assessment. In reviewing case studies, policy analysts (Zietsma and Vertinsky 1999) have inferred five frames which shape the way executives in businesses react:

- not the firm's responsibility;
- threat;
- technical issue;
- opportunity;
- societal duty.

In the end, most of the systematic approaches not only yield direct savings but confer a marketing advantage, often recognised by some quality assurance award, as a short-hand for attaching value. The Environmental Management Ethic, once in place, is the most reliable safeguard of all, but it demands we look after our communities, and involve everyone (not least in lifelong learning, both formal and informal).

When attaching values, and establishing which impacts are significant, we will need to exercise judgement. The 'we' should include (at least) representatives of (or proxies for) the population concerned. Traditionally, in Britain, decisions have been taken by representative democratic institutions. Local authorities have played a major role, but increasingly the advantages of augmenting these approaches with more participative techniques are being explored. In Britain a prestigious think-tank, the Royal Commission on Environmental Pollution, conducted a study of energy and the environment and, at the time of writing, is to study whether 'a radical reform of the planning systems . . . would help deliver a more effective, accountable and transparent way of protecting the environment within the context of sustainable development', in a report to be published in 2001. The five main themes of the study are: environmental sustainability, boundaries, integration or co-ordination, subsidiarity and democracy, and assessment approaches (http://www.rcep.org.uk/epissues.html).

Ideas for further study or work

1   Assess the different techniques currently available for measuring benefits and costs, and describe their respective advantages, disadvantages, and appropriate areas for application. Are there any techniques of general application, worthy of development?

2   Some benefits and costs are relatively easy to identify, others not. Reflecting on the previous two weeks, list ten leisure or recreation events of personal significance. How would you measure the benefits and costs of each, or otherwise set them in some order of priority? You may be among those who think this approach is absurd. If so, set down the arguments for, and describe, a different approach.

3   Which of the alternative approaches to attaching value or measuring performance has most relevance to you? Put together a package of measures to take account of all the benefits (including the difficult-to-measure benefits) of some leisure or recreation service, or facility, in which you have a special interest.

4   For your area of interest, make a mind map of the surveys in existence, (you may find an Internet search helpful to augment the few mentioned in this

chapter) which can provide helpful information. In your map, show how they relate to each other, recording what population each survey covers, as well as any relative strengths and weaknesses.

5   Put together the bones of a scheme to promote the Environmental Management Ethic in respect of leisure and recreation in your community. Having identified each audience, what motives and traits could you build on, and how?

Reading

Measuring

Economics lends itself more easily to (simple?) measurement than does psychology or sociology, but there are examples in the texts mentioned to point to different problems and assumptions. Huff sets out some of the problems in presenting statistics of any kind, in an account which is as amusing as it is informative. Part of the problem seems to be that expected results are accompanied by some that are unexpected. For those who want to know why, Magee and Gleick provide plenty of food for thought. To keep up to date with some of the alternative approaches in economics, the New Economics Foundation is a useful source.

Benefits and costs: from an economic viewpoint

Clawson and Knetsch is a classic text. There is more recent work by Walsh, and by Tribe, giving the conventional approach to the economics of recreation. Gratton and Taylor cover sport in their work.

Surveys

The surveys most helpful to you will depend on the focus of your interest, and the geographical area in which you operate. Many countries will have a similar range to that which applies in Britain. Explore the following:

◆   *General interest.* General Household Survey (Office for National Statistics); Time Use Surveys (Office for National Statistics); British Leisure Survey (Department for Culture, Media and Sport).

◆   *Arts. Cultural Trends*; Reading (a proxy is books sold), *The Bookseller*; music (compact discs and other recordings sold), video and 'games' – 'the charts'.

◆   *Countryside.* The UK Day Visits survey (National Centre for Social Research); All National Parks Survey (Centre for Leisure Research). Miscellaneous surveys conducted by owners and agencies (search through the Countryside Recreation Network, or see websites in Appendix 2, Forest Recreation Surveys (Forestry Commission) and for canals (British Waterways).

- ◆ *Sport*. GHS, and try the relevant Sports Council – Sport England, **sport**scotland, Sports Council for Wales and Sports Council for Northern Ireland.
- ◆ *Tourism*. UK Tourism Survey, International Passenger Survey. The easiest way in is through the website (http://www.staruk.org.uk) which includes reports and tables from many of the surveys, for all the UK Tourist Boards.

Environmental management and impact

Barrow (in this series) introduces the core principle and themes, and provides many references, and Harrop and Nixon focus on impact assessment. Through a European international project, Tribe *et al.* have reviewed a number of case studies and described how they take on the principles of an Environmental Management System. Try http://europa.eu.int/comm/environment/emas/ for information about EMAS and http://www.ieem.org.uk/ for the site of the Institute of Ecology and Environmental Management.

Chapter 6

Preparing to manage: planning, from research to design

Overview

Preparing to manage includes the three stages: survey, analysis and plan. At the start, we need to determine precisely what the information requirements are. Following a simple stepwise process allows us to identify what research is required. The information may already be in the library, on the Internet or in the minds of your staff. The focus of research is likely to involve the dual systems we have been living with: the bio-physical and the socio-economic, at its simplest, the site and the people. Sometimes differentiation between the two systems helps, at other times we need to integrate thinking. For the site, we may need to collect information about:

◆ *natural systems*, species present, distribution, type of soil, etc.;
◆ *cultural systems*, archaeology, historic buildings, legends; and
◆ *managerial systems*, leases, existing regimes, etc.

For the people, we will need to be sure whose plan it is, and who needs to be involved in the different stages, and some techniques are suggested.

Plans need to build on the past, and to be written and read within a context which will include the plans of others, and of predecessors. Early on, decisions will need to be taken about scale, in relation to space (area to be covered) and time (period to be covered) and focus (scope of recreation and leisure covered). The plan will refer to resources: land – the physical resources; labour – human resources, knowledge, skills, time and effort; and capital – accumulated wealth. To ensure success, an aim is required, together with objectives that are SMART: Specific, Measurable, Agreed, Realistic and Time bounded. Not everything that is worth pursuing can be so easily or SMARTly defined. For these less tangible aspects, we need to describe the expected outcomes or milestones. The focus will be on providing greater quality and responsiveness, greater quantity or frequency, and at less cost; more with less. A cluster of indicators may be helpful in channelling actions to the desired end, and choices made about their nature, whether they are pointing to economic, environmental or social costs and benefits.

Although there will be unintended and unexpected effects of our actions, we need to look as far ahead as makes sense. Some of the relationships are simple, even linear. Before making an attempt at forecasting, it is worth considering how sensitive the outcome may be to initial conditions. Writing scenarios can be a helpful way of exploring the possible outcomes. Other techniques worthy of exploration (brainstorm, photography, Delphi) are suggested.

The plan is a device to share information. Depending on the wishes of the stakeholders, the plan will provide for different ways of sharing in the development of the information. A matrix can be drawn up to map the level of involvement the stakeholders desire, and to plot how to take the project forward. To allocate responsibilities for the achievement of the aim and objectives, the plan must be communicated to all those involved. Many plans sit on shelves, because people simply find them boring to read. Plans can nowadays be put on the Internet, and made interactive. Even without electronic wizardry, there is no reason why we should not make full use of visual representations in the plan, and for that matter, poetry, song, rhyme, dance or any other art form we can think of.

The starting point in preparing to manage the leisure and recreation environment is to understand what kind of experience you are managing for. There are many variables present: settings, participants and expectations. One approach is to simplify these settings into types, and to segment individuals into different groups. The Recreation Opportunity Spectrum (ROS) devised in the United States is an example of an approach that could be more widely applied.

Recreation Opportunity Spectrum

Originally developed by the USDA Forest Service (Clark and Stankey, 1979; US Forest Service, 1982), this approach to recreation planning has been adopted by a number of different countries, each giving its own interpretation. It provides a way of mapping areas of the countryside according to the properties that, combined, help to meet the different expectations of visitors. The Recreation Opportunity Spectrum, or something very like it, has been used in the United States, Australia, Canada and

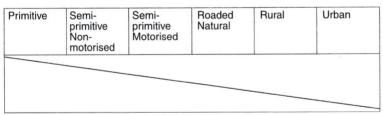

| Primitive | Semi-primitive Non-motorised | Semi-primitive Motorised | Roaded Natural | Rural | Urban |

(a)

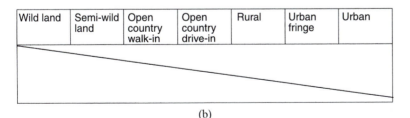

| Wild land | Semi-wild land | Open country walk-in | Open country drive-in | Rural | Urban fringe | Urban |

(b)

FIGURE 6.1 Recreation Opportunity Spectrum classes, or zones of expectation. (a) Classes. (In an earlier paper, Clark and Stankey (1979) recognised only four types of opportunity: modern, semi-modern, semi-primitive, primitive.) (b) Classes suggested for the United Kingdom
Sources: (a) US Department of Agriculture Forest service (1982); (b) Philpin (1996)

New Zealand in the attempt to impose some rational order on how wilderness is managed.

The underlying principles have a wider currency beyond wilderness, and could be used to good effect to impose discipline on the management of the countryside in the United Kingdom, for instance. In the United Kingdom there is virtually no land that has not felt the hand of man (to a far greater extent than is obvious when admiring wilderness in places such as the Rockies). The language is important, and there have been suggestions that a different order of classes is required in Britain (Philpin, 1996). The spectrum recognises that the Experience results from Activity in a given Setting. Simply put, we could caricature the two ends of the spectrum (Table 6.1). Of course, even on the enormous scale of the United States, the Forest Service recognises that what appears to be natural, e.g. in the forests of the south and east,

TABLE 6.1 Characteristics at each end of the Recreation Opportunity Spectrum

	Wilderness/wild lands	Urban
Activity	Walking, hiking, viewing scenery, camping (in tents), study, hunting, climbing, canoeing and other non-motorised water craft	Use of cars, motor bikes, team sports, commercial use, trains, buses, hotels, resort camps . . .
Setting	Essentially, unmodified and extensive natural environments	Substantially urbanised, with mass transport systems, and sights and sounds of people
Experience	Isolation from sights and sounds of people, close to nature, tranquillity, elements, challenge and risk	Mostly with people, with convenience, no need of special outdoor skills, as challenges of natural elements not presented
Remoteness	In the United States at least three miles from roads, railways and rails with motorised use	No distance criteria
Size	In the United States usually 5,000 acres (but less if contiguous with the next class)	No size criteria
Evidence of humans	Not noticeable to an observer wandering through	Built structures dominate
Social setting		
Managerial setting		

Source: US Department of Agriculture Forest Service (1982), Philpin (1996)

may well have been subjected to a great deal of human interference or management in the past. For this reason, the ROS guide encourages the use of the phrase 'natural-appearing' instead in these cases. In other words, it recognises that there is value in a relative spectrum for differentiating the classes, the settings, the activities and the experiences for which we are managing.

Zones are determined which will provide for these different experiences sought by visitors. The analysis determines the type of management necessary to ensure that the opportunity for such experience is safeguarded or enhanced, taking into account the Remoteness and Size of the area zoned, and the Evidence of humans. As originally propounded, the Social setting criteria is a crude reflection of perceptual carrying capacity, the degree to which individuals and groups are likely to come into contact with others. The appropriate figures for each class are set locally, to reflect the expect-ations of visitors. The Managerial setting is an expression of the degree to which visitors can be managed implicitly, by means of permits and regulations, at the wild end and, more explicitly, by means of pathways (or routes) and barriers, at the urban end.

When the Recreation Opportunity Spectrum was originally developed, the concept of carrying capacity (as evidenced by the use of People at Any One Time and Recreation Visitor Days) was very much to the fore. However, ROS is a very useful approach, which has subsequently been used in a number of different ways to shape the way outdoors recreation management is put into effect. Having determined what the zones are, different management interventions are considered for each, consistent with meeting the particular expectation. This has been used to match the measures taken to provide for the appropriate degrees of freedom from risk, challenge and safety expected by visitors, in Canada (Sparkes 1995a) and the United Kingdom (Ramsay 2001).

Surveys

Planning the research

The first task is to establish why you need the information, and precisely what the research problem is. Time spent being specific here can help save a great deal of time in later stages. The following has been suggested (Dewar 1998) as a series of steps to be taken in planning research:

1   Identify the problem.
2   Clarify the problem.
3   Formulate success criteria.
4   Seek advice on available solutions.
5   If no solution is readily available, then 6.
6   Assess costs and benefits of research; if favourable, then 7.
7   Invite research proposals.
8   Reassess costs and benefits; if favourable, then 9.
9   Commission research, and if successful, then 10.

10   Implement solution.
11   Monitor outcomes – against 3, and for unintended consequences.

Seeking advice on available solutions amounts to considering what information has already been collected and whether it is reliable and, if collected some time ago, whether it is still relevant. The local library and university are good places to start. The local authority or relevant national agency may also hold information. Increasingly web sites hold grey information, resulting from research projects and studies undertaken by university staff written up as working papers, and by students as part of their graduate or postgraduate courses. Some of it should carry a health warning, as not all has been subjected to peer review or necessarily had the same level of skill (or rigour) applied. It can, nevertheless, provide a very useful starting point. A quick search on the Internet can save a good deal of time (see Appendix 2 for a listing of some starter addresses). To ensure that the information you collect is comparable with other work, and that you can help build a larger data set, it is worth considering whether you can

TABLE 6.2 Planning for leisure: family or local authority

Stage		Family		Local authority
Survey	a	Discussion round the supper table to identify what the different members might wish to do; or	a	Household survey, or community survey carried out using the help of market researchers, or less formally using service providers: community workers, arts outreach teams, leisure managers or Rangers
	b	Confer with other families to explore what goes down well; or	b	Investigate what the best councils are providing (perhaps using benchmarking)
	c	A framework for a day out is suggested, round the supper table, and the views of different members sought	c	A proposal (which may have emanated from a group within the community, from council members, or officials) is advertised, and the views of the community sought through consultation
Analysis	d	Parents (or the driver at any rate!) weigh up the pros and cons of the different options, and	d	Officials conduct an investment appraisal, and/or environmental assessment to identify all the relevant costs and benefits in following the various options identified, revealing the best option, which

	e	May consider bouncing off the family the revised plan, before	e	May be replayed to the community, and/or, as a fully costed plan, may be
	f	Reaching a decision	f	Taken through the democratic process of the council: leisure and recreation committee, policy and resources committee, and full council to reach a decision
Plan	g	An entry is made in the diary, responsibilities for preparation are allocated to or assumed by different members of the family	g	The costed plan is converted into a project plan, identifying different tasks to be achieved by different departments, or under different contracts, from start to finish, and work begins
	h	During the trip, members of the family check with each other on how and to what degree expectations have been met, and make adjustments accordingly; and	h	From its inception, visitor monitoring is conducted either formally using surveys and other data gathering, or using field staff, and adjustments to the management made
	i	Decisions or recommendations are made about whether the exercise is worthy of repetition, and under what conditions, openly, and possibly by parents when children have gone to bed	i	From time to time, reviews might be carried out to establish the worth of the project, and to consider whether there are different ways of providing the same or a better service

use common questions in your surveys. The English Tourist Board, Northern Ireland Tourist Board, Scottish Tourist Board and Wales Tourist Board all co-sponsored a piece of work to derive a set of standard questions for use in market research (Scottish Tourist Board 1997) and the model is well worth replicating.

To help you decide how to research, plan, design and manage environments for leisure and recreation you will need to determine what you can afford, and what the value and cost of the information sought will be. It is not always necessary to collect all the information yourself, or even at your own expense. Often you will be able to enlist the support of specific stakeholder groups, who are often the people with the degree of specialist expertise to do a really good job. The USDA Forest Service, at the time of writing, has thirty-two forests looking for volunteers, and a further thirty forests recruiting volunteers to help with surveys next month: to interview visitors, set up and retrieve traffic counters, and to act as 'secret shoppers' to safeguard quality control. An advertisement is placed on the Forest Service's home page for Recreation (http: //www.fs.fed.us/recreation).

Surveying the site or the environment

Once you start collecting information there is always the temptation to continue collecting without checking back to the actual requirements specified at the outset. The information becomes an end in itself. At worst, its continuing collection post-pones the need for decisions, and the best becomes the enemy of the good. It may be more helpful to collect fairly coarse grained information to begin with (perhaps by using a SWOT analysis) and then fill in the gaps, as resources (whether time or money) allow. Working to a plan will help avoid this potential pitfall.

Having collected the data or information, it will be important to consider how to store the information so that it can be retrieved at a later date, and revised or added to. There are now some very sophisticated user-friendly packages which will enable

Strengths	Weaknesses

Natural environment
Well preserved heritage
People-friendly
Range of visitor attractions
Wide range of accommodation types
Excellent activity and special interest
  holiday facilities
Good road, rail and air links from Europe
	Expensive
	Accommodation standards variable
	Service standards variable.
	Absence of children's facilities
	Low on bad weather activities
	Poor package tour presentation
	Information for travellers often poor
	Some environmental problems (beaches
	  dirty, mountains eroded, fauna and
	  flora badly presented)
	Problems of seasonal imbalance
	Problems of regional imbalance

Opportunities	**Threats**

Extending the season
Development of holidays linked with
  culture and environment
Develop young outdoors market
Development of customised packages
Development of fauna, flora, culture
  products
Extending geographical base of tourism
  outwith Glasgow/Edinburgh
	Complacency
	Failure to keep up with competition
	Growth of competition in future
	  (e.g. Eastern European destinations)

FIGURE 6.2 SWOT analysis of Scotland
Source: Seaton and Hay (1998)

you to assemble databases which can then be drawn on to build a network of information within your own geographical information system (Blamey 1998). The information may not exist solely in terms of written data, names and inventories. Photography can be an extremely useful method of collecting information swiftly. With digitisation the images can be transferred to electronic form, and used in geographical information systems. The development and power of inexpensive computing hardware and software enables very large data sets to be articulated with spatial information, to render the results much more meaningful. A laptop computer can now do what could only be done using a mainframe computer in the 1970s.

When collecting baseline information it will be necessary to have some understanding of the ecological relationships, or the nature of the systems. We often need to collect data about a dynamic system. For this, we need to collect data over time, or be aware of probable changes. Even for relatively simple systems such as the maintenance of sports halls, we will need to know how temperatures vary, and so keep monitoring a number of basic indicators in selected stations. Examples of the data we may collect are:

◆ *Natural*
  Lists of species (present/absent)
  Distribution of species
  Topographical features – slope, aspect
  Geology – hard and soft
  Climate – macro, micro; temperature and humidity
  Hydrology

◆ *Cultural*
  Archaeology
  History (including social history)
  Myths, legends and stories

◆ *Managerial*
  Leases and other constraints
  Present uses
  Accessibility – physical, programme
  Risk assessments

Safety

Major catastrophes have stimulated work on sport and safety management (Frosdick and Walley 1997), and there are comprehensive guidelines in respect of safety at sports grounds (Department of National Heritage and Scottish Office 1997). Occasionally an accident in the outdoors (or the subsequent media attention), such as that accompanying the canoeing tragedy in Lyme Bay, when four schoolchildren drowned on a canoeing trip that went badly wrong, unreasonably threatens to change our whole approach to risk and safety (Driscoll 1995). We should, instead, take a careful, measured approach (Ball 1995), and learn the particular lessons from each event. People take part in outdoor activities, in part because of risk, and often for the challenge. In the 1990s Parks Canada developed a Visitor Risk Management system, which relied on matching the expectations of visitors with the appropriate degree of intervention (or management), melding the latest techniques of risk management with the ideas developed out of the Recreation Opportunity Spectrum (Etchell 1996c). Developing this system for use throughout Parks Canada required a system to compare the likelihood as well as the nature of any outcome. As a result of a visit of Jennie Sparkes, Parks Canada, to Britain, under the auspices of the Countryside Recreation Network (Sparkes 1995a, b), a working group of representatives from the leading countryside recreation agencies started work on developing an agreed set of principles to guide practice. The objective was to provide a framework suitable for Britain, to help in the management of visitors taking informal recreation out of doors. The principles seek to provide a framework to guide individual managers and to help inform judgement when issues of visitor safety are being considered.

When managing safety:

♦ Take account of conservation, recreation and landscape objectives.
♦ As far as possible, avoid compromising people's sense of freedom and adventure.
♦ Avoid restrictions on access.
♦ Avoid over-regulation. It is more important to strike a balance between user self reliance and management intervention.
♦ Ensure, as far as possible, that all risks are taken voluntarily.
♦ Assess risks and develop safety plans for individual sites.
♦ Risk control measures should be consistent.
♦ Inform and educate visitors about the nature and extent of the hazards, the risk control measures in place, and the precautions which visitors themselves should take.
♦ Recognise that people taking part in similar activities will accept different levels of risk.
♦ Recognise that risk control measures for one visitor group may create risks to others.
♦ Work with visitor groups to promote understanding and resolve conflict.
♦ It is reasonable to expect visitors to exercise responsibility for themselves.
♦ It is reasonable to expect visitors not to put others at risk.
♦ It is reasonable to expect parents, guardians and leaders to supervise people in their care.
♦ Monitor the behaviour and experiences of visitors to review visitor safety plans.
♦ Ensure work activities are undertaken so as to avoid exposing visitors to risk.

To ensure that the appropriate safeguards are taken to deal with risks which can reasonably have been foreseen, the methodology developed by the Visitor Safety in the Countryside Group proposes a number of simple steps.

Assess the risk
Assess risks and develop safety plans for individual sites.

1   Identify the hazards that are present, and the people who might have an accident.
2   Estimate the likelihood of accidents occurring, and how severe the injuries would be, taking account of existing controls.

Control the risk

3   Consider whether it is necessary to remove the hazard or do something more to lessen the risk. Risk control measures should be consistent. Inform and educate visitors about the nature and extent of hazards, the risk control measures in place, and the precautions which visitors themselves should take.
4   Record the findings.
5   Have a mechanism for review.

What actions should be taken will depend greatly on the likelihood of accident, and the severity of the consequences, and a matrix (Figure 6.3) helps in setting priorities.

To manage settings for leisure and recreation, we need to ensure that we have the appropriate information about people. Just how far you go will rest on the same principles of cost and value of information, as already mentioned. In the United States, the Outdoor Recreation Resources Review Commission started a major survey (the National Recreation Survey) in 1960, and there have been five surveys since, with last of that series in 1982. This has now been replaced by the National Survey on Recreation and the Environment held in 1994–5, with further survey work in train. Information is available on the web (www.srs.fs.fed.us/trends/nsre.html).

Surveying people is more complicated than recording information about the physical environment, and yet, at another level, 'staying close to your customers' (Peters and Waterman 1982) is the most natural action for a good manager. All you have to do is to ask questions, record answers and carry out some analysis. However, asking people questions is complicated by at least these six factors:

◆   the different meanings we each might attribute to the same phrase at any one time;
◆   our different approaches to revealing what we think;
◆   our inability to explain those actions in which we engage without too much thought, almost automatically;
◆   the gap between intentions and behaviour;
◆   choosing who to ask;
◆   our behaviour in groups or as individuals.

Recording is not so simple either. Closed questions require little more than a cross or a tick. Open questions require skilled (and still selective) listening or copious tape recordings to be made, which lead to more problems when conducting the analysis,

RISK CONTROL

User self reliance ←

Management intervention →

ZONE	 WILD	 RUGGED	 RURAL	 URBAN
Level of user's skill and self reliance	ADVANCED	MODERATE	MINOR	MINIMAL
Personal safety skills	Advanced skills, training and experience of first aid, leadership, personal safety and self reliance are essential	Skills and knowledge of basic first aid, personal safety and self reliance are important	An understanding of emergency first aid, personal safety and self reliance encouraged but not expected	Skills and experience of emergency first aid, personal safety and self reliance not expected
Level of expected support from land manager/owner	MINIMAL	MINOR	MODERATE	MAJOR
Terrain	Extremely rugged terrain. High level of fitness required. No access facilities for the less able	Rugged terrain, reasonable level of fitness required. Access facilities for the less able unlikely	Varied terrain, modest level of fitness required. Limited access for the less able	Easy terrain, accessible for all ages with full facilities for the less able
Hazard management	No management intervention	Minimal intervention, few warning signs. Limited use of physical safety measures	Modest management intervention, some advisory signs	Major management intervention, high profile signs, barriers, warnings, and welfare provision

FIGURE 6.3 Risk control matrix

Source: Ramsay (2001)

not least the time taken to replay tapes and listen. Why should analysis be easy? In our (natural) bid to make information more digestible and to simplify our model of the world, there is a danger that we reduce everything to mean values. As is often said, your head might be in the oven and your feet in the freezer, and *on average* your middle would be comfortable. Not in every case are means of great value, and they always hide pictures, often multi-modal curves, which would provide the manager with much more meaning. It is easier to collect quantitative information: but not necessarily more useful than collecting qualitative information, we need both. Which people should we survey anyhow? Decide and be clear about the implications, and which of these is the appropriate sample group:

◆ the population at large (but for which area, at what time, all of them or certain ages, etc.?);
◆ the community (bounded by what, over what time, all of them or certain ages, etc.?);
◆ the potential users (existing and those who, under some other conditions, might be users);
◆ the existing users (is it easy to define 'user'?);
◆ groups of users (selected on what basis: women, children, older people, frequency of use?);
◆ non-users.

Some surveys can be very simple, and there is good advice to be had (Love 1996; Tull and Hawkins 1984). Quantitative information concerning the number of tickets sold can be put together with information collected by visual inspection (a glance back from the stage) about the nature of the audience, and the level of appreciation (whether laughter or tears). Where money changes hands, or where cards can be swiped, the data capture can be almost automatic. One of the difficulties with leisure surveys is defining the object, the experience that you are seeking to know more about. People have very different understandings of what leisure is and therefore may unwittingly leave great gaps in reported leisure. A way round this is to survey time use, using leisure diaries (Gershuny and Fisher 1999; Glyptis *et al.* 1987; Robinson and Godbey 1997). The Office for National Statistics in Britain is conducting studies, in collaboration with European partners, to establish 'harmonised' data. Cross-cultural differences may limit the value of such comparisons (Cushman *et al.* 1996), although it will be useful to explore the similarities and differences using appropriate cross-referencing.

Focus groups are flexible, and, with the aid of a facilitator, enable you to test the reactions of your chosen audience to recent changes, or to changes proposed. Qualitative work of this kind can be used as the precursor to a programme of quantitative work. It is important not to fall into the trap of using such qualitative information alone, as if it was accurate quantitatively. Focus groups work well where recruits are alike. People are more prone to talk about their deeper feelings if they find themselves among like company and feel at ease. Often groups are recruited to investigate what particular segments of the market might think, such as women of a particular age, youngsters or people with a common ethnic background. One way of keeping in touch with your customers, and augmenting your knowledge of their likes and dislikes, is to establish a group of customers, or a regular panel, whom you can reach swiftly at any time you wish to test out ideas. This is a technique put into effect by retail businesses and other commercial organisations, and just as eagerly by not-for-profit organisations concerned with maximising social or environmental benefits. It is wise to use the technique to augment other information rather than rely solely on the output to guide your decisions. Even if the group was representative of your customer base at the time of recruitment, it is unlikely to remain so.

We survey consumers to find out about service provision. Another way to find out what is happening at the interface of service provision is to ask the managers and people at the sharp end. There are dangers here, but also some very positive advantages. You know a great deal about your staff. They will be consistent in the way they collect data, interpret and report it. Staff:

◆ accumulate great experience about service;
◆ feel part of the team, in sharing such information;
◆ will watch more carefully, in the knowledge that such information is valued.

On the other hand, managers can often perceive things rather differently from customers. We often presume that we know what people want. It is sometimes very instructive to ask. In a study which included surveys of operators as well as customers (KPMG Peat Marwick and the Tourism Company 1994), it was revealed that activity holiday operators with accommodation generally considered that in choosing a holiday, the standard of food and accommodation was the most important factor. The customers themselves identified safety and quality of tuition as the key factors in influencing choice of activity holiday. The location, too, was relatively unimportant. There were variations of opinion across the age groups, with the scenery becoming a more important factor for older participants. Younger age groups were least concerned about standards of accommodation and food.

One of the main themes of strategic plans for tourism in Scotland has been to work at spreading out activity throughout the year and away from the main destinations such as Edinburgh and Glasgow, into the hinterland, to help spread the benefits of tourism. Activity holidays are an important component, which matches some of the expected growth areas, such as the continuing trend to increasing numbers of short breaks. Something of the order of 425, 000 visits are generated through this market. The most popular activities identified as influencing the future choice of an activity holiday in the United Kingdom, were identified as:

Walking	47 per cent
Water sports (excluding sailing)	32 per cent
Riding	17 per cent
Climbing/caving	17 per cent
Golf	17 per cent
Cycling	16 per cent
Sailing	14 per cent

Simple brainstorming sessions with staff can be very cost efficient in terms of money and time. They can also help build morale, help to identify where people need extra training, and deliver useful ideas about how service can be improved. Carried out on a regular basis combined with action and feedback they can be highly motivating.

Working *with* (and not for) people

Consumers (and people in most situations) particularly enjoy responsive service. In some instances we can short-circuit the need to conduct research, to survey and to plan for the provision and management of a service, simply by working *with* people rather than for them. This can often reduce costs (surveys can be expensive), improve the quality of the information collected (by making sure it is more relevant) and ensure development is in harmony with what participants want. Therein lies one risk, that we may fail to include all the right people in the process, or even identify them as potential participants. Working with (rather than for) people is intuitively normal. The development of high-speed communications and transport has made it all too easy to overlook this natural step. We should consider using such techniques wherever we can. To involve people in reaching solutions that last:

◆   Identify the problem – e.g. accessibility.
◆   Identify stakeholders (or watch as people identify themselves), e.g. the local community, people who have problems with accessibility now.
◆   Refine the problem.
◆   Consider the resources and skills available.
◆   What are the likely trends that will affect such accessibility concerns – short, medium and long-term?
◆   What are the available options?
◆   Which gives greatest net benefit and can be acted on now?
◆   Monitor and review.

This may be very time-consuming, but it will often help to bring about lasting solutions and a build trust so that a mechanism exists to tackle other problems in the future. For many problems, it will be cost effective to give the resources to such groups rather than to teams surveying what problems are out there to solve.

Ideas and plans in context

Writing plans enables us to share information with others about how goals are going to be achieved. Others can see how what we propose will help or hinder what they intend to do, and where their actions may help (or hinder) what our plan proposes. Having written the plan, we can afford to be run over a by a bus, or just change jobs, be reorganised, be restructured or, better still, take a sabbatical. A good plan, then, provides a route to immortality, at least for ideas. Where we go is usually dependent

BOX 6.1 The Scottish Arts Council lottery funding strategy

The Council has a responsibility for distributing lottery funds in Scotland for the benefit of the arts. The National Lottery Act 1998 required the lottery fund distributors, of which the council is one, to produce strategies explaining priorities for distributing funds. Taking account of government policies; likely available funds; an assessment of needs at different scales; national priorities such as social inclusion, creating jobs, the needs of children and young people; and a quest for value for money, the council came up with eight priorities:

- Making the arts available to those who have had few or no opportunities to experience them (focusing on barriers: disability, socio-economic factors, cultural factors, and lack of local opportunities).
- Increasing arts activities for and by children and young people.
- Increasing the number and broadening the range of people enjoying and taking part in the arts.
- Making sure there is a fair geographical spread of buildings and activities throughout Scotland.
- Encouraging arts activity as part of the policies of non-arts agencies.
- Supporting film making and distribution.
- Helping arts organisations achieve lasting change.
- Developing the creative and technical skills of those who work in the arts.

This requirement in the Act followed much concern that most of the funds raised by the lottery were coming from people who had rather less, and were being spent on activities enjoyed by people who were mostly rather better off. Prior to the strategy, there was a requirement for the Council not to solicit applications, but now it can and does. The Council has already been instrumental in two major schemes being proposed for areas which otherwise would not have seen such developments, where provision should make a real difference. It produced the strategy drawing on the expertise of staff and council members; of advisers and working groups; annual consultation meetings with those who work in the arts; of local authorities and other agencies; and experience as a lottery distributor.

Source: Scottish Arts Council (1999)

on where we have been, and where we have come from. A good plan is most un-likely to follow past plans slavishly, unless they are very good. All plans will at least acknowledge past plans and seek to present new plans as building on the past.

The world is full of good intentions, and many of them are described in plans. In writing our own, we can gain a great deal by looking at the intentions of others and ensuring that ours are to some extent integrated with them, and set into a context. This context will have at least three important dimensions:

- *geographical scale:* global to local;
- *time scale:* millennial to minutes;
- *subject focus:* all leisure to one sport, for example.

While each plan and strategy will have its own particular structure, all should answer some fundamental questions. Plans will be especially important in complex situations, where the project involves many actors, where the project is to take place over a long time, or where conditions change quickly. But, in simpler situations, the plan is still important in communicating intentions. It allows others to marshal support, ideas and resources (Forestry Commission 1992). No matter how complex or simple the situation, any half decent plan should answer these questions:

- What's the aim?
- What do people want?
- What's the situation now?
- What needs to be done?
- What setting is required?
- What will be required in the future?
- What resources (including finance) are required?
- How are these resources to be obtained?
- Who is to be involved in drawing up the plan?
- How will you measure success?
- When will you review the plan?

Other people may have the same intentions as we do. It could be valuable to take account of this. Do we need to adjust our own plans immediately, or should we do so at the next revision? Can we persuade them to revise or adjust targets? Can we now divert our resources to some other cause, or make better headway in the same direction? Should we consider increasing the scope of the plan, or constructing a joint plan? On the other hand, other people may well have different intentions, sometimes in opposition, or in another direction altogether. This will be equally important to be aware of. Perhaps in our plan we have overlooked some important issue, or not taken some group's views into account. In the end, everything is joined up to everything else, and we are each affected by the plans of everybody else.

Only if others have committed their ideas to paper, recorded their intentions in some other way, or at least discussed them with us, can we take them into account. Even so, we must be aware of their existence and know where to find them. If you are searching for the plans of others, try the appropriate arm of government, local if your plans are at that scale, national or regional if your plans will be affected by large geographic scale considerations. Contact the appropriate government agency or non-government organisation (NGO) that serves as the umbrella organisation for the subject area of your plan, to find out what others exist. Use the Internet. If you prefer looking at 'hard copy', library shelves in all sorts of institutions are usually bowed under with the weight of such plans. In fact, we should reflect on this. Where do you wish the plan to repose?

Assembling resources and ideas

Ideas take shape in a variety of ways. Most often we will be reviewing the plans of others and seeking to construct something to cover a different set of circumstances. Some will arise through very careful and intricate analytical thinking, others will arise apparently spontaneously from intuition, from creative thinking. Many will be collected from other people, and bit by bit we can refine the structure until it becomes a plan to which all the contributors adhere. Ideas are easy to come by. Ideas that can be implemented are rather more elusive. People often say there are insufficient resources to carry out the plans. What resources are there to be applied? The classical factors of production, described by economists, are:

◆ *land*, in essence the physical resources;
◆ *labour*, the human resources, and expert services;
◆ *capital*, a proxy for accumulated wealth.

Identifying others with common objectives, who can contribute their time, energies, resources and ideas, will encourage success. These others may be able to achieve discrete parts of the plan, or may wish to contribute in some proportionate way to the plan as a whole. The key is in identifying who holds what view. To do this the overall objective and the expected outcomes must be clear.

The idea must be converted into reality. This takes two stages: design and implementation. The design (which takes account of balancing form, function and

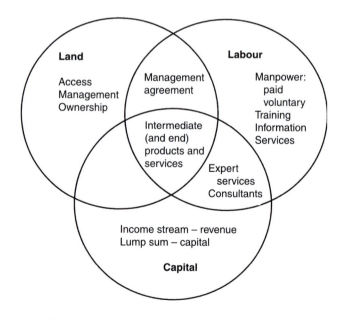

FIGURE 6.4 Identifying the possible resource mix
Source: Broadhurst (1989)

finance) takes the ideas, presents them (mostly) in visual form, and allows them to be tested. Often with computer-aided design, the level of testing can be quite sophisticated and allow a degree of simulation through virtual reality, to enable others to experience what the design will bring.

BOX 6.2 Design – balancing form, function and finance

Depending on the nature and scale of change you are intent on bringing about, you may have cause to use a number of professionals, to visualise and make real the ideas:

- architects, including landscape architects;
- designers, including graphic, interior, three-dimensional;
- engineers, including civil, electrical, mechanical, structural;
- quantity surveyors and cost accountants.

It is sufficient here to say that each has specialist skills to bring to bear in balancing form, function and finance.

Once made real, the responsibility for maintaining the balance is passed to a manager, and here too there are specialists who can help any particular aspect, whether landscape manager or accountant. The nature and scale of the project will determine how many such specialists it is worth involving. If you are unable to use professionals, or the scale of the project is too small for their involvement to be feasible, at least gain their advice, or think through each stage from their differing points of view, before making the decision whether to go ahead with the project.

In leisure and recreation, this stage involves looking at the probable sequence of experience that the subject will enjoy. Because choice is such an important part of recreation, the design often focuses on the design of a space, or a sequence of spaces, within which the relevant experiences take place. It contributes considerably to the discipline if the designer attempts to focus on the activity first, rather than the space or facility.

We do not have to redesign the physical characteristics of a building every time the particular form of recreation changes, but we do need to think our way through exactly what the subject requires. Adjusting the settings and the associated services should provide the solution. The good manager continually changes and responds to what is required, intuitively. We should remember to apply as much effort to the redesign process as is commensurate with the expected outcomes.

BOX 6.3 Designing the visit

To help sharpen up the approach to recreation in the forest, managers in the Forestry Commission are urged (Forestry Commission 1992) to focus on designing the visit, rather than facilities. Facilities are expensive, not always portable, and they need maintenance as well. Sometimes, of course, they are a necessary component of the visit, but by thinking your way through the stages of a visit it becomes easier to see just what is actually required and what may be desirable rather than essential.

- advance information and publicity
- journey to the forest
- a sense of welcome
- parking the car
- what's on offer
- picnic sites
- the walks
- signs and waymarks
- play areas
- leaving the forest

Of course, foresters are in the business of providing settings within which visitors choose what activities to engage in, so they must think about all the different settings which the many different users may need, in designing the forest, and balance this against the needs of the other stakeholders.

Source: Forestry Commission (1992)

Expected outcomes

The logical process we are describing requires us to think carefully about what results we expect from our actions, and to catalogue and describe the relevant expected outcomes. We need to have aims. As Straker (1997) writes, 'It has been said that if you aim at nothing, you will probably hit it.'

Aims can be quite general, and it often helps to be much more specific about objectives, and to describe them in SMART terms:

- *specific*, so that others will readily be able to identify exactly what is required;
- *measurable*, so that we can assess when the goal has been reached and then switch our resources or effort to some other goal;
- *achievable/agreed*, so that we take full advantage of the natural desire to reach the next goal, and avoid discouraging people by presenting too formidable a leap in one go;

- *realistic*, reaffirming the nature of achievable;
- *time bounded*, so we have a clear idea of when the goal is to be reached.

Not all the best things submit themselves easily to measurement, as we have discussed in Chapters 4 and 5. It will help if we try to think more clearly about all the outcomes which people seek or which we may expect to follow (or accompany) the leisure or recreation, and not just those that are easy to measure. The purpose of any planning is to share our ideas so that we can work with others to pursue an aim. Can we describe the outcome, no matter how intangible, in ways that others will understand? Can we describe it in such a way that we can replicate the success we aim for? What are some of these best things, which are difficult to measure? We have considered them in detail (in Chapter 4), but to specify what to measure we first need to develop a framework, to help work through the options in a more disciplined but pragmatic way. If it is difficult to specify measures for the outcome; it may be more helpful to consider the chain of actions and to select something more measurable, as a milestone along the way.

More and more people are searching for quality experiences (Alexander 1996: 51–70). This seems a natural consequence of the way in which travel and high-speed communications have had the effect of speeding up the rate at which we experience life, but not necessarily with enhanced intensity or depth of experience. The search for quality extends also to environments, goods and services that act as intermediaries in giving rise to experiences. The increasing speed of communication may also give rise to an overload of experiences. As an antidote, people value peace and tranquillity. Quality is measured only by the audience, and in their own terms (Buttle 1993). Managers must use proxies (with care) and avoid destroying the quality through attempts to introduce inappropriate measurement techniques, e.g. by intensive face-to-face interviewing immediately after an intensely emotional concert. Few of us would thank any researcher who started asking questions immediately after, or during, some climactic experience. Consider the use of instruments which participants themselves can use to report directly (Esteve *et al.* 1999). Even without such tools, there are ways in which we can describe more generally what quality is associated with. Quality experiences and recreation may share a number of characteristics:

- fit between what the individual wants and what is received or experienced;
- responsiveness of all those involved;
- goods and services which have been lovingly created, and show signs of that investment of human energies;
- richness, and complexity of experiences, integrated into a coherent whole.

The same trends which induce a quest for quality also drive people to seek more and more pleasurable experiences – more recreational activity and more leisure. The increased potential for leisure and recreation is certainly an expected outcome of most plans. The volume or quantity of recreation expected is easiest to measure in terms of frequency of events (Cope *et al.* 1999). This could, for example, be number of football games a month, or number of fishing trips per year. To be properly quantifiable there must be a shared understanding of what an event, a visit or other unit of recreation amounts to, or an understanding of the range of tolerance in the definition.

We touched on difficulties experienced in defining visits to the countryside in Chapter 5, whether by duration of time, or purpose. In the United States these difficulties have been resolved through accounting or recording outcomes in terms of days of recreation. In the United Kingdom, where everything is rather closer, this requires a certain amount of mental gymnastics. In a survey of Visitors to National Parks (Coalter *et al.* 1996), the estimate required six steps:

1   Estimate the number of vehicles on recreational day visits to the parks on surveyed roads during the six-hour survey period throughout July and August (proportion of surveyed traffic, and automatic traffic counters).

2   Estimate the number of vehicles on recreational day visits to the parks on surveyed roads during the twelve-hour survey period (8.00 a.m.–8.00 p.m.) throughout July and August (survey information about intended time of departure added to step 1).

3   Estimate the number of recreational day visitor days from outside the parks in the twelve-hour survey period on surveyed roads (average number of people in vehicles from surveys applied to the total number of recreational vehicles).

4   Estimate the minimum number of recreational day visitor days on all roads for the twelve-hour survey period on surveyed roads (proportion intending to leave on non-surveyed roads, from surveys, added to step 3).

5   Estimate the minimum number of all recreational visitor days (including holi-daymakers staying in the parks and resident day visits) for July and August (from survey, ratio of day visitor days : staying holiday visitor days, related to roadside estimate of day visitor days).

6   Estimate the total minimum number of recreational visitor days for the period January–December (supplementary data to calculate seasonal expansion factors to multiply up the result from step 5, such as traffic counts, information from visitor attractions, and carriers, bus and rail).

Other methods of aggregating recreation are just as tricky, and most seem to rely on a portfolio of measures of frequency and duration (e.g. hours of television viewing, and kinds of programme) to give an indication of levels of activity.

One of the concerns of any manager, user or citizen will be the financial cost of recreation. The expected outcome of recreation surely involves some costs. Costs may be of different kinds, and will accrue to different actors, and not just the user. The two kinds of cost that are usually the most visible are the cost met by the user in preparation for the recreation experience, which may include elements for travel, equipment, food, accommodation; and the price of admission or cost of using a facility to gain the recreation experience desired. (Pricing we return to when considering management issues in Chapter 8.) A desired outcome might be that the costs of recreation are kept at a minimum, or below a certain threshold, particularly if the cost (even a proportion) is being met from the public purse. Where there is a net (in the sense of identified costs, benefits usually being excluded as intangible and difficult to measure) cost to the public, it will often be expressed as a subsidy. Like most statistics, these are open to a certain amount of (mis-, even) interpretation. Where the expected outcome has as much to do with spreading wealth, there may be pressure to increase the costs of recreation to one group and ensure the revenue is spread throughout a target population. Leisure is political.

Recreation and leisure are also very personal. Individuals will have expectations that can be expressed in the targets which participants or managers set. Yet much activity depends on and revolves around the participation of other people, often in groups. Here different measures are required to be taken into account in specifying the expected outcomes. The measures are best set by the people, in their groups. In practice, many of the expected outcomes will be articulated by managers, albeit working closely with their audiences. Crudely, targets can be expressed in different ways to capture these different sorts of outcome:

◆   number of individuals taking part in the specified activity;
◆   number of families taking part in the activity;
◆   number of households taking part in the activity;
◆   proportion of the population taking part in the activity.

One of the important variables in recreation, and indeed in personality, is the extent to which people wish to be in close proximity to others. Most people are apt at some stage to wish to get away from it (or other people, that is, other than close family and friends, and sometimes those too) all, to disengage from their normal environment and to seek peace and quiet. Policy makers and managers concerned with countryside recreation (Countryside Group 1991) have pointed out that those who manage protected areas, National Parks and other countryside areas, need to decide just how much pressure each area can take. Carrying capacity is not rocket science. For that matter it is not, either, just science. Judgement is required. Managers have been working hard to limit the impacts of visitors on historic buildings and villages, to stop irreversible damage (Department of Employment 1991).

Wilderness sensations depend to a degree on perceived solitude, and the absence of signs that the area has been well visited. By managing visitors and encouraging people to spread themselves either across the one site or across several sites, the manager can help maintain the (perception of) wilderness values. The notion of carrying capacity was developed as a crude tool to help address this problem. The capacity can be described as the level of visits, beyond which further visits would cause erosion or other damage that could not be managed satisfactorily. Every visit, and event, will, of course, cause irreversible damage (or change) at one scale, unless we find a way of reversing the second law of thermodynamics, but we believe that in most cases the changes are manageable within the great scheme of things. In most cases the capacity we are concerned with is ecological carrying capacity. There has also been interest in perceptual carrying capacity, with the idea that a given area can be perceived as wild only if it is relatively rarely visited. Different habitats or environments can carry varying amounts of people without perceived crowding. Woodland absorbs the sight and sound of people and has a very high capacity, whereas open moorland where walkers can see one another from several miles away has a low capacity. The capacity can be increased if there are many indentations and irregularities in the contours of the ground, or rich complexity. This perceptual carrying capacity is properly associated with the visitor and the location, as different people have different perceptions of crowding. Studies have shown some interesting effects (an apparent increase of perceptual carrying capacity accompanying increased crowding). This can be explained by the displacement of crowd-averse by crowd-tolerant visitors (Chambers and Price 1986; Price and Chambers 2000).

Recreational carrying capacity is only a proxy measure for the damage with which managers are concerned, crudely reduced to the number of people managers consider the ground can carry. The concept derives from ecology, where it is used to describe the biological productivity of the land, in terms such as how many individuals can be supported without any degradation of the natural resource base (Barrow 1999: 139). The recreational version then is very crude, and is concerned with inputs and not outputs or outcomes. What we should be more concerned about is the impact of those people. It might be better to have a cluster of different indicators:

- *length of paths eroded* (beyond a certain width);
- *number of pieces of litter* visible from the path (per mile);
- *presence or absence of species* known to be sensitive to the presence of people.

These are exactly the kinds of indicators that emerge from the process devised by the US Forest Service, Limits of Acceptable Change (LAC), discussed in Chapter 8. In constructing indicators, be clear about their purpose:

- Is it to indicate costs?
- Is it to indicate benefits?
- Are these economic, environmental or social costs and benefits?
- Whose costs, and whose benefits?
- Who will use it?
- What will it cost to maintain?
- How many indicators do you need to give the answer at the required level of resolution?

BOX 6.4 Performance indicators

Leisure, at its simplest, is made in the mind of the person engaging in it. This defies simple measurement or performance targets. Any targets we are likely to devise will be extremely crude, and will be some far distance away from the leisure we are seeking to measure. However, having targets is one step better than not having them, and encourages us to think carefully about our particular roles in managing the leisure environment. It is usual to select indicators to describe different elements, at least one to deal with quantity and another to say something about quality, or use a cluster of indicators.

One of the six objectives of the Forestry Commission is to develop the recreational potential of woodland. Forest Enterprise, the agency of the Commission that manages the national forests, has performance indicators to help managers and to aid in reporting on progress. For quantity, they monitor the number of visits, by buying in to the UK Day Visits Survey, and by carrying out surveys in selected forests to give rather more information. In these forest surveys they also gather information to help measure quality, through levels of satisfaction. The results of the surveys can be viewed at their web site (http://www.forestry.gov.uk/).

In preparing to manage, you will need to marshal your arguments to win the necessary resources and support. To win over those concerned with budgets, you will need to convince, using terms which economists understand. The benefits, to the owner of the plan, need to outweigh the costs. The usual techniques for investment appraisal can be used, which make assumptions concerning payback period and interest rates for inclusion in discounted cash flows (discussed in Chapter 5). On its own economics is unlikely to yield the answer. Economics may win over minds, but sometimes you will need to win over hearts as well.

To win support from the politicians, the policy people and the public, you need to deploy an additional range of arguments, to explain why the project should go ahead rather than an alternative. This is not solely a matter of using rational arguments, but rather of looking for frameworks to reflect the other values that people hold.

For each of our intended consequences there will be consequences that we did not intend or that we failed to predict, as described by Popper (Magee 1985). These may have positive, value-free or negative properties. They may be long-term or short-term in nature, cover a wide area or focus on a point. They may affect an individual or whole populations. The fact that there will be any number of other acting forces ensures that the world remains a fascinating place. It is for these reasons that Environmental Management is gaining ground as a continuous process, and why monitoring is so important. The problem is made more interesting yet because the natural, social and economic environments themselves continue to change, in isolation from leisure, recreation and tourism. We are managing systems within systems, and we really have only the slightest idea of what the consequences of our actions will be. The further ahead we attempt to predict the less we can hope to be accurate. In many situations we make judgements knowing that there is a risk that the result of our intervention may not be quite as planned. If we were to wait until absolute certainty prevailed, the moment would have passed and no intervention would ever take place.

Forecasting impacts

In some cases, the amount of pleasure an individual gains from recreation is directly related to the amount of time he or she is engaged in the recreational activity. If for each hour the pleasure increased by a fixed amount, the relationship would be linear. The fixed amount of pleasure might increase in direct proportion to the time involved or be a function of the time. This simple relationship could be described mathematically in an equation or represented in graphical fashion. In such cases it would be easy to forecast how much pleasure could be gained if you knew how much time was involved. The phenomena we seek to study are rarely so simply related. Consider two cases:

♦   The first hour of recreational activity gives a fixed amount of pleasure, the second rather less and the third still less. This experience follows the law of diminishing returns. The extra pleasure can be described as marginal, and still adding to the total utility (see p. 122). If we know how the pleasure varies, we can still find a curve that describes the phenomenon.

♦   The first hour lays the foundation of more pleasure, and as each hour passes the pleasure builds as more and more is achieved, and the experience becomes enriched. In this case each extra hour might be more valuable than the last, up to some point when fatigue sets in. This relationship can still be described by a curve if we know how the pleasure builds, to add to total utility.

Drawing such a curve works rather better where the saying 'all other things being equal' really applies. Unfortunately, as already mentioned, this famous dictum very rarely does seem to apply. This explains why economists so often fall short of adequately explaining the way we behave. After all, their discipline seeks to simplify the world so that we may comprehend the result, not necessarily believe it. In the last twenty years or so we have been able to speed up the rate at which we do our sums enormously, thanks to the power of computers. This in turn has enabled us to explore relationships which we could not previously comprehend, and where there seemed to be no simple logic. First revealed through the study of population dynamics, there now appears to be a grouping of these events that do follow a simple relationship, but not one that immediately reveals itself. Often the result in these cases is extremely sensitive to initial conditions. This sensitivity can be mimicked using the simple dichotomous decision tree. The first answer determines on which side the eventual solution lies. Each subsequent decision makes rather less difference.

There is no need or reason for the relationships to be so blindingly obvious. What makes for pleasure in a concert of popular or classical music? What are the factors involved? The musical score, the brilliance of it, the complexities, the harmonies, the order, the new sounds or chords (even discords), the rhythm, the interpretation of the score, the skill of the performers, the sounds of the different instruments, the different sounds of each instrument, the sounds woven together of all the instruments, the appearance of the performers, the responsiveness of the performers, the physical surroundings, the colours, the materials, the temperature, the food, the drink, the social surroundings, chance encounters, meeting friends, good

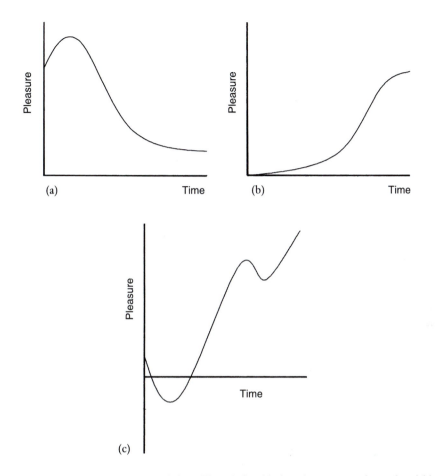

FIGURE 6.5 Leisure, pleasure and time. The relationship is at least as complex and variable as these charts suggest. (a) For a short while a game is very pleasurable and the pleasure increases until it becomes rather boring. (b) The rate of increase in pleasure rises, perhaps as skills increase or as friendships are made. (c) Pleasure drops as anxiety increases until it is overcome and pleasure increases. The variety of settings, activities and individuals will ensure that this remains a subject of study

conversation, shared pleasures, the immediate company, the crowd, the attendants, the memories? These are just some of the factors, so clearly it should not be surprising that some of the relationships are hardly linear or, if they are, hardly simple. Most of them are written as if they were positive, but as every performer knows a note played wrongly or the right note at the wrong time may spoil the entire concert for some of the audience. There are factors which (if present) inhibit value, and may even turn a would-be pleasant concert into a nightmare.

For social scientists, and especially economists, it is a small step from describing these factors to reducing such experiences to mathematical equations. By conducting surveys it is possible to put values against the different variables. Having done so, they can be loaded into equations and tested for their ability to explain the variation in different cases, and the extent to which the different variables explain the result. Regression analysis assumes (Walsh 1986: 148–53) that:

- The relationship between dependent and independent variables is linear or can be transformed into linear form, using quadratic equations, logarithms etc.
- The variation of the dependent variable is the same for all values of the independent variable.
- Values of a variable are independent of one another, and values of independent variables are independent of each other.
- Values of the dependent variable are normally distributed about their mean values.

An example was given earlier, in Chapter 5, when discussing the Travel Cost Method. With such assumptions, it should be no surprise that regression analysis has problems. First, correlation is not the same as causation, and even if it were, there is no direction given, a could cause b, or b could cause a. Extrapolation can be tricky, as the linear relationship for which regression works may not be linear beyond the range of data explored. Inevitably, these analyses are used to predict how populations behave and will therefore look for mean values, seeking to explain behaviour in the round. Like many forecasting tools, regression relies on looking back to look forward! In a straight-line world that might be fine. Even in a chaotic one, so long as we know what to measure and have identified the relationship, it might be fine. In many situations it is of no help whatever.

A photograph taken at any one instant gives us little information on which to make any forecasts. Strive as we may to make sense of our surroundings, we will not shrink from making all sorts of assumptions. Some people are better at it than others. The value of taking a snapshot, whether visual or recorded in some other way, is that it enables you to survey all that can be seen at that instant, freezing the instant for more detailed study. In real time there is a natural (and very important) tendency to focus, and to neglect what seems to be insignificant, peripheral or out of the picture frame altogether. Of course, it may become dominant in the next frame. Forecasting becomes rather simpler when two consecutive photographs are taken, especially when you know how much time has elapsed between the two, and assuming that you have an understanding of the overall relationships, and which came first. We are all tied to time, and think about different time scales. Sometimes, showing two photographs, before and after, enables one to test and explore futures. We are so used to making these judgements that we would think nothing of guessing at the outcome. This is the equivalent of drawing a line between two points on a graph and projecting the line into the future, and just as risky in predicting the outcome. A video or sequence of images gives more contextual information to enable a forecast to be made. Because it gives better information, it is easy to become duped into thinking that we know with greater certainty what the answer will be.

A simple party game experiment will show you how tricky the whole process is.

Show ten minutes of some video. Stop the showing, and then ask each member of the party audience what happens next, and compare the hilarious and different descriptions. Now complete the showing and surprise some, if not all, of the audience. Why are we so surprised that forecasting results from a clip, no matter how long, of a video, or in life, is rarely successful? This is surely the reason we enjoy travel, enjoy books, films and for that matter videos. It is precisely because we do not know what the ending will be, what the journey will bring. We can only guess.

If the relationship between the variables, or the nature of the variables, is unclear one way forward is to consider a number of scenarios, or cases. We can ask the question in terms of 'What if . . . ?' If we have an idea what the desired outcome is, we can compare the likely outcomes in the best-case and worst-case scenarios. We can, for example, think about the planning of leisure facilities for an area, assuming different mixes of a few simple variables:

- ◆ *time available*: more, same, less, in a different pattern;
- ◆ *age structure*: same, people living longer and therefore more older people;
- ◆ *structure of households*: same, or more single-person households;
- ◆ *growth patterns in the community*: shrinking as younger people leave, increasing as professionals move in to take advantage of rising house prices;
- ◆ *money available*: rising wealth, but only for particular groups.

In writing scenarios, we are not attempting to cover each and every variant, but attempting to show the range of possible outcomes, usually to focus people's thinking, and to give room for manoeuvre over time as it becomes clearer which scenarios are the more likely. Where the variables are simple, and can be measured or described in a similar way, they can be presented in matrix form. When writing scenarios, we need to be clear whom we are writing for, over what period, and under what conditions. You may wish to write scenarios, plots and sub-plots for groups within the population. You might, for example, write scenarios for certain life-cycle stages, or other segments of the population. Other social variables may be important to consider, for example ethnicity, gender and degree of integration. It is obviously easier to consider what may be the case within a year, more difficult within five years and extremely difficult beyond ten years. The relevant period obviously depends on the scale of changes and period of cycles involved. Looking at leisure opportunities in forests commonly demands consideration over periods of twenty years and upwards. What are some of the conditions or common variables that we should consider? Sometimes these will fall out of brainstorming sessions or daydreams, rather than a conscious thought process. Encouraging groups of people to discuss what might happen should throw up the appropriate and explanatory variables.

Different experts see different futures. We can collect different scenarios and futures and consider the likelihood of each. Another way of driving out the uncertainty or encouraging credibility is by checking and refining the predictions. Circulating the scenario among experts and asking them to amend them in a series of rounds or iterations, using the Delphi technique, has for a long while been an accepted route, or perhaps a way of sharing the blame.

BOX 6.5 Future-scoping

Time is money, they say. Being able to see the future allows you to plan and to gain an advantage, or edge, over your competitors. A technique, increasingly used, is to set up future groups of interested individuals who are in the key organisations concerned with the area of enterprise under consideration. The groups focus clearly on events that have caused some significant change, and search these events for a pattern and a context that may offer clues about what futures are possible. There are no fixed rules, but the consultant may collect and provide core abstracts, and group members will be encouraged to collect additional abstracts relating to their areas of expertise. Prior to the meeting, abstracts are gathered together and circulated to the group members. Members are allocated tasks, to ensure that the complete range of abstracts is scanned, and invited to put together clusters of abstracts that elucidate some point, or hold significance for them. Instructions are given to ensure that the complete range is covered. The abstracts might be cuttings from the national press, extracts from trade magazines, snippets from journals, clips from the media, downloaded articles from the Internet or remembered conversations. Individuals are given responsibility for some area, and encouraged to bring forward 'interpretations' at the meetings. The facilitator ensures that the discussion makes the most of building on the various interpretations, to help construct possible futures. In short, the steps are:

1   Identify a common area of interest, and form a group (fifteen to twenty-five people?) to carry out scanning.
2   Employ (or appoint from the group) a facilitator.
3   Identify reports of events where some particular change is crucial, and organise the collation and distribution of abstracts.
4   Arrange for members to digest a number of abstracts and come up with papers analysing these, and preparing their interpretations.
5   Arrange the meetings, and capture discussion to inform scenario writing.
6   Arrange future cycle of meetings and (at intervals) repeat the cycle.

Source: Mackay Consultants (1999)

In the first round the background is described, and a series of assumed trends described. This is circulated to a field of agreed experts, and different degrees of formality can be deployed in checking against each trend. Participants (better to call them experts, as vanity encourages continuing commitment) may be asked to use a semantic differential. For example:

♦   In the next ten years the population in this area is set to grow, at the expense of the surrounding rural areas.
♦   Do you agree strongly/agree/disagree/disagree strongly?

BOX 6.6 The UK Foresight programme

After a review of Government science, engineering, and technology policy, the UK Foresight programme was launched.

The purpose of Foresight is to:

◆ develop visions of the future – looking at possible future needs, opportunities and threats and deciding what should be done now to make sure we are ready for these challenges;
◆ build bridges between business, science and government, bringing together the knowledge and expertise of many people across all areas and activities; in order to
◆ increase national wealth and quality of life.

The Foresight programme arranged a conference on Opportunities in Sustainable Tourism, and the results (Department of Trade and Industry 1998) were fed into the consultation on government's tourism strategy for England (Department for Culture, Media and Sport 1999). For details of Foresight and a useful list of web-sites concerned with the future see: http://www.foresight.gov.uk/.

Of course, the risk is that you ask the wrong group of experts, or that the group is unbalanced. However, we usually rely on more than one type of forecasting to give us insights into possible futures. Certainly this is a very fertile area for consultancy. Anyone concerned with making money from providing or servicing leisure, or for that matter spending large sums of public money, should be concerned about what markets and society will want in the future.

All of us, intuitively, practise worst-case/best-case scenario writing. When we do, we are at least clear, or think we are, about what is worst and what is best. When forecasting for others we need to be rather more careful. All this approach does is to consider two extremes, as if the future could be considered to lie between two points, or poles. The discipline invoked by repeated use of this technique makes us think about all that can go right or wrong. Uncertainties about the values to explore can be reduced by working with the population under consideration to elicit from them, by survey or discussion, what the relevant extremes may be. The same difficulties exist about deciding whom to involve, or providing some framework that allows people to put themselves forward. Many of the factors which determine the future are beyond our control, but if we can think about likely trends and their effects, it will help us limit the negative effects and build on the positive.

Information is all

All our existence can be written in terms of information. Genetic information from two parents is combined to create the cell that divides until we have life. We are born with a brain, and our five senses, to make something intelligible of the world around us – of what we see, hear, touch, smell and taste. Managing is about arranging information in ways that achieve objectives set down in a plan. Information is central to every action we take. Some of the signals are happily sent while we sleep, or subconsciously while awake. When it comes to planning and managing, the same rules apply. Information and its transmission, its understanding and its effects are what will make the difference. Some of the best plans will rely on shared values. As Handy (1999) has said, 'Shared values are great, of course, if the values are great.' We will need to check these values periodically to be sure that they are shared, rather than risk assuming that they continue to be shared.

Preparing to manage, and even preparing to plan and design, requires a great deal of information gathering and processing. The people who have the information you need are to be found first in the community for whom you are intending to provide the service. They may not know that they have the information. They may have great difficulty in articulating clearly in a language that you understand what the crucial snippets of information are. Further, they may well resent the intrusion of an outsider, who has apparently come to deliver a plan. Some of the best leisure plans for communities are those that emanate from the community, which grow by some organic and natural process. In many countries they will be developed through people working for local authorities. Many of the best plans are automatic, and would never be referred to as plans by the people who conceive them.

The linkage between plan and action may be very direct. A household plans its leisure, and few families would consign to paper their plans for leisure, but plans they will have. The first stage is gathering information. This can be done in any number of ways: listening to what people say, discussing with them to probe what they want, observing, formal research studies to try to gain in-depth knowledge, surveys to reveal what the population as a whole may want. The objective is to give only the information people need to respond with what their wishes are, and to avoid imposing value systems that are not theirs. The plan must then be communicated to the people for whom it was written, and to the people who are to take the necessary action to bring the plan into being. It follows that this stage is well worth investing in, every bit as much as gathering or processing the information. The scope for misunderstandings is great. To avoid them, we have to make sure that everyone knows what to do, and what the plan requires. That is not to say that the plan has no life. It should be written with revision in mind, from the outset, so that when things stop being so equal we can adjust our actions to bring about the desired result.

What are the ways to inform people about what the plan is? Choose the most appropriate one from this list. You may yourself recognise the kinds of plan that work for you, and the ones that would end up sitting on a shelf unused. There are more ways than using the written report:

- conversation
- written report
- story
- play
- poem or song
- map/plan with text
- glossy leaflet with illustrations
- picture
- artist's impressions
- video
- presentation

Consulting implies rather more than just making information available. Informing suggests that the outcome of the plan is transmitted in some way. Consulting includes 'informing' but implies that there is room to take on board suggestions for change. It allows for (at least) another iteration of the plan. Of course, different people will have different meanings in mind when making reference to 'consulting'. Some would-be megalomaniacs may be using the term in order to advance their own ideas. In order to promote the ideas, they need to gain a degree of acceptance from the local community. This is the worst kind of tokenism. The purpose of consulting is usually to check back with the community concerned, to check that the planner or designer has fully understood the needs of the community and brought forward a solution with which the community agrees, and will continue to support.

One of the pitfalls of consultation is raising false hopes. Be open and honest from the outset about the scope of the plan and the room there is for variance from what has been shown. If it is a plan that has been drawn up for public expenditure or for the common good it is most unlikely that more funds can be magicked out of the sky to deliver more goods or services and, if this is explained, the consultees will usually agree. If consultees understand from the outset what the rules are, the information supplied is more likely to be helpful in refining your plans, to obtain the best fit to the circumstances, and the needs of the community.

The best way to go about consultation will depend on the people to be consulted, the plan and the planners. A complicated plan may need an extended period of consultation consisting of a number of stages. Consider:

- publishing the plan in the paper;
- putting articles in the press;
- arranging advertisements in the press or using radio;
- direct delivery through the postal service, or using the milkman;
- providing a display or exhibition in shops, school, library, gymnasium, public house;
- touring a roadshow with video;
- touring a roadshow with live (!) planners;
- sending the plan to community councils;
- arranging meetings to discuss the plan.

Meetings are probably the most dangerous, as they can be dominated by atypical members of the community, the articulate and extrovert, and indeed on occasion by extremists, and ideas tend to be drawn to opposite poles. The silent majority is usually just that. Meetings also need to be arranged with care so as to be held at times and places convenient to those being consulted (rather than just to those doing the consulting). The right atmosphere needs to be created. Any staff not usually involved with consultation will have to be reminded of some of the ground rules, and advised what to wear. In some instances (for instance, in rural Britain) grey suits will freeze an audience or build a solid barrier between planner and consultees.

4 Increasing awareness and participation, *for a better fit between benefits sought and involvement intended*				
<<<<<<<<<<<<<<<<<<<<<<<<<<<< Involvement >>>>>>>>>>>>>>>>>>>>>>>>>>>>>>>>>>>>>				
Benefits from people, trees and woods	a Inform	b Consult	c Involve	d Participate	e Adopt
1 Development (economic)					
2 Recreation and access (social)					
3 Quality of life (environmental)					

FIGURE 6.6 Framework of benefits sought and involvement intended: a matrix used to map or describe existing projects and to explore options. (a) Better information leads to better realisation of benefits. (b) Consultation allows communities a say in what happens. (c) Involvement suggests continuing interest of some sort in how the forest is managed. (d) Participation suggests more active involvement. (e) Adoption suggests that the community has taken control without necessarily taking ownership

Participation implies something more than mere information or mere consultation, and implies a direct contribution from the population under consideration. No longer is the plan being written for the community, but rather with or by the community. There have been many attempts to describe the scope for participation. One of the most popular is the ladder of participation attributed to Sherry Arnstein, and often transformed in some way (Forest Enterprise 2000; Wilcox 1994: 4). This describes the degree of control that people have over the process. It is often interpreted in terms that show, at the lowest level, the agency as having control and, at the highest level, the people are seen to have gained control. There are other models that endeavour to take a more neutral stance, providing a matrix, which can provide for different levels of engagement of different parties for different purposes (Forest Enterprise 2000). This is a useful device to map out what is happening and also what is sought, in an easy non-threatening way. Three examples of different kinds of participation will help to give colour and suggest ways that managers can develop the process:

◆ *Brainstorming*. With or without specialist facilitators, this can be a very cost-effective way of stimulating and collecting ideas in an enjoyable way. As described elsewhere, there are real benefits in making the session as enjoyable as possible, beyond the purely hedonistic reasons, because humour and original ideas are often mutually supportive, if not interdependent. Because atmosphere is so important, the brainstorm group has to be very carefully chosen. The members need to understand each other and have empathy. Usually the same rules which apply in recruiting focus groups will apply: same sex; similar age groups; similar background; and/or other strong commonality. Work groups can be mixed, and such activity is also very effective in team building. Usually the groups are restricted to between six and eight people.

◆ *Planning by a member of the community*. When people carry out skills audits of communities they are sometimes surprised at the breadth and depth of skills available. It is quite likely that, if someone acts as a catalyst, the community will be able to draft a masterly plan. Even if they are not able to do so at the outset, one option is for the community to grow those skills, as an alternative to working with a consultant. At any rate a consultant may have to spend a good deal of time picking up local knowledge, and overcoming suspicion. A plan drafted within the community is more likely to have its full ownership. This is an approach made possible where an agency will help the community to gain the skills.

◆ *Community or club plan.* Increasingly communities and groups themselves are coming forward with proposals and plans. In many countries, challenge funds are arranged by government and government agencies. Bids are encouraged from non-government organisations to provide a vehicle for achieving development in the heart of communities. This kind of development may be more sustainable. The problem posed for communities is that the capital for development is easy (perhaps sometimes too easy) to come by, but what is difficult is securing the revenue funding. With communities rather more mobile than in the past, it is also sometimes difficult to arrange or secure succession. But, such schemes do encourage participation and the principle of subsidiarity, that is, devolution to the lowest level of community organisation possible. The process of applying for funds inevitably requires a plan and a good deal of communication to arrange the necessary inputs from all concerned.

Ideas for further study or work

1 For a chosen form of leisure and recreation, and for a specific target audience, consider how you would start up a chat group to keep managers properly informed of changing needs, and improvements sought.

2 Select a particular managed leisure or recreation facility or service, and ask three different staff, independently, 'What are you trying to achieve here?' If the answers are different in kind or degree, attempt to write SMARTer objectives. Check them out in turn with the staff you asked, and then the manager, to see whether you were able to make their collected aspirations operational.

3   Find a charitable organisation (in the area where you work, study or live) which provides leisure or recreation opportunities, and help them develop their funding package.

4   Prepare outlines for different scenarios in the future which would help in the development of a plan for leisure and recreation, for any given target groups or market segments, in the short term and medium term.

5   Take an existing plan (or section of a broader plan) for leisure and recreation provision from the shelf and sketch out ideas for communicating the contents to different interested parties (e.g. children; under 18 year olds; and young adults, 18–25) effectively.

Reading

Research, survey and analysis

Market research texts, such as Tull and Hawkins, may help to inform on the full range of options, but for a short cut Veal covers a range of techniques specifically from the point of view of recreation.

Planning and design

Architecture, interior design, landscape architecture and design, and (town) planning are specialist fields, each worthy of exploration. For recreation design in the countryside read Bell. For planning and management of water sports read Peter Scott Planning Services. A good deal can be gained from reading good plans in the sphere of your choice, and your local university department, commercial practice or local authority is likely to be able to refer you to the best.

Managing the environment: the physical setting

Overview

In determining when and where to intervene, four principles should be borne in mind, that:

◆ We should err on the side of caution if we are uncertain about the effects of our actions.
◆ Everything is connected.
◆ There are always unexpected effects.
◆ We are personally and collectively responsible.

People have always sought to tame nature and manage land, although most of the objectives have been concerned with uses other than recreation, such as agriculture, forestry and the safeguarding of water supplies. Land managers are often encouraged to take account of recreation in their management. The manager of recreation and leisure can choose to intervene at different scales in managing the natural and the built environment.

In doing so, we need to remember that the natural environment is dynamic, made up of ongoing processes and systems. There are many natural forces at work that will often mask our efforts at rearranging or modifying our environment. We often notice most the visible trail left behind, whether by management intervention, or by recreation impact. A framework for studying the impacts of (and determining the interventions for) outdoor recreation is provided, to consider trampling, erosion, agents of change, and disturbance. The effects of noise are also considered before looking at a number of ways of increasing carrying capacity. There are opportunities in reclaiming damaged landscapes, or in changing land uses to provide more opportunities for recreation. The most attractive areas are usually edges, and it is here that most effort needs to be focused. Water is a feature that is paramount for recreation, and we touch on some aspects of its management before briefly considering the air. Greenways and other corridors that serve to join up the recreation opportunities are crucial.

In the built environment too, corridors are important for travel, and much recreation depends on travel. It is in buildings themselves that we have the greatest opportunity to manage and control. Hotels, restaurants, sports halls and the home are used as examples. A framework for looking at the impacts of recreation focuses our attention on physical, chemical and biological changes that are perceived and construed. Skateboarding is considered as an example of recreation that is difficult to manage, before considering problems associated with noise.

For each activity, the impacts will be different. They will be revealed only through careful mapping. The example of orienteering is used to show how impacts can be managed, and these impacts need to be seen in context, so that we concentrate our efforts where they will do most good. In some cases, it will be the direct effects that are significant. In other cases, the indirect effects may be of much greater importance, and 'By Car for a Walk' is quoted as evidence.

Principles of sustainable leisure are discussed, as part of the solution, along with the need to view the natural environment over time, as a series not just of stocks but of flows between those stocks. Landscape ecology provides a way of integrating this information over time. In concluding we recall that sustainable leisure has two important components which relate to ensuring that future generations have at least the options available to us now (the subject of this chapter) and equity among all individuals now (which is the preserve of the next chapter).

What is manageable and what is not

In our management of leisure, we should be considering whether the leisure we are advocating is itself sustainable, and also how it can contribute positively to sustainable development, that is, 'development that meets the needs of the present without compromising the ability of future generations to meet their own needs' (Brundtland Commission). We could be running cost–benefit analyses, implicit or explicit, on all our activities to determine their sustainability. We could then devise a framework which rewarded participants according to the sustainability of their actions, and penalised or taxed those whose actions were unsustainable. On the other hand, this might be rather irksome in itself, and wasteful of resources. For most, a strong system of shared values is surely all that is required. If only we could be as good at developing social issues and the practice of dealing with people as we are at developing technology.

Many peoples have had (and outside the developed Western countries still do have) codes of living which respected the land in a way that we are struggling to regain almost against the odds. The first nations in North America have been able to pass down some of their wisdom, which the rest of us would do well to take account of in our development of leisure today. If everyone upheld such values a system of sustainable leisure would naturally follow. Problems arise when individuals choose to follow their own desires and pursue their leisure no matter what the effect on others, or on the environment. Sustainability requires us to take into account the needs of communities (Agenda 21), and has two important components:

◆ equity in the future, ensuring that future generations have (at least) the options we have available now;
◆ equity among all individuals now.

At least these four principles are worth keeping in mind:

We should err on the side of caution. One of the tenets of nature conservation in the United Kingdom is based on the precautionary principle. The beauty of the principle is its simplicity. In essence, it requires that in considering any course of action in areas we have set aside for protection (such as National Parks), and where we are unsure of the outcome, we should err on the side of caution. We should ensure that we consider certain conservation, rather than risk any unknown effect. By definition, this is risk-averse behaviour, and one could argue that it does not take into account reasonable decision-making tools which we employ in many other areas of human endeavour.

The principle has been described (O'Riordan and Cameron 1994: 17–18) as being underpinned by the following basic concepts:

◆   prevention rather than cure;
◆   safeguarding ecological space;
◆   proportionality of response;
◆   onus of proof on those who propose change;
◆   intrinsic natural rights;
◆   paying for past ecological debt.

The principle may be more useful in the management of particular sites than in more pervasive conservation (wise use) practised across wider areas of land. Rigid adherence to the precautionary principle is seen by some as pedantic, stifling, and by others as anti-recreation. Those who were first on the scene (in many cases the conservationists) seem to have established their position and blocked any succession thereafter (Sidaway *et al.* 1986).

Everything is joined to everything else. Across space, every surface is in contact with another surface such that we are all joined together. It follows that every physical action will have some (albeit maybe difficult to measure) physical consequence elsewhere. The same could be said of each chemical reaction. We are connected with the moon and beyond. Across time, the picture is rather different. Again we are connected with everything else across time. We are living in a continuum, but the arrow of time means that this connection is in one direction only. We inherit from our ancestors and bequeath to our children.

Every expected effect has unexpected associated effects. When we put up coastal defences in concrete, we merely displace the effects of the sea to a different area (French 1997). Such knowledge is giving rise to an interest in working more closely with natural processes rather than trying to resist them. Much the same is true of putting up barriers against the wind.

We are personally and collectively responsible. Whether as individual consumers, or as citizens in a community, we have to take responsibility for our actions. This encourages us to make the most direct link between what we do and the environment; the 'polluter pays' principle. We also have a responsibility for previous impacts, as we reap the rewards of the technological developments that grew out of the endeavours of our forebears. There is an environmental debt that we should seek to repay before passing the world on to future generations.

What are the components that managers can work with? Crudely we can focus on the bio-physical or the socio-economic environment. Simply put, we can manage the resource or the visitor (Goldsmith and Munton 1971: 267), and extend that to considering the service we provide (Jubenville and Twight 1993). The following components are worth considering:

Managing the resource:

◆ elements: soil, water, air;
◆ land types: forest, mountain, moor, heath, farmed land, desert;
◆ water types: lakes, ponds, rivers, streams, sea;
◆ plants;
◆ animals;
◆ interventions: building, roads, tracks, paths, drainage.

Managing the visitors:

◆ elements: visitors, staff, volunteers;
◆ organisations: companies, not-for-profit organisations, church, state, school, college;
◆ communities of neighbourhood, and of interest;
◆ families;
◆ interventions: information to attract, advise, make aware, lead to understanding, channel, influence, prohibit;
◆ opportunities to take part in leisure and associated activity.

Managing systems:

◆ cycles: carbon, nitrogen, water;
◆ other processes: speed of change, chemical changes like fire, physical changes like wind, wave or avalanche;
◆ interventions: adjust stocks and flows between components of the system, at each and every stage of the process, minimising the disturbance or negative effects associated with leisure, and optimising the beneficial effects.

This may seem rather abstract for the humble manager who is concerned with providing and managing settings for leisure and recreation, but alternately focusing on global and local aspects is more likely to lead to the imaginative solutions required.

The built environment is the ultimate manifestation of our desire to control and bring order to our surroundings. The home, the school and the workplace are examples of built environments that have important components related to leisure and recreation already encoded. Think of your own neighbourhood. Think of all the places you have enjoyed leisure and recreation, and there will be a host of additional buildings: restaurants, public houses, bars, shops, football stadia, swimming pools, squash courts, gymnasia, cinemas, art galleries, libraries, concert halls and churches. Beyond the buildings, there are all the other structures which play their part: links such as bridges, roads, canals; and barriers like gates, fences, hedgerows, walls; and all the meeting places such as squares, steps, beside clocks, sculptures, the bus stop and so on. The task of the manager concerned with leisure and recreation is not usually construed to include architect, planner and designer, but must at least extend to an understanding of how people use these areas, and the effects of our interventions; the sensations we perceive (Armstrong 1962: 127–8). The components to manage include all those suggested above, and in addition those that this greater degree of control allows us: the propensity to provide for the stimulation of each of the senses; and controlling the settings and flows, within and between spaces, of substances and of people.

Travel is integral to much recreation. On land, the transport infrastructure often comprises linear routes. Providing fast routes in one direction necessarily affects the speed of travel in any other, and may well lead to the severance of other routes for wildlife or social severance for humans. Considerable areas of land are given up to roads and rail tracks, as well as the necessary junctions, stations and termini. Canals and rivers offer linear routes also, but on most water (and in the air) we may choose to define pathways, but the infrastructure is restricted to the airports and harbours. The beginning (and end) of journeys is often associated with delays or periods of time spent waiting; something which has not been lost on the merchandising managers, who have developed shopping villages at some airports. Profits from these shops can contribute significantly to the cost of (or profits arising out of) providing such infrastructure. Recreational travel and tourism are a significant part of many economies and the nature of the infrastructure has a bearing on the quality of recreational experiences. Providing all the essentials to meet the needs of passengers (for food, water and toilet requirements) is only the start.

Where usage is high, it may be desirable to surface the route along which people will move, whether it is a path, track or road. The temptation in the past has been to intervene by resorting immediately to hard surfaces such as asphalt or concrete. Any

TABLE 7.1 Stimulating the senses. Designers and managers can deploy a wide range of stimuli in providing or maintaining the desired ambience

Sense	Example
Touch	Smooth or rough; even or irregularly patterned or textured surfaces
Taste	Indirect, through foods and drinks, sometimes enhanced by use of spices or chemical enhancers, such as monosodium glutamate
Smell	Smells of new mown hay and of sea water are being used in airports to induce calm
Sight	Quality of light Colour: warm/cool; stimulating/relaxing
Hearing	White noise to provide scope for intimacy, music to attract or repel people
Heat and humidity	Values and changes = hot/cold; moist/dry

path will change patterns of drainage, and this needs to be taken into account at the design stage. In recent years severe flooding has been exacerbated (in parts of Britain, for example), by the sheer amount of hard surfacing, covering the earth, from which the water runs off much more quickly. Apart from the surface of the route, we need to determine the most appropriate alignment, to meet the needs of travellers and to have minimal impact on the natural functions of the landscape (Aitken 1984; Bayfield and Aitken 1992; Hamilton *et al.* 1999).

Hotels provide for a very large degree of control that can be exercised over the internal environment, enhanced by creating separate compartments, rooms and spaces. The amount of managerial effort we devote to providing and maintaining such settings for leisure and recreation will vary a great deal. Here we merely scratch the surface. To determine where effort should be expended we need to know something about what customers want. It is quite common for hotel managers to run customer satisfaction surveys. To enable the most appropriate management interventions, such surveys need to be calibrated against studies which point to the attributes which contribute most to the satisfaction of customers in each of the identified market segments. Inter-continental Hotels undertook a review of their hotels in 1991, and developed an environmental operating manual. Working with the Prince of Wales Business Leaders Forum, Inter-continental Hotels decided to share their manual with other hotels, which altogether accounted for more than a million hotel rooms (van Praag 1992). It was used as the basis of the first manual produced by the International Hotels Environment Initiative (1996), subsequently updated.

People are attracted to restaurants for a host of reasons beyond those concerned with the food and drink available. The initial ambience may be determined by the architect and interior designer. The task of the manager is usually restricted to providing and managing a series of settings within the restaurant, and maintaining (and adjusting) the specified environment. For some chain operators, the latitude will be further restricted by staff codes and manuals, which are intended to ensure that the customer receives the appropriate level of care, and in this case standard (standard is the word) level of service. Some restaurants may be catering for people in a hurry, but, for most people eating at leisure, speed of service is rather less important than some other attributes. The temperature, humidity, colour, decor, pictures, ornaments and other movables including plants can be used to create a variety of spaces with different qualities, and contribute to the atmosphere. Much of the atmosphere is dependent on the other customers. A few restaurants go so far as to have a dress code.

In sports halls, where the focus is more on physical activity, adequate space, surface, a well sprung floor, pitch markings, equipment, temperature, ventilation, access on to court, changing rooms, hot water and showers, water to drink are the attributes which are likely to be more important. For any activity we can specify certain requirements (John and Campbell 1993, 1995, 1996), without which the activity cannot proceed. Beyond that, a number of additional factors are important.

A great proportion of leisure time is spent within the home, and (in Britain at least) there are signs that more attention (and money) is being spent there. This space

---

BOX 7.1 Skateboarders: claiming spaces

The development of skateboards has enabled an urban subculture to develop in countries that feel the influence of the United States. Rather than the sophisticated skateparks that exist in many areas, it is the railings, steps and street furniture and fittings, including barriers, that provide the challenging surfaces for practitioners. Because the sport has developed among youngsters who are seeking excitement and freedom from the constraints usually placed on them by parents and others in authority, the activity flourishes outside the scope of normal management. Indeed, in some areas, considerable skill is required to provide sufficient management to allow the activity to proceed, safeguarding the interests of the youngsters as well as the safety of citizens going about their daily business.

---

above all others is subject to continuous management, and intervention. Spaces within the home are continually changing use as the occupants change life stages or roles. Some rooms may be very closely regulated according to rules imposed by the household (or keeper of the household), whereas individuals' rooms are usually regarded as the sovereign responsibility of the individual.

The scale at which we need to intervene is governed by the requirements of the processes that we seek to influence. The processes that are most relevant to manage, in turn, depend on the settings and the activities with which the individual or community is concerned. In managing an outdoor recreation area, for example a ski field, we may be concerned not so much with managing the weather as with choosing the areas where skiing should take place, and helping to safeguard the snow-holding properties, perhaps with the management of the forestry programme. We may be concerned with managing activities by zoning across a large area, such as the Hohe Tauern national park in Austria (Stadel 1996) where a core of more than 1,000 km^2 is reserved for nature conservation, some 500 km^2 where the preservation of cultural landscapes is given prominence, and a much smaller amount (less than 50 km^2) is set aside as special for some defined purpose. In some areas, e.g. where management is attempting to secure cultural landscapes, special incentives are given, and in others, activities are regulated. We seek the areas and the conditions that will give maximum pleasure with least damage to the environment, and the appropriate balance between risk and challenge. Of course, it has been pointed out that it is not always clear when damage is about to occur, because many of the effects are incremental and cumulative (Holden 1999). Similarly, we do not always know whether it is better to focus the activity on a specific area or to spread the load (Sidaway 1991: 21–2). We may also assist snow cover with artificial snow, using snow cannon where it is practicable and (financially) cost-effective for the operator. This may be easier to do in North America and the Alps than in Britain, where the atmosphere is generally too humid for good snow production (ASH Environmental Design Partnership 1986). Heavy use of machinery, whether to produce artificial snow or to groom pistes, may lead to long-term effects of compaction and erosion, as well as additional use of energy and fossil

TABLE 7.2 Relative impacts. Three examples are used, and only examples of marginal impacts are included, not the impacts associated with, for example, the cost of building sports centres. For that approach, Life Cycle Assessment would reveal the total impacts. Of course, it is true that there is an 'opportunity cost' of impact; you could be causing more impact with your activities elsewhere, doing something less economically, environmentally or socially friendly!

Activity	Economic	Environmental	Social
Squash – a game	Transport Sports centre – jobs Equipment suppliers – jobs	Energy costs of transport Energy costs of heat, light and showers Waste	Interaction with other players and with sports community
Art gallery – a visit	Transport Gallery – jobs Related services – shops, restaurants	Energy costs Waste	Interactions within visiting party and with arts community
Countryside – a walk	Transport (to starting point) Land managers – parts of jobs and grants for providing facilities and services Related services – shops, public houses	Energy costs Waste (exhaust)	Interactions within visiting party and chance meetings with other walkers, and people working in the countryside

fuels. Such activities may not be so cost effective (environmentally) over time. In this case the management effort must be directed at the scale of systems, and catchments.

Some impacts may be more transient but have far-reaching effects. There is a growing trend among some time-starved and money-rich skiers to go on holiday to the Rocky Mountains and be airlifted to the top. Then they ski downhill through virginal powder snow. On reaching the bottom, they are airlifted to the top of the mountain again, or to another mountain. This avoids time being 'lost' on conventional lifts and tows, and there is obviously a premium to be paid, considering the expense involved. It also means that areas which have so far remained relatively tranquil for wildlife, and for people, are now 'open season' (for very rich skiers anyway.)

Other local short-term events can have long-lasting and sometimes extensive effects too. An obvious example is wild camping overnight at a very specific location, where the human waste left behind, if not properly disposed of, may taint the water supply, and may in extreme cases lead to the spread of disease. Here the emphasis is on managing the resource or system through managing the visitor. Awareness-raising campaigns can help (Mountaineering Council of Scotland *c*. 1999), to encourage the development of safe behaviours.

BOX 7.2 A framework for studying and managing impacts

Biophysical impacts
- *Physical change*, e.g. trampling, erosion, accretion, addition, subtraction and restoration of materials.
- *Chemical change*, e.g. chemical reactions such as those that lead to the erosion of stonework, e.g. soap in waste water.
- *Biological change*, e.g. disturbance of ecology, effects on populations, relationships among wildlife, introduction, mutation or extinction of species.

Social and personal impacts (see also Chapter 8)
- *Perceived change*, e.g.:
  sight: visual stimulation, discord, over-excitation, light pollution, deprivation;
  sound: noise, discord, volume, pain;
  touch: unwanted (or desired) contact, harassment, violence, pain, affection, love;
  taste: unexpected, or unpleasant tastes, e.g. spoilt water; pleasant, new and exciting tastes;
  smell: unpleasant, strong, or inappropriate; pleasant, subtle or strong , or appropriate.
- *Construed change*, e.g. with changed meaning (arising out of the receipt, or reconfiguring of information).

Much natural activity is cyclical in nature. At the fundamental level, the carbon, nitrogen and water cycles need to be borne in mind. We should think the cycles through and identify where leisure activities may have a particular effect and seek to take into account actions that will mitigate the negative effects. The importance of reducing the production of greenhouse gases including carbon dioxide is well known. As managers we can encourage a reduction in the use of fossil fuels, and in many parts of the world the planting of more forests and woods. We can take more care in our use of fertilisers (and pesticides). Unthinking use of fertilisers in sensitive areas will build up nitrogen levels that may take many years to be washed out. In the meantime such concentrations may lead to eutrophication, especially in shallow waters, with consequent algal blooms (often of toxin-producing blue-green algae). In some parts of the world, tourism developments have put extreme demands on water supplies (Wall 1996). Even in the relatively moist British Isles, there are times when hosepipe bans are put into operation to ensure there is adequate water to maintain the natural functioning of rivers and basic ecology, and with climate change supplies may be less secure. At various points in these cycles we have the opportunity to encourage practices that will render our leisure more sustainable. The amount of rubbish we create and energy we use can often be reduced with just a little forethought (Watson 1997; International Hotels Environment Initiative 1996).

Our desire to do more and more, to live faster than our forebears, speeding up the rate at which we travel and communicate, is leading us inexorably towards a twenty-four-hour culture (Future Foundation 1998). Such a pattern of living is very hungry in terms of energy. We could, though, use leisure to recover natural rhythms and biological time, and work towards their restoration and sustainable development.

Few things stand still. When you think you are standing still you are actually moving faster than 1,000 m.p.h. (the speed at which the surface of the Earth is rotating in one direction, while rotating round the Sun at 66,600 m.p.h.). When we

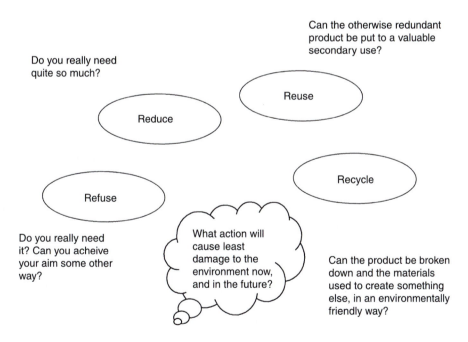

FIGURE 7.1 Ecological efficiency. In every action we should strive for ecological efficiency. Do we really need all the materials we are using? Do we need to use so much energy? Are there more efficient or effective alternatives? How can we keep the footprint small? We need to take account of the longer-term effects. Recycling is not necessarily the best option

BOX 7.3 Center Parcs

In 1967 a small company was set up by Piet Derksen as Sporthuis Centrum Recreatie, with thirty villas and an outdoor swimming pool, at De Lommerbergen in the Netherlands. By 1985 the company had grown substantially and was floated on the Amsterdam stock market. The launch of Center Parcs, as it then became known, was forty times oversubscribed.

In 1987 Center Parcs opened Sherwood Forest Holiday Village in England with 609 villas set within a secure environment within coniferous forest acquired from the Forestry Commission. The forty-acre development includes three miles of landscaped (and necessarily convoluted) water features, and a covered 'subtropical water paradise' under a geodetic dome, alongside shops, sports facilities and a café-style village square.

Management of the contract to build the village was extremely tight. Heavy penalties applied should there be any damage to trees. What were to become the linear water features acted as the basic infrastructure or road system to do the work from during construction. The result was extremely swift recovery, and an apparently mature setting. Great attention is paid to the needs of customers, with staff involved in customer care training and quality control.

This development was quickly followed up by another, at Elveden Forest, using a similar model to provide 850 villas. The short break holidaymakers spend almost all their time in the holiday village, although there are opportunities for cycling on trails into the neighbouring forest, and for pony trekking. The visitor spend is therefore also focused.

The same thorough management has been applied to recording and encouraging the ecological development of the water areas, and recently one of the sites was being considered for designation, for its value to nature conservation. Steps are taken to minimise the ecological footprint, through recycling water and conserving heat. Center Parcs collected awards, including the 'Come to Britain' trophy in 1987 and the English Tourist Board's inaugural award for green tourism.

Source: Gordon (1988), Allen (1993)

rearrange components, we are doing so within a context. The more we know about that context the more effective our actions will be. An example is trampling on the beach, which we generally ignore, as we know the incoming tide will wash away all signs of human activity, returning a beautiful virginal surface. The shore is a very useful example of a particularly dynamic piece of the land, because it is so obviously changing the whole time. Needless to say, all parts of the earth's surface are continually changing, but at different rates. Changes are viewed at a particular scale. If we were able to switch to fast forward, we would be continually surprised by new mountains, and coastlines. Whatever system we manage is itself within a context that is continually changing, and subject to greater than mere anthropogenic forces. This suggests that for economy of effort we should be working with these environmental forces and processes, rather than trying to stop them. Some of the most obvious (to us) impacts are those that leave a visual trace behind. But are they the most serious? We do not know whether it is better to focus impacts or spread them evenly across a wider area (Sidaway 1988, 1991). We need more research. Here we consider the more obvious direct impacts out of doors.

The passage of an individual across the landscape has an effect on the soil or vegetation underfoot. The nature of the effect will depend on the substrate, and on the vegetation under consideration. In many instances, after a few passes, the route becomes preferred, and the same individual or other visitors will follow in the foot-steps of the first pass. In alpine environments 200–500 passes may require a full year, or more, for recovery (Whinam and Chilcott 1999). In many cases, this will lead to damage to the rock, soil or plants underfoot, and even to the microbial populations (Zabinski and Gannon 1997). This is a natural consequence of the combined force of the movement of the human traversing the area and gravity.

Solutions are simple, but the effects require careful consideration. Visitors can be channelled. Erecting barriers to restrict movement, identifying preferred routes, and providing surfaces are some of the chosen methods of reducing the effects of trampling. In some cases management of the soil will be, or should be, integral to the management system in place (Jim 1998). In any event, early monitoring may prevent the necessity to adopt more intrusive management techniques later on (Buckley 1999). If the passes continue, the vegetation (if there is any) will degrade. Roots that provide a matrix of support for the soil will be removed. If the area concerned is soft or fragile rock, the rock may begin to split or crumble. In any event, further passes increase the risk of damage. This is then exacerbated by natural forces, as the slightly eroded areas form channels and pools into which water flows, making the area more susceptible to further and greater damage.

If visitors are to continue to use the area, some of the management options are to:

◆ drain the area;
◆ support and repair the surface by: providing support for developing roots systems, through the use of terrestrial fabrics (man-made plastics or natural fibres such as jute);
◆ plant more resistant species, recognising the possible effects on the ecology;
◆ harden the area, by: providing a water-bound path; providing a sealed surface (e.g. with asphalt); or providing a paved area.

Each will have effects that need careful consideration before intervention and need monitoring thereafter. The solutions need to be acceptable environmentally, eco-nomically and culturally. Often, therefore, what is a solution in one area may be seen as desecration elsewhere. It is as well to take advantage of locally produced guides and handbooks (Hamilton *et al.* 1999). Different sites will have quite different sensitiv-ities. Sites that are particularly sensitive, are often those which are dynamic, like dune systems, any involving water, and the uplands, which may have particularly thin soils.

Scale	Small area	Extensive area
Short term	e.g. concert: managing the promotion of the event, and the concert hall for the duration of the concert	e.g. transatlantic yacht race: managing the support logistics, and monitoring progress
Long term	e.g. statue: keeping it clean, and free of graffiti; guarding against erosion	e.g. National Park: managing the physical resource, the settings, and the behaviour of visitors

FIGURE 7.2 Managing at different scales? Management intervenes on many different scales. These are examples of settings of intervention on different scales of time and space. Intervention needs to be matched not only to the setting but also to the experience sought. This will affect the level of complexity to which the intervention is directed, a combination of time, space and intensity

Eight strategies (and thirty-seven tactics) have been identified by Leung and Marion (2000) for managing recreation impacts on resources and experiences. While these have been collected in their state-of-knowledge review focusing on the use of wilderness areas, many of the techniques could easily be transferred to other settings.

Agents of change

In the examples above, feet (assumed to be human at that) were considered as the primary agents (or triggers) of change. It will come as little surprise that the effect of horses' feet is somewhat more severe (Whinam and Chilcott 1999). The impact depends on (but is not necessarily proportionate to) the weight of the individual pressing on those feet. The angle and speed at which the feet are brought into contact with the substrate will play a part, as will the nature and shape of the foot. The weight can be spread according to the type of shoe, boot, snowshoe, ski or other device strapped to the foot. Choosing and encouraging the use of footwear and treatment appropriate to the conditions, and appropriate styles of movement, are interventions that could be used more often.

Other agents of change include vehicles, motorised and the simple mechanical. Mountain bikes have increased in popularity (among users!) enormously in recent years. The combination of fat tyre, strength and lightness of frame have combined to provide a means of transport for the outdoor enthusiast who can now strike deeper into the country in a shorter period of time. Because the wheels (generally!) tend to inscribe a continual mark, where the ground is soft or wet, erosion can develop swiftly.

There is a continual tension between exploration (or movement) and conservation. Four-wheel-drive vehicles enable drivers to strike off road into the country, where the damage created can match what must be the exhilaration of driving under such conditions. The outputs can include noise, erosion, mud, dust and disturbance. It has been estimated (Lovich and Bainbridge 1999) that the damage already caused

BOX 7.4 Principles of upland pathwork

Formulated by the Paths Industry Skills Group, based on a policy statement of the Mountaineering Council of Scotland, endorsed by the House of Commons Environment Select Committee, in 1995.

- Pathwork will be carried out within a coherent management framework, including a commitment to long-term maintenance. It will integrate with other management objectives.
- An understanding of the underpinning philosophy and practice of path improvement is required of managing and funding agencies.
- Pathwork will be generated by area survey and prioritisation.
- Priority will be given to curtailing and restoring environmental damage, while also enhancing visitor experience.
- Environmental sensitivities will be given stringent regard, particularly in sites of outstanding landscape and or/natural heritage quality.
- Management of the path will be informed by suitable consultation with interested parties.
- The purpose of the path and its expected use will be defined and the path built to fit this purpose.
- Pathwork will be of the highest standard of design and implementation, preferably using locally sourced materials in harmony with the site.
- Good environmental practice will be paramount. No material won in works will be wasted. Techniques used will protect existing vegetation and cultural remains and the site will be left in as natural a state as is practicable.
- Those involved in the design, implementation and supervision of pathwork should be demonstrably professionally, and technically, competent.
- All work will be carried out in accordance with legal obligations and the requirements of Health and Safety legislation.

Source: Hamilton *et al.* (1999)

by off-road vehicles in the Mojave and Sonoran deserts of California is of the order of $1 billion and that to re-establish vegetation to pre-disturbance levels may take 50–300 years. For systems to recover fully may take 3,000 years, and restoration techniques do not show great success. The advice offered is to limit the extent and intensity of disturbance. Choosing (zoning) appropriate areas, training in less aggressive (but just as enjoyable and more skilful) driving techniques and laying out trails that are less damaging are some of the options available. Managers need to provide opportunities (or substitutes) for such activities, because, if not provided for in managed situations, the activity will continue unmanaged. The trick is to manage without

TABLE 7.3 New sites for old? Restoring the land through recreation. Leisure, recreation and tourism can restore or add value to sites and settings previously damaged, or otherwise impoverished

Old use	New use
Environmentally damaged land	
Coal mines	Country park and lake: Lochore Meadows, Scotland
	Lakes and woods: Cologne, Germany
Silver mines, lead mines, copper mines	Tourist attractions with interpretation: Wales, guided tours with ex-miners; caving and potholing opportunities
Salt mines	Tourist attractions: Salzburg, Austria
Gravel pits	Nature reserve for wildfowl, or water sports centres
Brownfield sites	Temporary green space or playing field
Scenes of violence and destruction	
Historic battlefields (beyond living memory)	Culloden, Flodden and Bannockburn, Scotland; Gettysburg, Pennsylvania
Battlefields	Pilgrimage/history tourism: France, battle of the Somme, trenches, war graves
Concentration camps	Family/social history: Auschwitz, Poland; Belsen and Dachau, Germany
Bombed sites	Pilgrimage/tourism: Thriepval, Normandy; Hiroshima, Japan
Natural disaster sites	
Shipwrecks	Diving attractions: Scapa Flow, Orkney and Mediterranean
Volcanoes	Tourist attraction: Pompeii, Italy; El Teide, Tenerife, Canary Islands; Mount Hood, Washington
Fires	Tourist attractions
	US National Parks
Earthquakes or lightning strikes	Photographic attractions, albeit brief

intruding unnecessarily (perceptibly) on the freedom that drivers seek, and to develop among them a sense of responsibility (Range Rover of North America 1990). In some parts of the world, the motorised sledge or skidoo has similarly provided the recreational traveller with opportunities to strike further and deeper into countryside than others would wish. The activity is best managed by a combination of techniques, as at Kananaskis, Alberta.

BOX 7.5 Tread Lightly!

The US Forest Service and the US Bureau of Land Management joined forces with fifty founding members (manufacturers, clubs and media among them) to inaugurate the 'Tread Lightly!' programme in October 1990. The aim was to encourage safe and environmentally responsible off-road wheeled recreation. The target was principally four-wheel drives, skidoos and motor cycles, but the programme extended to mountain bikes also, and included a pledge:

> I Pledge to TREAD LIGHTLY by:
> Traveling only where mountain bikes are permitted;
> Respecting the rights of hikers, campers, horseback riders, and others to enjoy their activities undisturbed;
> Educating myself by obtaining maps and regulations from public agencies, complying with signs and barriers, and asking owners' permission to cross private property;
> Avoiding streams, lake shores, meadows, muddy roads and trails, steep hillsides, and wildlife and livestock;
> Driving responsibly to protect the environment and preserve opportunities to enjoy my mountain bike on wild lands.

Source: US Department of Agriculture Forest Service and Bureau of Land Management (1990)

BOX 7.6 Zoning the problem out

Often the easiest way to resolve conflicts is to allocate the activity to a spatial or temporal area of its own. It follows that, to do the job properly, it is important first to identify all the possible conflicts or competing interests (including, for example, wildlife). Zoning is the way we allocate space, and programming is the way we allocate time. Naturally, there are occasions when the two techniques are used together.

A reservoir complex with three water areas, formerly used for supplying drinking water, was converted into a country park, Monikie (Jamieson 1984). Whereas there had been certain activities allowed whilst it was still a reservoir, when opened up the demands from some of the newer sports grew. There was a good deal of conflict at the outset of opening. Some areas (notably Island Pond) were equally popular for fishing and sailing, but on closer examination and after negotiation it transpired that the peaks in popularity could be used to programme the activities across the zones, to create greater capacity.

What became clear during the process was that the real zones (in this case) were of expectation and were to do with leisure styles rather than physical requirements. In determining the carrying capacities that should be set, the

---

effects (and their reversibility) need to be taken into account. Ballinger (1996) gives some useful examples of zoning water areas where the interests of recreation and wildlife have been taken into account.

The visible trace of human movement is most often erosion, but the disturbance effects on other species, particularly animals, are more difficult to gauge. The single pass through a glade in a wood may be sufficient to alter the networks of routes taken by small rodents. The use of skidoos may affect the migration patterns of deer, and the presence of humans near nesting birds (Sidaway 1988, 1991; van der Zande and Vos 1984) or near mammals rearing young may affect their breeding success. These disturbance effects can be reduced. Those taking their recreation need to be more aware of the animals and their behaviour patterns. Mostly, they will then be willing to modify their approach to recreation (British Mountaineering Council 1996; Scottish Canoe Association n.d.).

In the sea the changes take place, for the most part, unobserved by others. Sometimes the consequences are dire, as for example with manatees injured by water craft (Mignucci-Giannoni *et al.* 2000) or the erosion of large areas of the Great Barrier Reef, where education (supported by enforcement) is required (Alder 1996; Inglis *et al.* 1999) to prevent the loss of this vast ecosystem. In some areas feeding by visitors has grossly changed the behaviour of species. The example of brown bears in the parks of North America is well recorded. There are also cases where the diurnal rhythms and diets of sea creatures have been substantially altered, as in the case of 'Stingray City', off the Cayman Islands, where on a busy day 500 divers may be observed feeding and stroking rays (Shackley 1998). Incorporating education and interpretation is the long-term answer, as discussed in Chapter 3. In the short term, zoning activity across time and space, judiciously tempered with marketing, may offer the best hope.

A world of sound

Absolute silence is not a common feature of our world. We are for the most part surrounded by sound. Some sounds are universally appreciated, and others are universally loathed. In between the two extremes much depends on tone, pitch, volume, rhythm, duration, timing, personal preference, personal mood and social setting. What is fantastic in a concert hall or over the radio is not necessarily welcome when watching wildlife in a wilderness setting. Similarly, a tiger growling would be perceived rather differently depending on whether the listener was outside or inside the enclosure. Personal preferences will have been shaped by experience. People who have flown light aircraft can identify with the pilots of small aircraft who, to others, may just be producing an irritating whining noise. A good deal depends on expectations.

Whatever we may think of sound or noise, we recognise the need to manage it. Acoustic engineers and designers need to consider the quality of buildings and how different surfaces will reflect or absorb sound. Indoors we will often be concerned to insulate rooms or areas to ensure that unwanted sound is contained. In a large void where the noise echoes, putting up fabric or filling the room with people will muffle the sound. Sometimes silence can be intrusive and the manager can introduce background noise (white noise). In other areas, music can be used to soothe people – even outdoors (Jenkins 1999). The therapeutic effect of music is well known. Traffic noise is one of the most intrusive sounds for people attempting to relax in cities, close to busy roads. Planted material, hedges, and trees can be used to absorb the sound, and push it up into the sky, so reducing its effect on those adjacent to the roads.

BOX 7.7 Tranquil Areas Mapping

Tranquillity is said to have declined from 63 per cent to 50 per cent from the early 1960s to early 1990s in an area close to London, according to the Council for the Protection of Rural England. This is not surprising, as road noise from motorways can travel 2–5 km. The process of mapping has been developed by the ASH Consulting Group, and widely reported (Rendel 1996).

Tranquil areas (so defined) are at least:

◆ 4 km from the largest power stations;
◆ 3 km from heavily trafficked roads such as the M1 and M6 and from large towns such as Leicester and from major industrial areas;
◆ 2 km from other motorways and major trunk roads (such as the M40 and A1) and from the edge of smaller towns;
◆ 1 km from medium disturbance roads, i.e. roads which are difficult to cross in peak hours (about 10,000 vehicles or more per day) and some main railway lines.

Tranquil areas are drawn with a minimum radius of 1 km, and lie beyond military or civil airfield noise lozenges, and beyond extensive open cast mining.

Within tranquil areas, zones of weakness are recognised where there are linear elements, creating lower levels of disturbance:

◆ low disturbance roads;
◆ 400 kV and 275 kV power lines;
◆ some busy railway lines

Within tranquil areas sites such as large mining or processing sites, wind farms and settlements of more than 2,500 may create zones of weakness.

Tranquil Area Mapping identifies undisturbed countryside. The process is not objective, but is a pragmatic approach that has pointed to areas of significant change, which may be of concern.

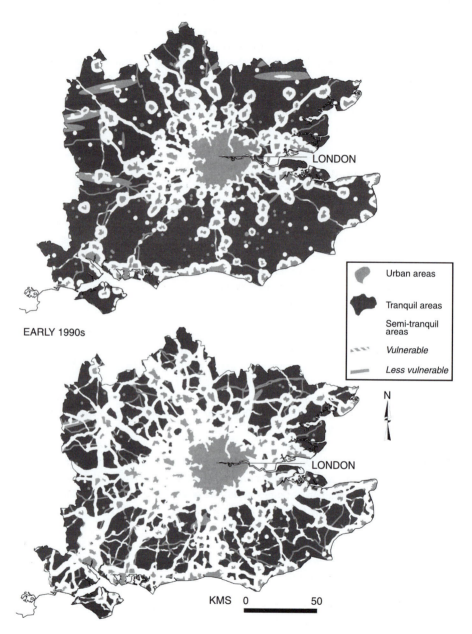

EARLY 1960s

EARLY 1990s

LONDON

LONDON

Urban areas

Tranquil areas

Semi-tranquil areas

Vulnerable

Less vulnerable

N

KMS 0   50

FIGURE 7.3 Tranquil Areas, unspoilt by urban influence, south-east England, early 1960s (upper) to early 1990s (lower)
Source: Tranquil Area Maps, Campaign for the Protection of Rural England and Countryside Commission (1995)

It is sometimes hard to create sufficient personal space in towns and cities. The noise generated by traffic provides a background hum. Against this background hum, machines in factories create their own special noises. General background noise is something to which individuals become 'acclimatised'. It is harder to become immune to sudden, intermittent or other noises that occur without warning. The same is true of sounds that produce irregular patterns, or permeate domestic quarters. Older buildings tend to have thicker walls or to have deadening (in Scotland, often ash from coal fires) between boards, which keeps noise down. New houses should have sufficient insulation to contain most domestically generated noise. Unfortunately the difference between noise and sound relates to values which the listener attaches to the sounds. Unwanted sound becomes noise, and wanted sounds become music or recall pleasant memory trails.

A survey of households in Britain in 1999 (Department of Environment, Transport and the Regions 1999) revealed a rising number of complaints about noise created by neighbours in urban areas. Top of the list was annoyance at couples arguing late at night. Other complaints included the playing of music too loud or too late. Noise does not have to be loud to be injurious to health. Studies of noise close to airports (Bond 1996) indicate that young people affected may have diminished attentions spans, and that complex tasks become more difficult. Increased blood pressure and levels of adrenalin have also indicated that this level of exposure induces stress. Insulation ensures that sounds generated by households are contained or kept to reasonable levels.

BOX 7.8 In something of a spin

The Isle of Skye is rightly famous throughout the world, for its song, for its whisky, for its mist and for its breathtaking scenery, where the Cuillin mountains reach the sea. In 1997 there was a proposal from a London-based firm to run tourist helicopter flights over the mountains. After a storm of protest the proposal was modified. The storm persisted, and the proposal was eventually dropped (*Scottish Wild Land News* 1998). At the time the British Helicopter Advisory Board had guidelines for flying over National Parks in England and Wales. Because of different histories and legislation there were at the time no National Parks designated in Scotland. The Advisory Board has since amended its guidelines.

Operating helicopters in National Parks [in England and Wales] *and* [in] *National Scenic Areas in Scotland*
Helicopter operations in National Parks and places of outstanding beauty often cause annoyance and irritation to those seeking quiet and tranquillity. Pilots should follow these Guidelines:

1   Before undertaking an operation in a National Park, make a detailed study of the map: note where the key beauty spots are located and the routes leading to them.

2   Plan your operation to ensure that your chosen route avoids sensitive areas, e.g. beauty spots, built-up areas.

3   Spend as little time in the National Park as possible and always fly as high as possible consistent with safety.

4   Operations prior to 0700 and after 1800 hours are to be avoided.

5   Seek advice from and inform National Parks office about your intended operation giving as much notice as possible.

6   Avoid crossing National Parks if suitable alternative routes are available. When crossing National Parks always fly as high as possible unless safety requirements, e.g. weather conditions, require lower levels. Follow key line features, e.g. railway lines, motorways, etc.

7   Helicopter training flights should not take place.

8   There should be no flights on Christmas Day.

Source: British Helicopter Advisory Board (1999)

The development of the personal stereo has enabled people to take their favourite music (sound, or noise) wherever they go. Inevitably the facility was taken up early by young people. With wrap-around headphones the music seems literally to enfold the listener and create a world within which only it can be heard. Depending on the shape of a listener's ears and the headphones purchased, the stereo player will play to the individual, excluding all external noise. If the match is less than perfect, it relays an all too familiar rhythmic susurration, sshhh-sshhh, to others round about, to their considerable annoyance. Few places are immune from the rash of 'personal' stereos. One case reported in the press revealed a (well-prepared) commuter on a train so irritated by the sound that he withdrew a pair of scissors from his pocket and cut the wire leading from the player to the earphones!

Reclaiming the land

When we run out of land to play on, we can always start reclaiming it, from the sea perhaps? The people of the Netherlands have done so in grand style, by creating banks, allowing the areas to silt up, and eventually planting to stabilise them, with effects elsewhere that will no doubt become apparent in time. Is there scope to create extra recreational areas through reclamation schemes? The construction of Chek Lap Kok airport in the Hong Kong Special Administrative Region of China and the steady northward advance of the shoreline on Hong Kong island give an indication of what is possible, on a large scale. The scope for recreation contributing to the rehabilitation of land is enormous, as Table 7.4 suggests.

Land is continually on the move, as tectonic plates shift, giving rise to seismic activity. Most such activity occurs over a longer time, such as the growth of mountains. Just sometimes, natural forces move the land about at a rate we can perceive. Forests have been established in parts of Britain to freeze geomorphological processes,

TABLE 7.4 Strategies and tactics for managing recreation impacts on resources or recreation experiences

1 *Reduce use of the entire area*
Limit number of visitors in the entire area
Limit length of stay in the entire area
Encourage use of other areas
Require certain skills and/or equipment
Charge a flat visitor fee
Make access more difficult throughout the entire area

2 *Reduce use of problem areas*
Inform potential visitors of the disadvantages of problem areas and/or advantages of
    alternative areas
Discourage or prohibit use of problem areas
Limit number of visitors in problem areas
Encourage or require a length-of-stay limit in problem areas
Make access to problem areas more difficult and/or improve access to alternative areas
Eliminate facilities or attractions in problem areas and/or improve facilities or
    attractions in alternative areas
Encourage off-trail travel
Establish differential skill and/or equipment requirements
Charge differential visitor fees

3 *Modify the location of use within problem areas*
Discourage or prohibit camping and/or stock use on certain campsites and/or
    locations
Encourage or prohibit camping and/or stock use only on certain campsites and/or
    locations
Locate facilities on durable sites
Concentrate use on sites through facility design and/or information
Discourage or prohibit off-trail travel
Segregate different types of visitors

4 *Modify the timing of use*
Encourage use outside of peak use periods
Discourage or prohibit use when impact potential is high
Charge fees during periods of high use and/or high-impact potential

5 *Modify type of use and visitor behaviour*
Discourage or prohibit particularly damaging practices and/or equipment
Encourage or require certain behaviour, skills and/or equipment
Teach a wilderness ethic
Encourage or require a party size and/or stock limit
Discourage or prohibit stock
Discourage or prohibit pets
Discourage or prohibit overnight use

6   *Modify visitor expectations*
Inform visitors about appropriate uses
Inform visitors about conditions they may encounter

7   *Increase the resistance of the resource*
Shield the site from impact
Strengthen the site

8   *Maintain or rehabilitate the resource*
Remove problems
Maintain or rehabilitate impacted locations

Source: Leung and Marion (2000)

in Thetford in East Anglia and at Culbin on the coast of the Moray Firth, where sands threatened to engulf the community. These forests were not planted for recreational purposes although both are now extremely popular for recreation.

Areas of land that have been subjected to disturbance through mining and quarrying have provided landscapes which, with a modicum of lateral thinking, or serendipity, can provide superb sites for leisure, recreation and tourism. Such areas are usually close to centres of population, where people were first attracted by the presence of minerals, and where other open land may now be scarce. The processes often leave large depressions which, when flooded, provide a haven for water-loving people and wildlife. There are old open-cast mining areas close to Cologne in Germany which are now pleasing lakes surrounded by forests and walks. Although some of the tips in the past have proved less sightly and even dangerous (Aberfan), with forethought the extractive processes can be planned to provide shaped tips or mounds, and the designs included in the landscape plan for the after-use.

Planning authorities in many countries now require a detailed plan to show how the site will be used after the minerals have been extracted. The assumption has usually been that it should be returned to its former use, often agriculture. Increasingly we should question such assumptions. Water areas can provide rich opportunities for wildlife and for people, especially if linked with a mixture of landscape types, some grassland, some forest and some water. One of the problems has been in the difficulty of treating ground which has been contaminated. Often the contamination has been caused by bringing to the surface a cocktail of toxic chemicals associated with the mineral targeted for mining.

In parts of North Wales the hills were rich in minerals, but these areas have largely been mined out. There are the remains of copper, silver and lead mines, as well as slate quarries. Some of these mines and quarries are now operating as tourist attractions. In some cases retired miners have been encouraged to take work (or serve as volunteers) telling stories about how the mines operated in days gone by. Such interpretation enriches the visitors' experience immeasurably. Interpretation, oral history and living history combine to generate pleasure, meaning and wealth.

Before man started cutting and burning trees in the boreal north, to clear space on which to live and grow crops, forests abounded. They have always been great venues for recreation, because of their generous distribution and the rich variety of settings that they provide. Some countries have managed to maintain their forest cover better than others. Sometimes, this has been because the land was found to be too steep to cultivate, or because crops would not grow easily. In any event countries like Britain, which for one reason or another allowed their forests to decline to dangerously low levels (only 4 per cent in 1919), are now replanting and creating new forests. Once again, these forests are multi-purpose. The revival of interest in urban forestry has been largely due to the recreational values of these forests (Whiteman and Sinclair 1994).

There is also enormous interest in the environmental values of forests, and great concern about the plight facing the great forests of the world, in the boreal north and in the tropical rain forests of Amazonia and in Indonesia. This serves to remind us that the value of forests and other environmental goods extend to non-use values that can be enjoyed at considerably more than arm's length. Tree planting is an activity in which communities can take part easily, and which encourages great community spirit, providing long-lasting links between a community and its environment, and of course among members of the community. Through leisure, people can rediscover or learn how much enjoyment can be obtained from working together to improve the local environment (see list of web sites). From a recreational point of view, new forests provide a way to create settings of many different shapes and kinds for all the different activities. Although woodlands cannot be created overnight, planting trees and shrubs can provide a very significant change in the landscape within three to eight years, and a real sense of woodedness.

Managing water

It is to edges that we are most attracted, and none more so than the interface of land and water, whether fresh (or sweet) water or the sea. For many, recreation on water is the goal they strive for, and we are now able to supply opportunities close to where people live which mimic those available naturally in other places. The management can be at many scales: polders in the Netherlands; rivers, lochs and dams in hydro schemes; artificial lakes and ponds; artificial white water; swimming and boating pools; and all the way down to fountains and water clocks (http://www. jackson.u-net.com.htp). In reclaimed landscapes, water has an enormous part to play. It requires management, to ensure that it becomes integrated into natural cycles without contaminating natural systems, ensuring no leaching out of any of the noxious chemicals that may have been brought to the surface in the previous industrial life of the landscape.

Holme Pierrepoint, England's national water sports centre, is Britain's first purpose-built water sports centre and provides the highest standard of facilities for rowing, canoeing and water skiing, It provides for other water sports too. A central feature is the regatta lake, which provides standard conditions for competitors, within a pleasing rural setting. In 1986 an artificial canoe slalom was opened, modelled on

---

BOX 7.9 Thorpe Park

Situated on the outskirts of London, this park has been created from a very large sand and gravel quarry. In 1970 the site was still an active quarry, which at the usual rates would have taken something like twenty years to restore, or at least complete the infilling. Thanks to a little lateral thinking, the site has instead been restored and developed as a water park, following renegotiation of the planning agreements. Instead of twenty years, the site was been restored and opened, after just eight years, as a theme park in 1979 – a park which attracts more than a million visitors a year.

The site extends to about 500 acres, half of which is water. Originally the central theme was 'The History of the British People as a Maritime Nation'. Each year themes are added or varied. As well as exciting historical exhibits, attractions include something for all ages, including the animals on the farm. In the early to mid 1980s more thrill-seeking rides were introduced. Typically, in developing the components something like £2.5 million or so would be spent every other year. As the publicity points out, Thorpe Park started as more of an exhibition-style theme park and has moved to a highly successful family leisure attraction.

---

and improved from a course built at Augsburg for the 1971 Olympics. The centre is set within a country park, and the two elements are managed as a joint venture between Nottinghamshire County Council and Sport England.

Water is an extraordinary substance. Its properties ensure that it is involved in transport systems of all kinds, anything from oceangoing cruises (as the major component of the sea) to carrying substances around our bodies (as the major component of blood). Most of our towns and cities sprang up beside rivers. Many would have been settled from rivers. We used rivers for transport, to provide fresh water and to carry waste away. More recently we have treated the waste first. We have a tendency, certainly in temperate countries, to take water for granted. Pressures and tensions about water as a resource are increasing around the world. Recreation can have many different impacts on water systems:

♦   *Extraction and consumption.* Water may be taken to serve the needs of tourists. In places like the Mediterranean, when a major part of the population of Europe heads south in holiday time, this can lead to problems, particularly for local communities. Elsewhere, there are reports that water pressure may be reduced for local populations to allow sufficient water for tourists in nearby resorts to have showers (Wall 1996).

♦   *Physical disturbance or disruption.* To meet the needs of sport, groynes and piers may be built which will have an effect on flow patterns (that is their intent), but this has knock-on effects (literally) downstream or further along the shore (French 1997: 75). People or accompanying dogs paddling in a river, or along a lake shore, will disturb the system—as will the wash, turbulence and the dragging of anchors by boats (Liddle 1997: 19–20).

◆ *Biological disturbance or disruption*. The situation on and in water is little different from the situation on land. Trampling on banks, and disturbance among life forms, and ecological patterns, may give way to serious or longer-term damage. In water, changes may be swift and extensive, as transport and dispersal through water are effective. Recreational fishing seems unlikely to have led directly to the extinction of any species, but in its pursuit there have certainly been ecological disasters, with the introduction and spread of species such as rainbow trout throughout Europe, displacing local populations of brown trout (Liddle 1997) and changing whole ecologies. Recreational boating among the Florida Keys does seem likely to have contributed to the injuries inflicted on the manatees. In other cases, the introduction of species may have greatly changed local ecologies.

◆ *Chemical disturbance or pollution*. There are yellow plastic ducks stuck in the most unlikely places, as a result of fund-raising duck races, carried by currents that circulate water around the globe (MacLeish 1989). It takes a very long time for all the water to be mixed, because of the great depth and volume involved. Even a few drops of phenolic compounds are traceable many miles from the incidence of pollution. The former use of antifouling paints on yachts to prevent the build-up of barnacles and weed, unsurprisingly, had an effect on sea life through which the yachts passed (Goodhead *et al.* 1996).

The remedies are usually very simple. They rely on using less water, on using it more wisely and ensuring that toxic chemicals and pollutants are contained and not released. At least one service is now offered at Yarmouth, Isle of Wight, for scrubbing yachts, at about 50p per foot, which, allowing for five scrubs a year, is apparently equivalent to the cost of antifouling paint (Sjoberg 2000). One of the greatest difficulties is the charm with which the water washes nasty things (apparently) away, and lulls us into thinking the problem is solved (out of sight, out of mind).

Users of wet bikes, or personal water craft, are popular among water skiers for having usurped their place as the least loved water users. There are undoubtedly problems associated with impacts from these craft, as they can reach into areas where others cannot, causing disturbance to fragile eco-communities. So hated are they by non-users that their significance may be scored by respondents in surveys above other probably more significant pollution effects and changes (Burger 1998). There are problems in relation to safety, and compatibility with other activities. Personal water craft and swimming do not mix well. Conflicts also arise between groups holding different values. It will help in managing any water area to consider the different groups, and what they will be expecting:

◆ *personal water craft users* – freedom to use the water to explore the fun, exhilaration and possibly noise associated with the speed and manoeuvrability of motorised personal water craft;

◆ *swimmers* – usually family groups or friends seeking the pleasure of swimming,

paddling and splashing about in reasonably shallow water, in one another's
company, with some members seeking a swim for a rather longer distance;

◆ *windsurfers, or board sailors* – usually youngish people, seeking mastery over
techniques which will enable them to sail on all points and to race across waves,
with acrobatic tendencies, and possible starts from the shore;

◆ *sailors* – usually seeking close contact with the elements, peace and tranquillity,
perhaps recreational fundamentalists.

Not only do personal water craft create machine noise (which arguably could be
reduced by the manufacturers), but they are driven mostly by young people who are
striving for freedom, and who may not yet always have matured into bearing in mind
what other users want. These are of course stereotypes, and may or may not resemble
what actually happens, but thinking through what the different intentions maybe
helps management. A joint working group has produced a very useful guide for
local and harbour authorities on managing personal water craft, which includes
a step-by-step plan for putting together a local scheme (Eardley and Mawby 1999:
17–19).

Beyond any disturbance to others, powered craft disturb the water, and most
notably their wash can cause or assist bank erosion. Not all the impacts are necessarily
bad. It is more helpful to think of the different impacts, their effects, and any knock-
on effects, rather than to jump to conclusions. For example, propellers will cause
turbulence, which (depending on the propellers) may encourage more mixing of
water with air, encouraging oxygenation, which may be helpful on some canals. The
same motion will clearly affect any algae on the surface. Whether this is good or bad
will depend on the outcome sought. Certainly the presence of powered craft will, by
definition, displace those species sensitive to powered craft, and encourage others.
Damage from wash can be severe, but is related to the power applied and the speed of
the craft. In most situations, therefore, the effects can be relatively easily controlled,
by controlling the speed of the craft.

**BOX 7.10 The British Balloon and Airship Club Code of Conduct for
Pilots and Farmers**

The club has agreed this code (British Balloon and Airship Club 1996) with the
representatives of farmers (National Farmers Union) and landowners in England and Wales (Country Landowners' Association) to avoid conflicts between
balloonists and farmers as far as possible. The code sets out advice to pilots
under seven headings: insurance, flight planning, the take-off, in the air, the
landing, record and retrieve, and post-flight. The advice ensures that would-be
pilots are aware of the possible effects of their activity, especially the risk of
disturbing livestock. The advice to farmers emphasises that farmers should be
reasonable in enabling balloonists to retrieve their craft, and sets down a suggested level of recovery fee. The code does little more than point out some
salient points of the law (to both parties), give advice as to what is perceived by

---

each group as courteous, and advise on how to avoid problems. The code gives the minimum information to alert the pilots as to what the farmers' fears are. Where problems do occur, a procedure is suggested to enable either side to gain access to the appropriate contacts. The added value is in having the code agreed (and revised) by the representatives of both groups.

A good deal of recreation is founded upon passage or travelling at ease with no strict timetable or obligations to others. In many cultures, canal systems have been devised to link rivers and lakes forming a network of waterways. Such networks allow travel between cities, and across from one river system to the next, and have been enjoyed by millions. Many writers have told of the pleasure to be obtained on such journeys (Stevenson 1896a; Jerome 1957; Jones 1987). What links many of the stories is the chain of unexpected events, the serendipity attached to journeys. Managers need to avoid interventions that might interrupt the natural flow of experiences or erode local characteristics and the distinctiveness on the way which marks passage along these routes.

It is not just in diverting and canalising rivers that mankind has sought to interfere with waterways. Sometimes man has created lakes through the damming and flooding of valleys, for hydro-electric schemes or reservoirs. Occasionally artificial lakes have been created for ornamental reasons. That the Serpentine (in Hyde Park, London) resembles a stretch of river is not surprising, as it was created by damming the Westbourne river in 1730 and designed into Hyde Park, in the centre of the West End. Creating lakes, purely for ornamental reasons, is an expensive business, but there is no doubt of the magnetic properties of water, and this is as true of artificial bodies of water as it is of natural ones. To maximise the value of such areas, it is important to vary the treatment (Forestry Commission *Wildlife Rangers' Handbook*). This will (necessarily) then appeal to a variety of creatures and plants and so encourage diversity from the start.

Managing recreation in the air

We need air for life processes, breathing for a start! Storage is less of a problem, although for certain recreation activities where air is in short supply – under the sea, or on the highest mountain tops – we bottle it. We may also choose different locations that naturally manage the air, such as cliffs that provide thermals to give gliders and paragliders lift. We also use every conceivable kind of sail to capture flowing air or use its force to keep us sailing across water, or flying through the air.

Despite designs as early as those by Leonardo (1452–1519) depicting aircraft (Taylor 1960: pl. 12–13), it is only relatively recently that man has taken to the air for recreation, but the range of contraptions in which recreation is taken is already diverse and still segmenting: man-lifting kites, balloon, airship, parapenting, paragliding, hang gliding, gliding; fixed wing powered flight, microlite, autogyros, helicopters.

The impacts of taking to the air can be of several different kinds and the problems can often be solved through better communication. Balloons suddenly appearing over hedgerows can startle cattle, and people, and the noise of powered flight can carry for miles and reach into and disturb tranquillity. Many governing bodies produce codes that suggest behaviours designed to minimise the negative impacts of each sport.

Corridors into greenspace

Providing links between where people live and the countryside through green corridors intuitively feels right, providing as it does networks for other animals to use also. Large mammals often use the marginal land alongside railway lines for migration routes. (In Moscow, elk use railway lines to move from the woodlands to the north of the city down to Elk Island – a wood on the eastern outskirts).

To make green space as attractive as possible to all species there should be plenty of variety of cover, of herbs, shrubs and trees. People want greenways to be not just routes to attractive countryside but attractive in their own right (Countryside Commission 1998). Ideally, these networks should have links with water systems, and should provide access to food and water. How wide they should be will in part depend on cost, but ideally there would be routes wide enough for all the different uses involved, from pram pushing to horse riding (Cotton and Grimshaw 2000, Ove Arup & Partners 1997; Sustrans 1994), with sufficient open areas to afford women security where they might otherwise feel threatened (Burgess 1995). Large areas of green space in the centre of towns and cities provide enormous opportunities for people to enjoy a neutral space out of doors where they can meet and make friends. Parks are managed to provide for a range of activities to cater for all ages, and all abilities. The best parks are responsive to the customers they serve, and function with unobtrusive management. The harder they are used, the more resilient the surfaces have to be, but a great deal can be achieved through sustainable management techniques to reduce the cost of maintaining these heavy-duty areas.

BOX 7.11 Five simple aids to sustainable park management

- ◆ Stop cutting grass and introduce grazing regimes.
- ◆ Remove old signs, and think twice about replacing them.
- ◆ Where possible (and if you have to have barriers) use natural barriers – plants or water – or those made of renewable materials such as wood, rather than iron, steel or plastic.
- ◆ Use less energy derived from fossil fuels.
- ◆ Consider working with (rather than against) natural cycles and processes, but take care!

Relative impacts of different activities

Not all activities have an impact of the same magnitude, although there is no simple scale or barometer we can use to describe what is going on. There is no short cut to thinking through the impact in each case (or measuring, but still thinking).

BOX 7.12 Orienteering – Sweden

Orienteering originated in Scandinavia. It is a sport that can attract hundreds, and on occasion even thousands, of competitors. Competitors are divided into age classes for each sex. Each class is set a particular course, which demands that competitors have to go to a number of stations in a particular order. The competition is timed and allows individuals to compete on running speed and navigation. While everyone goes through the start and the finish, the level of visitation at each of the other stations is in the gift of the individual who sets the course.

Following years of experience, the clubs are able to manage events so that the wildlife suffers minimal disturbance as far as can be told. Certainly, there is evidence that the deer return to the areas after competitions. Course planners can also avoid setting courses or timing events to disturb ground nesting or sensitive species of birds.

Source: Broadhurst (1987)

Negative impacts

In Britain there has been concern about the possible negative impact of leisure activities in the countryside (House of Commons Environment Committee 1995) and of tourism (Department of Employment 1991) on historic monuments, villages and the countryside. It is easy to see how concern has arisen. In retrospect, it is extraordinary that such concerns should have been seen in isolation from the effects of other activities. In relation to the countryside, some authors (Sidaway and Thompson 1991) have sought to draw attention to the rather greater changes being wrought by land uses other than recreation. In the past, impacts resulting from work-related activities have been brushed aside as the legitimate consequence of creating wealth, and therefore of a different order.

Typically, we may have detailed information about the effects of a recreational activity on a particular area but rather less information on some of the wider effects. Eco-tourism is an oft quoted example. An environmentally aware holidaymaker may take every precaution to avoid negative impacts on the natural environment whilst on holiday, and ensure that the host community is taken fully into account. But to reach the holiday the journey may have involved a few car trips, and even a flight or two. The impact can be substantial. Using recent survey results, it has been estimated that

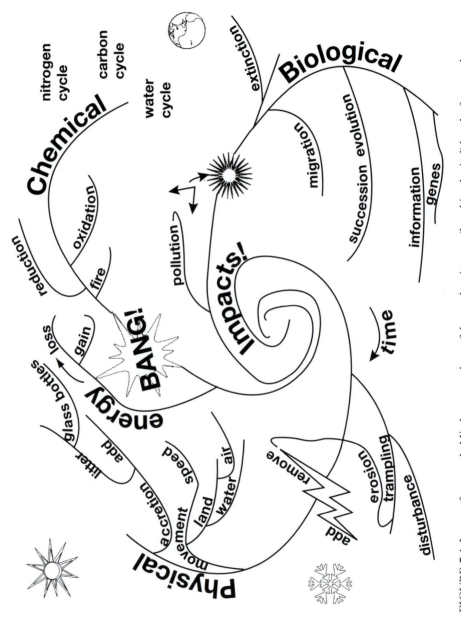

FIGURE 7.4 Impacts framework. Mind maps can be useful to explore impacts (here bio-physical) but the framework to use will depend largely on the nature of the site and the scale of the investigation

---

BOX 7.13 By car for a walk

Professor Jost Krippendorf quoted from an essay by Claus Mampell at a conference on tourism and the environment in London:

> There is really nothing like walking in the country. It is just a pity that it takes so long to reach the place where I like to walk because there is so much traffic on the main road. I had to travel by car for one hour each way to go for one hour's walk there, but I accept this because I find walking so relaxing and I get so agitated after the journey there in the terrible traffic that I need the relaxation of a walk to strengthen me for the journey home. The extra-ordinary density of the traffic on this road is caused by the fact that so many people are going to the same spot for a walk. The road is sometimes com-pletely blocked, so there are plans to widen it so that everyone will be able to enjoy the experience of being there, which is a good thing. It is a pity that more parking has to be created for the increasing number of cars, and this land has to be taken from the countryside. If there were to be room for everyone's car who wanted to walk there, the whole area would have to be turned into a car park, and perhaps that would be a good idea because then less people would want to go there and the problem would solve itself.

Source: Department of Employment (1991)

---

a return flight from Britain to North America contributes as much carbon to the atmosphere, per passenger, as a whole year's motoring by the average British motorist (Friends of the Earth 2000). We need to understand the full impact of our decisions. To be able to do this, we must endeavour to gather all the relevant data, and not repress uncomfortable data.

Golf: an example

A good deal has transpired since Schofield (1987) described some of the opportunities that exist for maximising wildlife on golf courses. The game has continued to grow in popularity, and the golfing establishment, like all others, has become more aware of its environmental responsibilities. *Enact*, one of English Nature's newsletters, carries news of developing awareness and practice (Newlands and Roworth 2000; Simpson 2000), and the introduction of mowing regimes which favour particular target species and prevent succession disturbing the nature conservation plans. Meanwhile the Society of Australian Golf Course Architects has produced a brochure describing the environmental contribution which well designed golf courses can make. There are apparently some 1,500 courses in Australia comprising approximately 100,000 ha of land, much of which is in urban environments. It is clear that there are enormous opportunities to manage golf courses to enhance opportunities for conservation, for

---

BOX 7.14 Principles of sustainable tourism

1   The environment has an intrinsic value, which outweighs its value as a tourism asset. Its enjoyment by future generations and its long-term survival must not be prejudiced by short-term considerations.

2   Tourism should be recognised as a positive activity, with the potential to benefit the community and the place as well as the visitor.

3   The relationship between tourism and the environment must be managed so that the environment is sustainable in the long term. Tourism must not be allowed to damage the resource, prejudice its future enjoyment, or bring unacceptable impacts.

4   Tourism activities and developments should respect the scale, nature, and character of the place in which they are sited.

5   In any location, harmony must be sought between the needs of the visitor, the place, and the host community.

6   In a dynamic world, some change is inevitable and change can often be beneficial. Adaptation to change, however, should not be at the expense of any of these principles.

7   The tourism industry, local authorities and environmental agencies all have a duty to respect the above principles and to work together to achieve their practical realisation.

Source: Department of Employment (1991)

---

example by using the course to filter run–off from roads and stormwater, and even to treat sewage. Equally, if the course is in an area where water is scarce, keeping the greens that colour can be an environmentally expensive option, and in some cases there may be a risk of inappropriate use of pesticides and fertilisers.

Landscape ecology

In developing their multidisciplinary approach to design and analysis, an ecologist and a landscape architect (Diaz and Apostol 1993) have been instrumental in developing a novel approach to analysis and design. They happen to be working at the forest scale, but because the approach could be much more widely applied, the word 'forest' has been dropped in describing how it works. The stereotype architect thinks in terms of structures and the stereotype ecologist thinks in terms of flows. Working together, with the support of colleagues representing many disciplines (botany, plant ecology, fire/fuels, fisheries/riparian, hydrology, landscape architect/visual quality, recreation, silviculture, soils, transport, wildlife biology), Diaz and Apostol have developed a sequence of steps which integrates these two approaches. If they had been working in another area they might well have included specialists in the cultural area too: archaeologists, historians, sociologists and even economists.

The approach recognises the fundamentals of ecology, that we can view the setting at any number of different scales, and consider the structures, the functions and the interactions between them. As well as the interactions within any system, there are interactions between systems to take account of. There are eight steps:

Analysis

1   Landscape elements, or structures. Identify and map the structures: patches, matrix and corridors.
2   Landscape flows. Identify and map the flows of interest.
3   Relation between landscape elements and flows, or interactions. Describe the interactions to help understanding of the functional aspects of the area.
4   Natural disturbances and succession. Define the natural systems, and succession, referring to natural disturbances – e.g. wind, fire or flood; population dynamics – and how these systems may be affected by the landscape patterns (the setting).
5   Linkages. Describe the linkages with other areas.

Design phase: establish objectives

6   Landscape patterns from the plan. Determine what landscape patterns already exist, from the plan.
7   Landscape pattern objectives. Develop statements that describe the 'target' landscape pattern (shapes, sizes, arrangements) using steps 1–6, and any specific local objectives.

Design phase: spatial design

8   Landscape design. Map the areas of landscape within which a particular approach is required, as described in the step 7 statements.

This approach could be developed to provide a very useful way of integrating all the different disciplines and applied in the planning, design and management of environments and settings for leisure and recreation.

Ideas for further study or work

1   Water is one of the most precious of resources. Choose a particular leisure or recreation centre, and identify the impacts that the activity has on the water cycle, and thereafter suggest a number of steps to render the activity more water-friendly.
2   In your favourite outdoor recreation area, identify and measure (or describe) the impacts on the ground, and draw up management prescriptions to reduce the impacts, and particularly any erosion occurring.

3   Select a few of your favourite spaces, inside and outside (close to where you work, study or live) and make your own map showing the distribution of noisy and peaceful areas. Draw up prescriptions for reducing unwanted noise and for securing tranquillity in the peaceful areas in the longer term.

4   The car is greatly used by people to reach their chosen settings for leisure and recreation, with the natural consequence that a greater share of the resources is spent on providing for the reception of the motor car. Identify 100 measures (as quickly as possible) for the chosen setting which would make it more attractive to those arriving on foot or by cycle, and save money or other resources.

5   Keep (and encourage colleagues to keep) a diary for a week, recording environmental impacts, on an hourly basis. These should include: energy consumption, water use, production of waste, production of greenhouse gases, irreversible changes, consumption of materials. (If you have sufficient time and (personal) energy, you could include the whole life costs involved – described if not measured.)

Reading

Managing natural systems

Every branch of land management has its standard texts, and each is no doubt worth exploring. For an insight into managing natural environments, and taking into account many different objectives, read Hibberd, who provides an introductory survey of forestry. Foresters have to manage many things: the trees, and the spaces in between: open space, water, wildlife, and visitors.

Managing the built environment

For maintaining the necessary conditions in such specialist recreation locations as swimming pools or leisure centres, there are guides produced by the Sports Councils. In relation to transport infrastructure, read Buchanan, and publications produced by Sustrans. It is worth also looking through alternative literature, such as the books kept by the Centre for Alternative Technology, in Machynlleth, Wales.

The House of Commons Environment Select Committee report on the impact of leisure activities provides a useful starting point, but is restricted to consideration of the more immediate effects. Liddle provides a definitive text on ecological implications, and Leung and Marion a state-of-knowledge review of impacts in the wilderness and how they can be managed. Journals in the area of leisure studies, recreation and tourism have articles on the subject, and so to do those related to environmental management.

Managing for people:
the socio-economic
setting

Overview

People are complex and individual. How they behave depends on the information that reaches them, in many different ways, in their genetic make-up, from newspapers and by word of mouth. People are also tied into groups. The manager needs to think of the political, economic, sociological and technological environments. The task of the manager is to achieve management objectives, and match expectations, without compromising perceived freedoms. Through leisure, recreation and tourism many exchanges occur beyond words and wealth. Managers need to be sensitive to the changing backcloth, the development of ideas, and the possible effects. Markets on their own will not safeguard the authentic or what is significant to people, but including people in the decision-making processes will help.

Working with (rather than for) people demands the provision of the relevant information to allow them to make choices. Existing structures strongly influence information flows, and we may need to employ techniques such as participatory appraisal to check that we are acting in people's best interests. 'Limits of Acceptable Change' is an example of a technique that recognises political realities and is highly adaptable.

Providing goods and services to support leisure and recreation requires material resources, and if we are to use these wisely we must use them effectively, by doing the right thing; efficiently, by using them as directly as possible; and, with economy, by using them frugally. Use of resources demands payment, in the form of prices or taxes, depending on who is the resource holder and who the beneficiaries are. Where costs can be passed on, we need to set the right price. Several techniques are explored: token, cost-plus and going rate. Taking into account the psychological dimensions and benchmarking will accelerate the chances of success. A logical sequence is set out to help managers looking at different ways to generate revenue, before considering the contribution that leisure and recreation events can make to our economy.

Advantages and disadvantages of partnerships in order to integrate service (vertically and horizontally) are described. Hidden agendas often contribute to the way in which projects grow to exceed their budgets. In any partnership, or team, we need to build in exit strategies, and strategies for succession. Waves of political change challenge long held views, and encourage experimentation, but not all experiments end in success.

In achieving best fit, we need to consider what happens, and what needs to happen: before, on the way to, during, on the way back and after each event. There will be a host of considerations about information to consider. Risk assessment and visitor safety management can be used to remove unacceptable risks and, coincidentally, reduce the liability of the manager. We need also to provide for memories, to extend the pleasure. To ensure that effects are positive, we need to encourage leisure, recreation and tourism that take into account all the exchanges, personal as well as material, and not just those mediated through money.

Where experiences depend on intermediaries, these require special inter-personal skills. People in the front line (paid or volunteer) have much in common. Development of skills is a strong motivator. The adoption of shared values that reflect concern for people is likely to reap enormous rewards. Unfortunately, for this reason, there has been widespread adoption of insincere customer care programmes that have cheapened the whole approach. Sincerity works. For that we need to choose the right people, and train them. Developing the full potential of people requires encouraging the exchange of information through appropriate networks, and the development of a strong personal morality or ethic.

What is manageable and what is not

Managing people is about achieving aims through working with other people. People are complex, each individual quite different from the next. How people behave is governed to a large extent by the information they receive: through their genetic complement, through experience, through the influences of parents, relatives and other referent groups, through the media, through expert sources and from every source on earth. How we react to a given stimulus will also depend on our personality and will differ on each occasion. Each occasion is by definition unique, and the result of any recreation activity will be sensitive to initial conditions (Gleick 1988: 20–3) although we may expect a particular outcome. Given that most major life decisions are made at leisure, or when away from work, leisure deserves a greater degree of respect.

As has been said, 'environmental issues are more than just physical. They are also inescapably philosophical, ethical, political and cultural' (Grove-White 1997: 109). Leisure, recreation and tourism, and the consequent impacts, are not all physical either. Managing for the socio–economic environment brings a focus on equity among people now. It implies that we are intent on changing current patterns. We can seek to address the inequities which exist and which we find disturbing (Dower 1993), but only if we weave in ethical values, and then only in defined areas. Neither is it just a matter of sharing out leisure, but rather of sharing out the benefits of leisure, which paradoxically may be employment for others. The socio–economic impacts that pro-vide benefits may also entail costs. New employment opportunities may distort exist-ing labour markets. New (and welcome) social contact may lead to a change in culture, and the erosion of previous values. Finding out what hosts think about tourism and the various interactions is a useful start (Brunt and Courtney 1999; Butler 1999) and it is often a more complicated picture than imagined, not least because some people gain while others lose (Fredline and Faulkner 2000). On occasion, a group of people recognise that they may need to lose a little, if the community is to gain overall, thereby displaying 'altruistic surplus' (Faulkner and Tideswell 1997).

The socio–economic effects can be extensive, changing the way that residents view and make their art (Cohen 1992) and sometimes leading to stratification of the local society (Nepal 2000). Arguably, we should be incorporating ethics in education and training for leisure and recreation (Hultsman 1995), to ensure that these effects are foreseen, and the ethical issues explored beforehand. Managing is about achieving aims, by rearranging things and working through others. This sits uneasily with the notion of freedom, so fundamental to leisure. That is why leisure management requires such care. The best management will be imperceptible to the participant. Delivery of service will be the visible manifestation of management, and there is some evidence that performance quality (against criteria that managers can measure) is a good indicator of whether participants will find it satisfying (Baker and Crompton 2000). People enjoying leisure should feel free from unwanted control.

It follows that Environmental Management systems need to extend beyond taking into account the crude social and economic impacts, and really seek to integrate all the significant effects. The question of significance to whom is best addressed through the inclusion or representation of the different stakeholders in the process. Good practice also requires attention from initiation to operation, and through out-puts and outcomes (Margerum 1999). As an aid to identifying what can be managed and when, the Life Cycle Assessment approach is helpful, and individual areas such as waste minimisation can become the focus of programmes that make substantial economic savings as well as environmental gains (Cummings 1997). Identifying sig-nificant impacts here requires assessment of extent – frequency (volume) and severity (value), as well as the degree to which management can bring about change (Tribe *et al.* 2000).

There are at least two dominant systems in which leisure is provided: through free markets, and through the public sector. In reality, there is a diverse range of opportunities, with elements of each. The 'free market' deals in goods and services for which a demand is evident, and where benefits are transferable to individuals. Over time, currencies have evolved to replace barter and facilitate trade, within free market economies, which are based on the notion of perfect knowledge. Leisure goods can be public goods (those that cannot be traded) or market goods (those which can). Public goods are therefore literally priceless. Experiences that rely on public goods may be seen as unsullied by commercial concerns, and in a different world, somehow closer to spiritual experiences. On the other hand, what were previously considered as public goods can be presented as commodities, and by association become sullied (Cloke 1993). A swim in the sea is worth more, not just because it is free (of charge) but because it belongs to a realm of palpably natural activities. Of course, it is still entirely possible to have a market regulation approach to the pricing of parking spaces by the beach or other related services, but it must be carefully arranged if it is not to spoil the experience. To charge for the swim itself, though, would destroy the freedom (from management). Determine which elements or components of a recreation event belong to which realm before deciding where to charge and where not, and how to charge. What is best for the economy may not be what is best for the environment (Augustyn 1998), in the short term. Some commentators suggest that the private sector rarely creates people-friendly environments (Tibbalds 1992: 5).

In most Western countries, public sector intervention is usually restricted to areas where there is market failure, or where, in the interests of equity, society wishes to make leisure available to all, or to discriminate positively in favour of a particular group. In between 'free markets' and 'public sector intervention' there are all sorts of opportunities to increase the likelihood of provision, or enhance quality of service through incentives and regulation. In Britain, the government uses a standard approach to investment appraisal (Glass 1997) to determine whether on balance the benefits of intervention outweigh the costs, and whether it is the only way of achieving the stated aim, and there are ample texts to give the background (Bailey 1995). When the public sector intervenes, it does so with the mandate of the electorate. To some degree it has had the involvement of individuals, probably in a rather indirect system of consultation, through representative democracy. The articulate tend to have their say, the disadvantaged rather less so. An alternative is to find out what people want by using more inclusive and participative techniques.

In a free market, agents provide information about the goods and services on offer, and consumers choose. Preferences are revealed through spending patterns and market research. In public sector intervention, community leaders often take the decisions on what goods and services should be provided. Any consultation is usually skewed. It is the most articulate who will make their case known.

Under certain circumstances, advantages will accrue to society if an organisation continues to exist, or a service is provided, even if the individuals do not themselves wish to support it at that time. Our public services and forms of government have ensured that many services are provided by government, if not national, then municipal or local. There have always been tensions between national and local government, and between adjacent local governments, and concerns about the growth of organisations at too great a cost to the population. Successive political waves sweep across the world, establishing different ways of arranging services provided for the public good. Right-wing governments have been experimenting with the degree to which services traditionally provided by government can be provided by the private sector, whilst still specified by the public sector. Left-wing governments have been exploring how to enable wider participation in decision making, through different forms of representation or participation. Comparing real costs is difficult because externalities come into play. It is in our nature to be selfish, and some would argue it is more efficient to provide incentives to reflect personal motivation. In many societies, changing demographic characteristics, more older people (living longer) and fewer working people supporting the economy (e.g. in Britain), mean that cost-effective solutions, and ways to economise are actively being sought. This is causing something of a revolution, with left-wing governments apparently just as interested to see how costs can be contained, and just as willing to see what different methods will deliver services most effectively, efficiently and with greatest economy.

There are situations when individual motivations will drive people to behave in a way that harms the collective and usually longer-term objectives of the community, as in the 'tragedy of the commons'. This essay by Garrett Hardin, written in 1968, described the scenario where commoners would compete for the benefits available from the grazing, mindful that their neighbours would otherwise gain advantage. The result, of course, is overgrazing of the common resource and a net decline in its longer-term value to the community and to individuals (Deadman 1999; Russel 1996: 269).

BOX 8.1 Limits of Acceptable Change

Developed by the US Forest Service (Stankey *et al.* 1985), this approach to setting limits and prescriptions as to what happens when the limits are breached represents a significant advance over traditional approaches to carrying capacity (Sidaway 1994a). The procedure, as designed, consists of nine steps:

♦ Identify area issues and concerns.
♦ Define and describe wilderness recreation opportunity classes.
♦ Select indicators of resource and social conditions.
♦ Inventory existing resource and social conditions.
♦ Specify standards for resource and social conditions in each opportunity class.
♦ Identify alternative opportunity class allocations reflecting area-wide issues and concerns and existing resource and social conditions.
♦ Identify management actions for each alternative.
♦ Evaluate and select a preferred alternative.
♦ Implement actions and monitor conditions.

In simple terms, Limits of Acceptable Change (LAC) recognises the value of focusing on outcomes rather than inputs. It is not the number of little feet that matter, nor even the weight upon those feet, but what impact they actually have, and what follows. As has been said, 'Our understanding of the use–impact relationship now tells us that it is anything but direct, invariant and linear' (Cole 1987). LAC requires participants to be specific about both quantity and quality. Its real strength comes from the involvement of all the interested parties in sharing the debate and decision making and in setting the limits and the prescribed management actions, should limits be breached. The debate requires, and contributes to, a shared understanding of everyone's needs.

The ASH Environmental Design Partnership (1986) recommended the use of LAC in the design of ski areas, in a report to the Countryside Commission for Scotland. Shortly afterwards the same partnership was engaged to prepare a development plan for a skifield site at Aonach Mor. The resulting planning permission required the preparation of a management and safety plan, and the management plan incorporated the LAC system.

From the outset the planning authority set up a management committee of the six organisations with the greatest interest:

♦ the developer – Nevis Range Development Company;
♦ local authorities – Lochaber District Council and Highland Regional Council;
♦ landowners affected – British Alcan and Forestry Commission;
♦ the Nature Conservancy Council.

Advised by the Institute for Terrestrial Ecology, which had carried out the inventory, and ASH, the committee agreed (before the development was in place) Limits of Acceptable Change and the management responses which would be implemented should the limits be breached. Thereafter an annual cycle allowed for monitoring and an annual work programme (Figure 8.1).

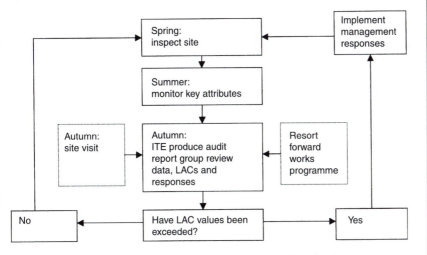

FIGURE 8.1 Limits of Acceptable Change: annual cycle at Aonach Mor
Source: Bayfield and McGowan (1995)

The targets are adjusted each year, to suit the changing circumstances. This avoids many difficulties, but will not answer arguments over what may be short-term gains leading to long-term losses, especially if the gains and losses are in different columns of the balance sheet. There is great potential to apply the principles and the process to many recreational projects where there may be differences of opinion and where different values need to be taken account of to safeguard the environment. It could just as well be applied to cultural as to natural resources. English Heritage has been using the technique in a partnership involving, among others, the Countryside Agency, Northumbria National Park, the local authority and Northumbria Tourist Board. The project will help monitor and shape the recreational use of Hadrian's Wall. The same technique could be applied at any scale, wherever the results outweigh the costs involved.

There is now a growing body of work which will inform those who seek to be wholly inclusive and meet the needs of everyone in the community, as well as the articulate, with techniques which can be matched to the particular circumstance (Agyeman 1996; Baines 1995; Bishop and Rose 1994; Bramwell and Sharman 1999; Brent Ritchie 1998; Carr *et al.* 1998; Chambers 1983, 1992; Etchell 1996a; Hockings *et al.* 1998; Jamal and Getz 1999; Lane 1994; Lankford *et al.* 1994; Loikkannen *et al.* 1999; McPhillimy and Guy 1998; Timothy 1998; USDA Forest Service 1995; Wilcox

1994). There is also ample evidence that more consultation is required (Jackson 1995). These same tools can be used to work with younger people to find out what it is that they want. The techniques for Participatory Appraisal are most helpful because they can reach a broader range of people, and help engage the community in determining their own programmes. There are also some important principles to apply in order to avoid the pitfalls, notably the possibility of raising false expectations:

- The techniques work best where the manager, facilitator and group are aware at the outset of what limits there are to the exercise (Nelkin 1982).
- The group should command (or at least be able to call on) the resources necessary to implement the solution even if through some intermediary.
- Each member of the group should have equal access to the process, and must therefore have a common understanding of the problem.

This also requires the development of skills and building of capacity (Victurine 2000). Care has to be taken to ensure that ideas come forward where the need exists, and not just from communities where the need is less pressing but the spokespeople are more eloquent and articulate (Curry 2000). Where conflict (or a dispute) arises, a different set of techniques can be deployed (Moore and Lee 1999; Sidaway 1998; Stewart, S. 1998).

Our beliefs shape our attitudes, intentions and behaviour (pp. 76–7; Ajzen 1991). To help guide others, most societies have developed codes and laws, so that there is a clear understanding of where the boundaries are: what is permitted and what is not. Education is the most effective tool we have in encouraging responsible behaviour. In some cases we have lost ground to make up. In a study of eco-tourism operators, personal morality or ethics was the only factor which explained the considerable variance in behaviours (Sirakaya 1997). In another study (of how regulations affected the behaviour of people camping) the introduction of regulations led to a change in the population attracted to the site and a consequent improvement in behaviour (Vorkinn 1998).

Some laws, like the Gaelic law tracts, existed for hundreds of years before being committed to paper, being passed down orally. Many of these ancient laws governed movement and access to natural resources, and the degree to which people might have rights over them, to eke out a living. Leisure and work were not then separately defined. Nowadays those with the greatest interest in ensuring that people behave with responsibility are the participants, and as well as rights participants take on responsibilities (Smith 1988). Many governing bodies produce codes which describe responsible behaviour, and increasingly there is co-operative working between environmental and recreational interests, such as between the British Mountaineering Council and the Royal Society for the Protection of Birds, to provide advice on how to avoid or minimise impacts on cliff nesting birds (British Mountaineering Council 1996).

At a larger scale, in 1994, at the Antarctic Treaty Consultative Meeting, reference was included to circulating 'Guidance for visitors to the Antarctic' of the following kind:

- ◆ Protect Antarctic wildlife (do not use motorised transport or contraptions to disturb, etc.; do not feed, etc.; do not damage, e.g. walking on lichen-covered scree slopes; no guns, explosives; keep noise minimal).
- ◆ Respect protected areas.
- ◆ Respect scientific research.
- ◆ Be safe.
- ◆ Keep Antarctica pristine.

BOX 8.2 Greater freedom of access

When the Labour administration of 1997 came into power, they did so with a manifesto that expressed the desire for people to have greater freedom of access to the countryside. Combined with the effects of devolution, this provided for different approaches within Britain. In Scotland, the Draft Land Reform Bill (Scottish Executive 2001a) is out for consultation (at the time of writing). This would establish a new right for individuals of access across land or water, for recreation or passage, subject to responsible exercise of that right. The proposals have been prepared within the Scottish Executive's programme of land reform (Scottish Natural Heritage 1998; Scottish Executive 1999). The law will be part of a package that includes an Outdoor Access Code, a programme of education, the development of a network of paths and the provision of more ranger services. The intention is also to encourage individuals to take greater responsibility for their actions. The Countryside and Rights of Way Act 2000 is now in place for England and Wales, providing for access to open ground.

The Executive had asked its agency concerned with conservation and recreation in the countryside, Scottish Natural Heritage, to review the arrangements for access, to see how greater freedom could be extended. The work was conducted making use of the Access Forum, a tripartite group comprising representatives of landowners, recreational users and public bodies. As each of the groups tends to take a somewhat different line, balance has been seen as very important, and in each forum, and sub-group, attempts have been made to achieve balance, necessary for developing a robust consensus. The Access Forum had earlier drafted a Concordat for Access to the Open Hill (Access Forum 1996), which to a great extent resolved the conflict between deerstalking (perhaps game management generally) and walking in the hills. (In Scotland the term 'hill' includes mountains.)

Having reported that it would be easier to write a new law than to unravel and make sense of the past, a second stage involved drafting a Scottish Outdoor Access Code. This code will describe what amounts to responsible behaviour, and make plain what behaviours are required of each group. It would be used much as the Highway Code is in relation to driving. It would be strongly evidential. For information about access in Scotland try www.snh.gov.uk.

It had also been recommended, by the Antarctic Treaty, that there should be designated areas for tourism, in effect Areas of Special Tourist Interest, to contain the impacts. At a quite different location, on the Great Barrier Reef, work has shown that education is a much more cost-effective tool than enforcement in encouraging responsible behaviour (Alder 1996).

Ryan (1991) charts the impact of tourism, but eco-tourism is often cited as a way of encouraging people to value the environment. Difficulties of definition enable unscrupulous activities to carry similar labelling to the most scrupulous, and questions have been asked as to whether eco-tourism is sustainable (Wall 1997), and whether the label 'alternative tourism' is simply disingenuous (Wheeller 1992). There are others who clearly foresee a sustainable tourism (Middleton with Hawkins 1998). Governments have also identified opportunities for sustainable tourism (Department of Trade and Industry 1998) and seek to deploy the language in strategies (Department of Culture, Media and Sport 1999). Whilst policies, programmes and schemes abound for minimising waste and energy use whilst on holiday (which also save money for the operator) there is less emphasis on minimising the use of transport (Watson 1997). The development of leisure, recreation and especially tourism is set within broader economic, societal and political changes (Shaw and Williams 1992).

Managing funds

Much happiness is independent of wealth (Carlson 1999; HH Dalai Lama and Cutler 1998) and seems more related to variations in mood. The presence or absence of funds does not, on its own, regulate the amount of happiness. This will be some comfort to many of us. On the other hand, as managers, our ability to provide services and facilities related to leisure, and to travel quickly, is clearly affected by the availability of funds. Travel on foot, cycle or horseback need not be very expensive, but it is not quick. Where time is money, this becomes important. If we are to use our resources wisely, we need to be sure to monitor for effectiveness, efficiency and economy.

Effectiveness is 'doing the right thing'. This may be the hardest of the three areas to monitor. It requires an evaluation of whether the issue identified was the relevant one, and whether the prescribed course of action had the desired effect, and all this in the judgement of those for whom the service is provided. The manager will usually have to make a judgement, without policy evaluation, although the degree of evaluation required will depend on the overall cost of the measure.

Efficiency is 'doing the thing right', and this is a matter of making sure that, between the input and the output, as little of the resource (the funds) as possible is dissipated on other (especially unwanted) effects.

Economy is achieving the result with the least funds possible. Arguably, efficiency and economy are both easier to measure than effectiveness. The rating is often enhanced by conveniently overlooking some externalities, some effects downstream that other programmes will have to pick up. This is one of the difficulties of single-minded pursuit of targets without adequate knowledge of the effects of interventions.

Marketing is concerned with meeting the needs of consumers profitably, and the discipline of marketing (Kotler 1984; Kotler *et al.* 1998) can help focus attention on the importance of some specific areas:

- *product*, which has been described in Chapters 1 and 4;
- *price*, which we need to consider here;
- *promotion*, which is all about encouraging flows of information (at least a two-way process), dealt with in Chapter 3;
- *place*, which really describes the channel through which the needs are met, in this book described as the web, Chapter 3.

In the simplest of cases, it is the person (or entity) who receives the benefits who will pay, and the person (or entity) who supplies the benefits who receives payment. This basic dichotomy suggests that if we are dealing with private goods, it should be the consumer or user who pays, and if we are dealing with public goods, it should be the citizen, i.e. taxpayer. As only those citizens who use the public good will have obtained direct benefit, the taxing route will confer disadvantages on those who choose not to, or who cannot afford to use the public good, perhaps because of associated travel costs. For this reason (if managing public goods) the answer may be to charge where feasible, and to look at ways of targeting those groups who would otherwise be at a disadvantage, using schemes like leisure passports (Collins and Kennet 1999). Where there is an advantage to the community or society for people to engage in leisure and recreation, as with savings in the health care bill, then care will be required to ensure that any charges do not act as a disincentive to the target groups. There are three good reasons for pricing recreation: to raise funds, to ration use and to determine preferences, so providing more of what people wish. This may address social inequities among consumers, but there are difficulties on the supplier side also. In the case of a host community on a tropical island that receives tourists, there is sometimes no suitable intermediate mechanism to enable a transfer of wealth to local people, and 'distributional inequalities favour external operators and urban gateway residents' (Walpole and Goodwin 2000). In some of the debt-ridden countries of Africa goods and services can be provided to build tourism, to help tackle poverty and conservation (Brown 1998).

If we are to introduce a charge, we need to determine the level to set it at. Once this charge has been set, it is not always easy to make radical changes and get them accepted. There are a number of methods for determing the charge:

- *cost plus*, where, having established the cost of providing the service, something is added, for administration or profit;
- *going rate*, where a similar price is set to that being charged by competitors or for similar services. It may bear little resemblance to costs.
- *what the market will bear*, where the price is set to take advantage of what people will pay. This may involve market skimming, where the price is set high initially to take advantage of those who can afford the goods or service and those who will pay a premium to obtain the goods or the service before others. This usually applies to innovations of one sort or another. When that segment of the

market has been saturated, the price can be lowered to take advantage of the next group, and so on.

Whichever technique is used in setting the price, it will be helpful to benchmark against others providing commensurate services. Inevitably, there may be reasons why it is desirable to raise the price. Sudden price rises can be disastrous, and it is usually much better to raise the price little and often. The truth is that the price cannot be successfully raised above the perceived value of the good or service. The perceived value can be increased by unbundling the benefits or changing the positioning of the service. The manager who wishes to practise the art of pricing in respect of leisure should read as much about the psychological aspects as about economics (Crompton 1982). In the public sector, pricing policy is likely to exercise the intellect rather more, not least because of the strong resistance from public sector managers.

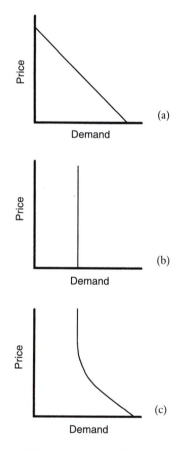

FIGURE 8.2 Price elasticity. (a) Price elasticity = 1. (b) Price inelastic. (c) Some elasticity. Price and demand have different relationships for different goods. Necessities tend to exhibit no elasticity but other goods vary

BOX 8.3 Forest of Dean parking charges

The Forestry Commission has some very popular sites in the Forest of Dean which in the 1980s became tired and overvisited. In times of stringency management had sometimes been constrained, with the net result that these areas badly needed a facelift. Traffic had also increased and more cars were visiting particular sites than the landscape could cope with. By sensitively developing car parking alongside refurbished and improved facilities, new toilets or a visitor centre, the quality of the overall experience was enhanced and a charge (for parking, emphatically not for access) was introduced. There was some initial resistance, especially among local people, who were concerned that the area might be promoted for tourism. In time, local people came to recognise the value of collecting money from tourists, where it would be reinvested in the forest. The charge is now £1.50, and the monies collected make a healthy contribution to the running costs of the recreation service in the forest. Local visitors can take advantage of a season ticket for £10, which provides for parking in a number of sites. Research has revealed that, in respect of any charges related to access to public land, it is social trust (over what the agency will do with the fee income) which is the key determinant of willingness to pay, rather than age, ethnicity or income (Winter and Palucki 1999). Here the Forestry Commission is charging not for access but for car parking. Increasingly people expect to pay for parking, and the original resistance among (most) local people has receded as they have appreciated the investments that have been made, in locations like Symonds Yat and Beechenhurst.

As a number of commentators have pointed out (Clawson and Knetsch 1971, Coopers & Lybrand 1981), the resistance from managers is stronger than any resistance the public may have. Where there are a number of different groups being served, differing pricing strategies can be applied to each. Sometimes it may be desirable to price complementary goods or services, because of the infeasibility of pricing for the targeted good or service.

Corporate entities sometimes become involved in programmes connected with leisure through an interest in community or environment. This might, rather uncharitably, be seen as a transformation of the 'polluter pays' principle into an imperative to sponsor projects to compensate for environmental losses elsewhere, attributed (whether justifiably or not) to the actions of that entity. This amounts to greening the corporate image, but there are programmes reflecting the same principle applied to social costs, such as those relating to health sponsored or paid for by revenues from tobacco and alcohol (Galbally 1996). In Britain, the Directory of Social Change publishes a wide range of books useful to the fundraiser (Clarke and Norton 1997; Norton and Eastwood 1997), whether targeting companies (Brown and Smyth 1988) or charitable trusts (FitzHerbert, Addison et al. 1999), or understanding the effects of the lottery (FitzHerbert, Rahman et al. 1999).

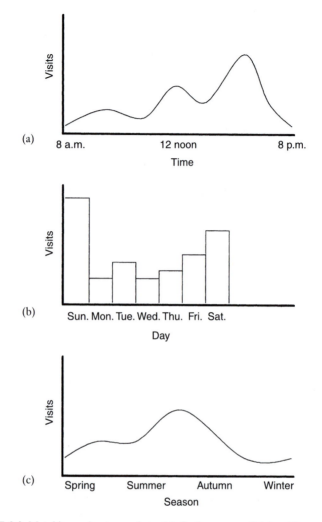

FIGURE 8.3 Matching price to markets. (a) *Daily pattern*. Pricing ideas: children go free (after school), off-peak morning specials, off-peak events. (b) *Weekly pattern*. Pricing ideas: midweek cheap tickets, encourage schools on Mondays and Wednesdays, increase the price on Sundays. (c) *Seasonal pattern*. Pricing ideas: spring and winter events, season tickets for local people in spring and winter, early summer specials

There are myriad ways of raising revenue through the provision of complementary goods and services, and these may generate more revenue than the primary activity (Tasks *et al.* 1999). Many recreation sites sell or hire out equipment related to the recreation activities that are popular there. This makes good sense and is just part of responding to the needs of the market, of participants. Following a logical sequence of steps may help explore the scope for revenue generation:

◆ Does the activity require any fundamental goods or services to enable participants to engage in or continue the activity?

◆ Can the activity be engaged in for different periods of time, and equipment hired out or provided over time?

◆ Are there any related goods or services that would enhance the experience, that could be sold or hired out?

◆ Are there any related experiences that could be sold or hired out to the existing customers, or their companions?

◆ Are there aspects of the service or work that cannot be priced, but that are valued, and for which a donation could be sought?

◆ Are there any aspects of the service that others would wish to support, in kind, by secondment, in cash or by any other form of sponsorship?

◆ Are there exclusive associated rights that could be sold: stories, interviews, film or television rights?

Partly this is a matter of determining exactly what the leisure or recreation experience sought comprises. Somewhere it should be possible to recover the costs, or make a profit. There are techniques to apply to make funds go further (Broadhurst 1989; Crompton 1999, 1987; Howard and Crompton 1980) and to encourage visitors to give voluntarily (Denman and Ashcroft 1997), but if they are unsuccessful you may have to suffer a loss, give the project away or let somebody else supply the service!

A landowner may own a forest, and within it there may be a visitor centre owned by a local authority, a hostel owned by a charity or a shop let to the private sector. Each could be managed separately, or with different degrees of integration. National Parks are examples of large areas where a number of different teams work together to provide a joint product. Often different parts of the leisure experience can be supplied in partnership. The package holiday is an example where a carrier teams up with an accommodation provider and adds a courier. The courier's job is to provide a seamless holiday, and resolve any difficulty (such as a mismatch between expectations and reality) that may arise. Managing joint projects is often more complicated than expected, not least because there are more stakeholders to satisfy. Given that it is more complicated, there must be clear advantages for joint working to make it happen, or to continue the venture, and usually:

◆ There must be greater benefits available to the stakeholders working together, than there would be with isolated working. Such benefits might include: access to larger markets; more secure access to markets; more secure access to raw materials, and services; access to greater economies of scale; complementary development (or matching) of skills and resources, i.e. synergies.

◆ There must be a clear understanding about how responsibilities are allocated and shared, over time.

◆ Provision has to be made for succession, and what happens under a number of different scenarios which are rarely envisaged (but which would cause great difficulty if they occurred).

> **BOX 8.4 Sporting events in the United Kingdom**
>
> Perceived wisdom a few years ago would have had cities competing fiercely for high-profile events, some of which may bring great benefits. Not all the competition in the past has been helpful. There needs to be a balance between competition and co-operation. There have been a number of moves (House of Commons Culture, Media and Sport Committee 1999) encouraging people to work together in attracting and hosting events. UK Sport has published advice (1999) to help guide decisions, and urges people to think about the wider effects of sports development, the environmental impact, traffic congestion, the loss of earnings, job creation, marketing benefits, and the long-term effects.
>
> The experience of the World Badminton Championships and Sudirma Cup, held in Glasgow 19 May–1 June 1997, resulted in £1.9 million spend in the city. The total additional spend was £2.2 million, with expenditure by governing bodies adding £445,000, equivalent to thirty-six full-time equivalents in work. Successful as it was in attracting spectators, some 21,642 visitors, they accounted for only 31 per cent of the money spent. It was the 1,632 people who stayed rather longer who accounted for the 69 per cent: the 812 competitors, 603 officials and 217 media representatives, (Leisure Industries Research Centre 1997; Gratton *et al.* 2000).
>
> Similarly, the 2,000 competitors, with their families, who take part in six-day orienteering events in relatively remote areas such as the Highlands of Scotland can have an enormous direct economic impact, and can give rise to repeat tourism visits also.

A festival is an example where the whole community can gain, although the spoils would not necessarily be shared out equally. The economic effects of large festivals are well documented (Weston 1996: 114–16) and make fascinating reading. The Albuquerque balloon festivals grew from thirteen balloons in 1972 to 626 by 1996. With a direct impact of $23.4 million on site, and an indirect impact of $62.6 million the effect on local and national taxes alone was $1.4 million. The Edinburgh Festival (Box 3.4) contributed something like £72 million in 1992 to the Scottish economy. There are actually several festivals – International, book, film, jazz, fringe and science among them – which together contribute something like £120 million (Scottish Arts Council 1999). To use tourism to spread the benefits throughout the country, into rural areas, which may have fewer options for diversification, requires a good deal of work. Some areas have worked hard at encouraging the development of local events – agricultural shows, Highland Games, walking festivals, festivals of traditional music and song, the Eisteddfod in Wales, and the Gaelic Mod – building on local distinctiveness and culture. These events require working together.

Most joint ventures that come unstuck seem to do so because the parties all have different expectations, which is why clear communication and recording of what is understood (by drawing up a plan) are so important. The plan obviously has to be owned by all parties to the venture, and it will usually have to run through several iterations before it adequately reflects the different interests.

The expectations may also change over time. Inevitably, the fortunes of all parties are likely to take different trajectories, and continued joint investment desired by one party may not be possible for another. It is worth thinking very carefully through different scenarios from the outset. There is a temptation to see only the advantages ahead, and to talk the venture up. Make use of your cynical colleagues, real or imaginary (Black 1996; de Bono 1990a), in testing out some of the less desirable outcomes, and work out how you would resolve the situation before you meet it. The expectations are even more likely to change when the individuals change, whether these are the parties' representatives or the owners themselves. It takes a very good plan to outlive such changes. Nobody should enter a partnership without an exit strategy.

No project, or organisation, should fail because a single team member leaves. There should be some rudimentary plan for filling any post should it become vacant. In particular, it is important to know how the leader will be replaced. In large recreation organisations, in both the public and the private sector, there will be all kinds of formal personnel procedures and rules governing selection and recruitment. In the voluntary (not for profit) sector it is not always so clear how arrangements will be handled. The kind of person required at different stages of the evolution of an organisation or project may be entirely different. Often it takes a very special, dynamic and creative person to start an organisation, and a rather different kind of individual (just as special!), rather more disciplined and analytical, to keep an organisation running well. The first kind may be more of an innovator, an explorer/assessor rather than the latter, who is more of a concluder/controller, at least in the terminology of Margerison–McCann Team Management Systems (Margerison and McCann 1991). The Margerison–McCann system analyses the degree to which

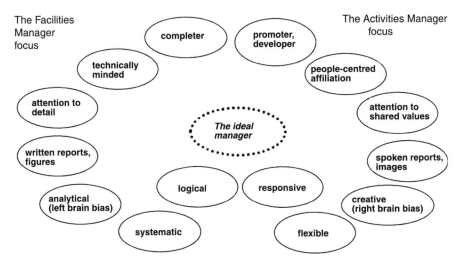

FIGURE 8.4 More than one kind of manager: suggested stereotypical view

---

BOX 8.5 Overspent?

When did you last hear of a project coming in under budget? There are frequent stories of projects being severely overspent. Why? Although we use budgets to keep our plans on track, the fact is that many of our decisions are not made on a single shared rational basis. Each member of the team will typically have his or her own approach to the budget, with a different set of objectives to be met by the project. There are distinct phases, and you may recognise behaviours that are likely to occur during each.

1 *Before the green light*. At the very start, the individual or group of individuals who generated the idea have not yet gained access to the resources necessary to see the project through. There is a natural tendency to overvalue ideas generated within the group, which gets the project off to a bad start. To make their case there is a tendency to underplay the costs and overemphasise the benefits. Having now gained the support of the champion, or person mostly likely to secure key funding, the argument goes through a further iteration, and the benefit – cost ratio is exaggerated still further.

2 *Drawing up plans*. The team has been enlarged, and almost everybody is infected with enthusiasm to introduce the very best scheme ever. The architect, landscape architect, interior designer and recreation manager can see how improvements and little tweaks can be made to improve the plans still further. The architect dutifully takes these amendments on board, and the project climbs well over budget. In a further meeting cost-cutting measures are agreed, but obviously everyone is at pains to limit the cuts. The project plans and the project budget are agreed.

3 *The project is installed*. The project is built or installed. During the process the client recognises elements of the project that must be put right now (as it would be much cheaper than doing the work later on), and architects' instructions are issued to the contractor. Additional costs appear, and anyway the potential savings previously identified had been overstated. The final costs climb still higher, and well over the budget.

4 *Cuts have to be made*. Additional funding has been made available, even though some heads may have to roll, but to remain politically acceptable, a number of components which were previously thought of as essential, or integral, have to be dropped. The eventual out-turn is still well over budget. The moral of the tale is that internally generated ideas get far too much support. Managers should build in sufficient external scrutiny at the earliest stages possible, in screening ideas, and developing the specification.

---

people are extrovert or introvert, practical or creative, analytical or apt to rely on beliefs and structured or flexible.

The results of the analysis allow team members to identify their strengths and to discuss how best to work with others, to best effect. If working on your own, it is quite possible to develop the complementary skills, or styles of working (de Bono 1990a), with the application of a little discipline, or theatrical skill, as you first think in one style (say, supercritical) and then in another (say, entirely lateral).

The individual who starts or develops an organisation may, even sub-consciously, be reticent about handing over the reins to another. In the worst case, he or she may have selected people who will never develop to fill the role (Ogilvy 1983). Wailing and gnashing of teeth may be a natural reaction to observers, but the awful truth is that all organisations have life cycles (Kotler 1984). The characteristics of each stage will be dependent on the objectives of the organisation. Some organisa-tions that have achieved their objectives, and others which have outlived their useful-ness, need to be closed down. This ensures room for new organisations and a pool of enthusiastic individuals to support them.

In the late 1960s and early 1970s the Western economies were booming and, in this period of relative wealth, substantial investments were made in public recreation facilities. In Britain, buildings were erected to house sports centres, swimming pools and theatres. Country parks were created close to cities. The world recession in the 1970s focused the minds of managers on how services could continue to be provided. The problem never seemed to be finding capital, but securing revenue funding. The logic is simple. The banks and other lenders gain from supplying capital, and this capital is accounted for quite separately from revenue. Revenue would have to come from charges or taxes. Generating income has been one approach to ensuring viability. Income dependent on charges is in turn dependent on prices, volume of service or throughput, competition, the disposable income of consumers, and consumer satis-faction. Where income is generated by ticket sales the time horizon is barely moments away. The horizon can be pushed further away by making an agreement with a group of people who are equally desirous of ensuring that the service is secure for a longer period. Parks services in parts of North America have been encouraging groups of people to take over the management of small parts of the park.

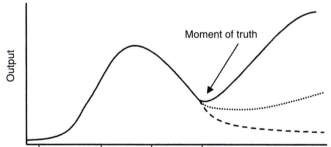

FIGURE 8.5 Organisations have their life cycle too

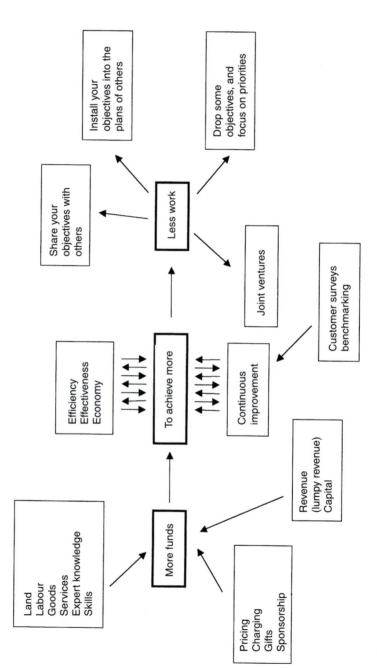

FIGURE 8.6 Ways to achieve more

BOX 8.6 Adopt-a-park

As the costs of maintaining leisure facilities belonging to the community rise, there are limited ways to bridge the gap between income and expenditure. Increasing the amount raised through taxes is the obvious way. This was often seen as the best way to help those who were not so well off. It would be an efficient way if patterns of participation were not so skewed. It is generally people who are better off who make most use of municipal leisure facilities. Raising taxes spreads the cost among all taxpayers, whether or not they take part, but leads to benefits for the better off. Methods of raising funds need to be more sophisticated, to ensure that the benefits are distributed in the way that society, or communities wish.

Another way to solve the problem is to encourage a greater sense of ownership, and Adopt-a-park schemes have done just that, enabling parks administrations to spread their resources, and to provide communities with an opportunity to give direction to what happens in their midst. In Berkeley, California, citizens have set up the Berkeley Partners for Parks organisation, which encourages many initiatives, such as the Friends of Sixty-third Street Mini-park. This group of volunteers worked with the city authorities to renovate an area which was being frequented more and more by drug users, and less and less by children playing. The result is more play provision, costs shared and less drug activity, at least in the Sixty-third Street Mini-park (http://www.bpfp.org/AdoptAParkGroups/Sixty-thirdStreet/). Cities throughout North America have set up similar programmes (http://seeh.spl.org/homepages/Adopt-a-Park.htm and http://www.ci.san-jose.ca.us/prns/parkst.htm).

There are also examples where local authorities have worked hard to restore relations with younger citizens, to encourage them to value their surroundings, and one such is 'Kids for Parks', in Anaheim, California (Crompton 1987). This approach is not restricted to parks, and may take a rather more financial line, where organisations or individuals are asked to contribute funds, rather than what the North Americans call 'sweat equity'. This may extend to sponsoring new holes in a golf course extension, or bricks in a path, also discussed by Crompton (1982).

Managing experiences and visits . . . for everyone

Before describing how we might intervene or manage the leisure experience, it is worth while reflecting on what precisely comprises the experience. We may consider life as being an almost endless coil of experiences, or one long experience. Consequently, we can decide to study the coil at any particular scale of resolution to determine where intervention would have beneficial effects, for the participant and for society. We could then explore how best to supply the information, using which channel – expert, media, personal and through the experience itself.

Determining what amounts to a sensible unit (of leisure) for management is the first and all-important step. The easiest way to be reasonably sure that the appropriate steps have been taken is for the manager or student to consider the experience from the point of view of the would-be participant. Start from some natural position of rest, perhaps sitting at home, pondering what recreational activity in which to engage, and think it through from then on.

BOX 8.7 Accessibility standards

There are guides to help architects, designers and managers to fully consider the needs of people with disabilities, for example in the arts (Pearson 1988), and more general introductions (Bullock and Mahon 1997). Disability Discrimination legislation should (hopefully) encourage more to be done, and there are some very useful checklists available (Adaptive Environments Center and Barrier Free Environments 1992). When it comes to the wild outdoors it becomes harder. In providing new facilities, it is simple to provide access for everyone, although surprisingly often there are problems with existing facilities. Managers on Forestry Commission recreation courses are encouraged to test (by taking to a wheelchair) the accessibility of picnic grounds, ramps and toilets labelled as being accessible to those in wheelchairs. The process is instructive, as the simplest of obstacles can prove insurmountable.

There are at least two ways to apply standards of accessibility:

◆   *Work with people who have a disability*. Ask them to give a rating of accessibility. More commonly, managers make judgements on their behalf. The disadvantage is that this cannot take account of the wide range of disabilities, and differing circumstances, and may put people off visiting because they will not know how the measure was applied.

◆   Provide a detailed and objective description of the characteristics of the site (or programme). This allows the potential visitor to judge whether the site (or programme) is (personally) accessible.

An imaginative project, which brought together the national agencies concerned with countryside recreation and a range of organisations representing the interests of people with disabilities, has resulted in a good practice guide with a difference. The Fieldfare Trust provided the secretariat to the group and worked to develop the *BT Countryside for All Standards and Guidelines* (1997). At the heart of the guide are standards of accessibility, to which the agencies signed up. The approach borrows something from the spirit of the Recreation Opportunity Spectrum. The idea is that in the more urban settings (and closer to built facilities) there should be an expectation of greater accessibility, and in

---

the wilder areas there will often be less intervention, but still some idea of precisely what standards (e.g. for the slope and camber, surface, width of a path) would need to be applied before something is labelled as accessible. Of course, while physical accessibility may have to vary, programme accessibility does not. Leaflets, interpretation, transport and the supporting infrastructure need to be accessible to people with disabilities too. Just as a path which is wheelchair-accessible becomes accessible to the large section of the population who push prams and buggies as well as to those who find walking on slopes difficult or unappealing, so too will widening programme access open up the opportunities for many more people to enjoy the countryside. This approach is a special case of designing the visit, and not the facility, and can be applied in all settings, not just the countryside.

When does any recreation event start? For the sake of pragmatism, we can start with the journey, between the home and the location of the event, and often between the exterior of the home and the exterior of the location. It is instructive to dissect each and every stage from the point and location of decision, through the experience itself. Such an approach helps to ensure that we do not overlook such simple things as steps. A single step can put an experience beyond the reach, of some. By examining this level of detail, we can identify critical points where action is necessary.

Having arrived at the threshold of the experience, there is still plenty of scope for intervention. We can break down the elements of the experience on site to explore the critical points for intervention. Usually there are a number of people involved all the way through. The experience is about physical (motor) activity as well as mental activity (cognitive and affective) that will affect the experiences of others at the location. Consequently, we may benefit by exploring the different experiences as time lines, to see how they interact and where intervention may be helpful (and it may be just as important to determine where intervention would be unhelpful). We have already agreed that the best management in recreation and leisure will either be perceived in terms of warm human relations, or be imperceptible.

In addition to a warm reception, most visitors like to be safe – more and more so, it seems. Why is not so clear. One reason may be the desire to achieve ever more within the time available (Future Foundation 1998). On the other hand, almost any-one (with enough money) can now sign up to travel on expeditions in the Himalayas, even for ascents on Everest. In managing the safety, the difficulty becomes a matter of providing information in such a way that the would-be participants can build it into their model of reality, and recognise the values on which to make a sound judge-ment. To make such a judgement in the past would have required learning over a considerable period, whereas now much of the preparation can be fitted into a holiday.

People can very easily find themselves in risky situations, and in parts of the Western world society is becoming more litigious, whereas in the past people would have taken responsibility for their own actions. There is a Latin phrase used by lawyers to reflect that climbers and other participants (in Britain, anyway) take part in their sport voluntarily, and therefore cannot pass liability on to others. It is *volenti non fit injuria*. People are now more aware that many accidents are caused by the coincidence of many actions (by many people), and liability may be apportioned or allocated out. In Britain the landowner, the occupier and the business operator have a duty of care, which amounts to taking all reasonable precautions against actions that could reasonably have been foreseen.

Providing appropriate information for the intended audience is the crux. It becomes particularly crucial in sport (e.g. boxing, motor racing) and outdoor recreations (e.g. skiing, climbing and diving). Much recreation is extremely peaceable, and it is not an area where participants often consider there to be any unacceptable risk. Going to the cinema, to a dance, the theatre, a football match or similar spectator event is not in itself usually conceived of as hazardous. Safety is taken for granted. Managers conduct risk assessments, consider the hazards, and take appropriate action.

The pleasure derived from sport is not a simple concept to map or disentangle (Lee 1981). Certainly, a key component in risk sports and outdoor pursuits is the sensation associated with the accompanying adrenalin rush, the heightened awareness and intense pleasure. The release of adrenalin can be induced by fear (in sport, termed excitement!) when confronted with a challenge. As managers we would not wish to remove this excitement. However, we may need to ensure that people are confronted only with doses of excitement that relate comfortably to their level of skill and ability. Csikzentmihalyi's model (p. 93) is helpful here, to show how we need to keep the challenge in proportion to the skills.

Children need recreation and play (cognitive, motor and social) to develop fully as individuals and to take their place as full members of society (Timothy Cochrane Associates 1984). Some of these situations need to present challenges which are (usually) overcome, and from which the child learns what the consequences are, and what action is required to meet the given challenge. Through this process the individual grows. We need to ensure that challenges are presented in meaningful circumstances, otherwise the individual will not mature fully, and will be unable to accept risk, or distinguish between a challenge which can be overcome and one which cannot. Organisations, in Britain, such as the Scouts' Association, Girl Guides, the Duke of Edinburgh's Award and Outward Bound provide an institutional setting for exposing youngsters to challenges of this kind, in a Britain that increasingly seeks bland security with a lack of opportunity to meet (let alone overcome) challenges. Media pressure encourages kneejerk reactions that sometimes serve to reduce the opportunities offered to youngsters to seize the challenge, as well as the day.

There is evidence that children gaining a large amount of their experience vicariously through television may develop a false sense of security. In Britain at least there is an increasing tendency for children to spend more time in the home, and to spend more time on their own, with video games, television and personal computers (*Independent*, 19 March 1999).

People taking part in recreation pay more attention to the end of each experience than most managers do. A manager is generally more concerned with welcoming the next customer, than with 'wasting' time on saying farewell to people who have had their experience (and probably paid for it!). Participants pay more attention to this phase because for them it is a transition to another experience. It is not an end, full stop, but merely a transition, and the beginning of another experience. People may be looking for reassurance that their experience was worth while, and good managers will ensure that people go away happy, to spread the news by word of mouth, and to return on another occasion or to a related place or activity.

BOX 8.8 Disney magic

What are the key factors behind Disney's legendary success?

In a world of its own. Once through the gates, you are in a world of its (Disney's) own, where many more things are managed than will be apparent to you, and managed very carefully. The magic and fantasy that fuelled the Disney studios are recreated to provide a whole series of particular experiences, and with the same painstaking attention to detail that made Disney films so successful.

In a class of its own. Everything appears effortless, and you should be virtually unaware of all the hard work that has gone into preparing for your stay. There is something for everyone, because they have planned very carefully, and taken good account of how visitors behave, and what their preferences are. The attention to detail is staggering. American management (perhaps more than managements elsewhere) work according to manuals, and certainly Disney is well equipped in this respect.

At a standard of its own. Despite the quality of what was on offer, the development of Paris Disneyland (its latest title) was far from a runaway success, making spectacular losses in the early 1990s which some have ascribed to the management regime and overheads being applied by the parent company (Taylor and Stevens 1995). The weather may have played a part, the edge of Paris providing a rather different backdrop from Florida. But, by 1996, Disneyland Paris had become France's most popular attraction, with 11.7 million visitors (Warren 1999). The French (and Europeans) tend to value, support and cherish regional differences and distinctiveness, rather than pursue standard guaranteed quality. The clash in cultures has been, and remains, an issue which undoubtedly played a part in the difficulties experienced right from the start. Disney attention to detail could hardly be surpassed, but focus at the level of community or culture was perhaps less than perfect (Weston 1996: 46–52). The attention to detail in the management of fun by the Disney organisation, in customer care and every aspect of management, is legendary, but as many questions are prompted by the Disney approach as are answered.

Managers can create a climate in which participants remember, reflect and relive their experience. Participants need to take away clear memories. They can be helped with photographs, videos, journals, sketches or other records. Alternatively, visitors may wish to purchase or take a 'souvenir'. If we are concerned to ensure that impacts are positive, we need to ensure that as far as possible the manufacture of the souvenirs does not contribute to negative impacts. What the next beginning, or next event, may be is mostly up to the participant. The manager can help provide leads and information to allow the participant to see what the possibilities may be.

For the impacts or effects of leisure and recreation to be positive, we need to ensure that the effects on the bio–physical environment do not have unduly harmful effects. The effects should not harm our interests now or in the future. In relation to the social and economic environment, we also have to ensure that the effects are positive. Again, making value judgements about all the effects may be extremely difficult. In the economic environment, leisure and recreation redistribute resources. Not always are the benefits and costs distributed evenly or in any sense fairly. Tourism, in particular, provides a means to encourage the exchange of money for goods, services and experience. There is much work to be done to encourage the spread of tourism of the kind that makes the greatest contribution to economies without undue damage to the natural ecology or to the host culture (Krippendorf 1987).

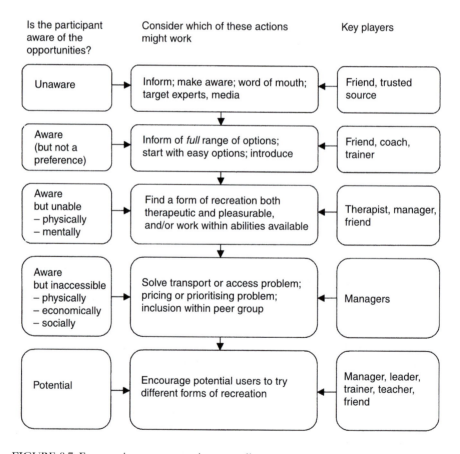

FIGURE 8.7 Encouraging more recreation, according to current awareness levels

One area where more study might reveal other ways of solving these problems is the consideration of how we could use currencies in a rather more broad-minded way. In the best of leisure and recreation, it is not just a simple matter of exchanging money for recreational goods and services (at least, not just material service). A great deal of recreation is about human interaction, about the exchange of much more than money for material service. The notion of personal service, so crucial to customer care, deals in the exchange of signals, language, respect, concern, affiliation, emotion, feelings, esteem and affection. We should look at transaction analysis and social exchange to help us understand the fundamentals.

Managing people: staff and volunteers

In cases where leisure occurs with the help of intermediaries or agents, whether staff or volunteers, the nature of the relationship between participant and agent, and between agent and owner/supplier or resource holder, is crucial to the outcome. The best intermediaries tend to:

◆   know their subject area well;
◆   be comfortable with their level of skill;
◆   enjoy meeting different people;

and to be good at:

◆   recognising the needs and moods of others;
◆   empathising with the situation of others;
◆   communicating verbally, and often in writing too;
◆   communicating non-verbally;
◆   reflecting the position and stance of others;
◆   making others feel comfortable;
◆   knowing when to use humour.

Although knowledge and skill in the subject area are important, the ability to communicate well is essential, as intermediaries are at the interface between all parties. All individuals vary in relation to their motivations for work, and over time (McCormick and Ilgen 1981: 260–300), and there is likely to be similar variation among volunteers. The economic motive is taken out of the equation, but there still remain a range of reasons sufficient to encourage volunteers to give of their time, their abilities, their ideas and their energies. These include the:

◆   *work itself*, which may be intrinsically interesting and satisfying;
◆   *physical situation*, which may offer related pleasures – e.g. working on a trail in the mountains;
◆   *cultural situation*, in a library with priceless fragments of history;
◆   *personal development*, with potential to enter into employment;

◆ *social setting*, surrounded by people to whom the volunteers can relate, with positive feelings, and a sense that they are contributing to and valued by that group and (ultimately) society.

The occupational psychologist would reconfigure these as working conditions, relations with others, advancement, responsibility, the work itself, recognition and achievement, although each theory would put it rather differently. As people live longer and enjoy better health, the range of volunteers available expands from fit youngsters of school age to older people who have acquired a great range of skills, knowledge and wisdom. This resource is significant and growing (Handy 1988).

BOX 8.9 The Appalachian Trail Conference

Benton MacKaye proposed the trail first in 1921. In 1925 the Appalachian Trail Conference was established, comprising a group of like-minded souls. The trail was completed as a continuous footpath by 1937. In 1948 one Earl Shaffer walked the trail from end to end in one season, becoming the first reported 'thru-hiker'. None of this would be remarkable but for the fact that the trail is 2,160 miles long, passing through fourteen states, 'eight national forests, six units of the National Park system and some sixty state park, forest, or game lands'. It was a team of volunteers that built the trail. Although it is strictly a unit of the National Park system, being the first of the National Scenic Trails, it is still maintained and managed by volunteers. The Appalachian Trail Conference consists of a network of more than thirty trail maintaining organisations. All this from the enthusiasm of one Benton Mackaye, see http://www.atconf.org/.

Rangers are frequently at the interface between visitors and the countryside, at the interface between visitors and the host community, and at the interface between land managers and visitor managers. The job is really about detecting the appropriate signals and passing the information on in the language of the recipient. Not surprisingly, the job can also be very taxing but is extremely rich in experience (Frost *et al.* 1998). Recruits come from all walks of life. Some are fresh from an academic training, while others are on their second or third career. For example, the Peak National Park, in England, has just over twenty full-time, and about 200 part-time and volunteer rangers to cover the 555 square miles of the National Park, which receives some 22 million visits each year. The training package they developed for National Park rangers received approval from Oxford University in 1997 (http://www.peakdistrict.org/).

The word 'ranger' has a long pedigree, appearing in common English usage in the fourteenth century to describe people who looked after a given area, the range. The tasks then were more often about protecting private property, the deer and game. Now the job is rather more varied. When the Countryside Commission for Scotland started up in the late 1960s it looked to the United States for inspiration in developing

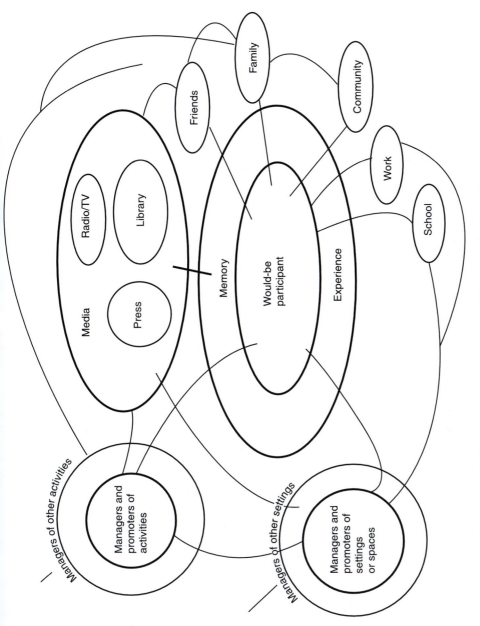

FIGURE 8.8 Information and influence. A starting point from which to consider the options

its Ranger Service. In Scotland, though, the rangers are not operated by a single government department, but by a host of land managers: public, private and voluntary organisations. As has been outlined above, there was a good deal of valuable experience to draw on in England too. Two of the most influential members of staff then on the Commission, the Director, John Foster, and one of the Assistant Directors, Don Aldridge, had previously worked in the Peak National Park. To develop a common approach, the Commission developed a grant support package for land managers employing rangers, an identity for rangers, and a common training programme that recognised the scope of the job. The first national ranger training programme ran in 1974.

Scottish Natural Heritage (the successor body to the Countryside Commission for Scotland) conducted a review of ranger services in preparing its policy statement (Scottish Natural Heritage 1997). In this, it is recognised that rangers could be doing many things, but four essential aims are identified:

◆   to ensure a welcome for visitors to the countryside through contributing to well managed informal recreation facilities and access to the countryside, and providing good information;
◆   to mediate between public use of land and water and other rural land uses, including the conservation of the natural heritage;
◆   to promote awareness and understanding of the countryside and, through this, to encourage its responsible use;
◆   to care for and enhance Scotland's natural heritage, enjoyed by visitors.

Employees are individuals and every bit as variable as volunteers. Do people who work in areas of recreation differ from people who work in areas not connected with leisure, say in banking and commerce? Probably, and people in the different sectors of the leisure industry probably vary just as much. Are public sector employees any different from those in the private sector? Public sector employees are often stereotyped as risk-averse and bound by rules (including ethics), whereas private sector employees are considered to be avid risk takers with little concern for anything but profits. The truth must lie between these extremes. Individuals in the public, private and voluntary sectors are more likely to have really important attributes in common, as against those that distinguish between groups. The variation within any of the three groups is likely to be substantial. What is clear is that all individuals have a raft of reasons for wishing to serve and achieve. For paid employees, pay is clearly important, although rarely the most important reward (McCormick and Ilgen 1981: 292). Pay in the leisure industry is notoriously low in relative terms among developed countries, when compared with other sectors. Often the jobs require work at times which some would consider antisocial, as clearly people are giving service when most people (and certainly others) are at leisure. Some of the jobs are worked as split shifts with insufficient time between shifts to use in meeting obligations at home, or for adequate recreation. Pay is important, not just for economic reasons, but also as recognition of skill, and level of achievement. Most people are acutely interested in equity among colleagues in similar employment, where the

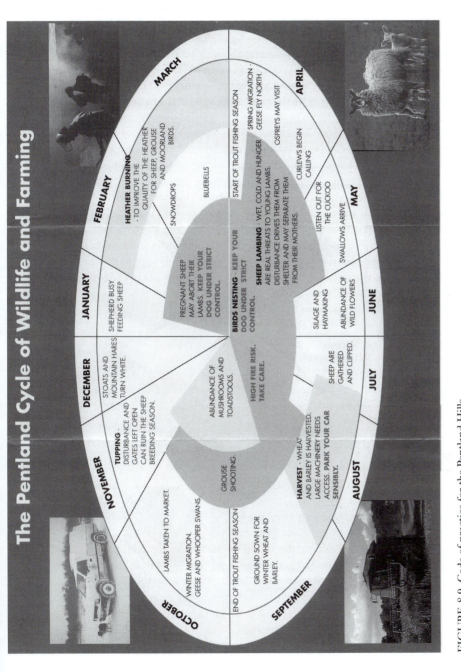

The Pentland Cycle of Wildlife and Farming

MARCH

FEBRUARY

HEATHER BURNING - TO IMPROVE THE QUALITY OF THE HEATHER FOR SHEEP, GROUSE AND MOORLAND BIRDS.

SNOWDROPS

BLUEBELLS

APRIL

SPRING MIGRATION - GEESE FLY NORTH.

OSPREYS MAY VISIT

START OF TROUT FISHING SEASON

CURLEWS BEGIN CALLING

JANUARY

SHEPHERD BUSY FEEDING SHEEP

PREGNANT SHEEP MAY ABORT THEIR LAMBS. KEEP YOUR DOG UNDER STRICT CONTROL.

BIRDS NESTING - KEEP YOUR DOG UNDER STRICT CONTROL.

SHEEP LAMBING - WET, COLD AND HUNGER ARE REAL THREATS TO YOUNG LAMBS. DISTURBANCE DRIVES THEM FROM SHELTER AND MAY SEPARATE THEM FROM THEIR MOTHERS.

LISTEN OUT FOR THE CUCKOO

SWALLOWS ARRIVE

MAY

DECEMBER

STOATS AND MOUNTAIN HARES TURN WHITE.

ABUNDANCE OF MUSHROOMS AND TOADSTOOLS.

HIGH FIRE RISK. TAKE CARE.

SILAGE AND HAYMAKING

ABUNDANCE OF WILD FLOWERS

JUNE

NOVEMBER

TUPPING - DISTURBANCE AND GATES LEFT OPEN CAN RUIN THE SHEEP BREEDING SEASON.

LAMBS TAKEN TO MARKET.

WINTER MIGRATION. GEESE AND WHOOPER SWANS.

GROUSE SHOOTING

SHEEP ARE GATHERED AND CLIPPED.

JULY

HARVEST - WHEAT AND BARLEY IS HARVESTED. LARGE MACHINERY NEEDS ACCESS. **PARK YOUR CAR SENSIBLY.**

AUGUST

OCTOBER

END OF TROUT FISHING SEASON

GROUND SOWN FOR WINTER WHEAT AND BARLEY.

SEPTEMBER

FIGURE 8.9 Code of practice for the Pentland Hills

Source: Pentland Hills Regional Park, © City of Edinburgh Recreational department

rewards are seen to be related to performance or potential. Pay is as much a signal about status and level of achievement as it is about economic need.

From whatever sector, people have a common desire to make the most of their abilities, and to use them to better their own and usually everybody else's situation. To enable people to feel at ease and confident in their situation, there is a responsibility for those supervising to match ability, skills, personality and the situation, whether for work or volunteering. We need to ensure that people have the right information, to be in the right situation, and to have the right skills to be able to achieve what is required. Training is vitally important to ensure that this is so. Managers can have a great effect on the environment in which leisure occurs: the physical and social environment. The physical environment can be selected or modified and then maintained. In just the same way the social environment can be selected, modified and maintained.

One of the greatest advances in training in the field of leisure, beyond specialist skill training, has been the advent of 'customer care' training. What seems so curious in retrospect is that the application of such common sense can make so big a difference, but it does. The essence of such training is that people who serve can only give their best when they feel good about themselves. The focus shifts from being solely concerned with how service is given to the participant in leisure to how interpersonal relationships are managed generally. Individuals learning these skills take them up more readily and can see the significance more if they are practised, as much at home as at work.

Some tips:

◆ Smile as much as you can.
◆ Laugh when you can.
◆ Enjoy physical contact (a hug at least) with someone who means a lot to you.
◆ Relax for some period.
◆ Be active for some periods.
◆ Enjoy some recreational activity of your choice.
◆ Eat well and, ideally, in the company of friends or family.
◆ Avoid fast food.
◆ Leave some time to think, to reflect on the day.
◆ Avoid stress at work.
◆ Do not continually attempt too much.
◆ Plan your day, your week and your month.
◆ Set realistic goals, achieve them, and record having achieved them.
◆ Have a reminder at work of activities or people that matter to you.
◆ Avoid aggression, but be assertive.
◆ Seek out those whose company you enjoy.
◆ Avoid stress at home.
◆ Listen attentively to those around you.
◆ Think about what gives rise to stress, and solve the problems not the symptoms.

Those who are dealing with the sharp end need to have leaders or managers who really care about them. At every link in the chain, individuals, whether caring for

---

BOX 8.10 Customer care, or just being human?

The basic building blocks that make up customer care are very simple indeed. Have we really forgotten what it is to be human, to give each other reassurance and praise when it is needed? The essence of customer care is sincerity. On the face of it, any training that seeks to change behaviour from what appears natural to trainees is unlikely to be successful. Even though selection is important, there is a place for training, in helping to legitimate natural behaviours that have been suppressed. We should remember that speaking or dealing with someone is necessarily a PERSONAL business. Each of us should:

◆ Pay full attention, make
◆ Eye-to-eye contact; show
◆ Respect.
◆ Smile (but naturally, and sincerely).
◆ Only use language with which both parties are comfortable, and use the personal
◆ Name, where appropriate.
◆ Accept differences, and
◆ Learn from each and every encounter.

Mostly we do this naturally. It's just being human.

---

colleagues or customers outside the organisation, should apply the same principles. If people inside the organisation know one another better, they will more swiftly discern what response is appropriate. Often internal organisational structures and climates serve to obscure the real character of individuals, but a sincere programme of customer care, a people-centred approach, will win through eventually. This requires focusing on effective communication, and especially listening. Much communication goes beyond words.

◆ Pick up as many clues as you can, as the person approaches, or as you approach the person (by looking but not staring) to gauge their mood.
◆ Greet the person.
◆ Smile when appropriate.
◆ Use eye contact when appropriate.

Then, make the person (and yourself) feel at ease by reflecting the position, style and stance of the person with whom you are engaging:

◆ Listen attentively to every word the person may say; do not interrupt.
◆ If dealing with a group, remember to encourage any reticent people who might otherwise go unnoticed.

- ◆   Use language (or other means of communicating) with which you and the other person are comfortable.
- ◆   Nod and give any other signals of reassurance that you are following and under-stand what the other person is saying (and if you do not understand, seek clarification).
- ◆   Respect each individual for their individuality.
- ◆   Use physical contact – hand on a shoulder, handshake – as appropriate.
- ◆   At the end of the meeting, close with an appropriate gesture of farewell, so that the other person understands the meeting is complete.

Only occasionally (let's hope) staff will have to deal with people who are unhappy, upset or very angry. Under these circumstances, reflective interviewing can do a great deal to reduce the temperature and reveal the underlying cause of the problem. Great care has to be taken, as, used inappropriately, it helps fuel frustration. The key is to remember the objective.

You cannot turn everyone into a good service provider. Managers have realised that, in many cases, it is better to expend more effort on selecting appropriate people – people who like dealing with people – than to expend the same effort on attempting to change the behaviour of someone who would really rather not work with people at all. This is not rocket science, but it is very powerful. 'Exchanging and spreading information to develop best policy and practice' is the strap-line that has been used by the Countryside Recreation Network (2000) for a number of years. It neatly sums up one of the principal aims of training generally, to share ideas, stop reinventing wheels and bring each other up to date. People in leisure, recreation and tourism businesses share many of the needs of other entrepreneurs in other businesses (McKercher and Robbins 1998). Development is an altogether different approach that seeks to encour-age the revelation and development of skills and aptitudes among individuals for their own personal benefit, whether at work, at home or anywhere else. Whether we sub-scribe to the idea of a travel career ladder (reflecting Maslow's hierarchy) or join others (Ryan 1998) in questioning it, we should recognise that people working and volunteering in this field will have their own personal motivations. We should also remember to practise a little of what we preach and encourage some playfulness at work, if we want people to be creative (Lieberman 1977). With the increasing power of communication and information technology today, there are many ways in which we can improve our skills and the way we work. Increasingly there are networks within organisations (intranets), and between people in different organisations who could usefully share information (extranets and discussion groups) as well as any number of sites across the internet. In Britain there are a number of key web sites (see Appendix 2) that will provide a way in. If you have a particular problem, say the expanding use of off-road or mountain bikes, it should be easy enough to set up a clearing house on a web page, with problems and fixes listed (Schuetti 1997).

BOX 8.11 Controlling anger, and the awkward customer

Non-reflective approach 1
Customer. Look here, I'm absolutely *** !
Staff. I'm sorry about that, but it really isn't our fault, you know.
Customer. Well, whose fault do you think it is, then, for heaven's sake? Eh?

Result so far? The customer is seething with fury. The member of staff is seething, and knows absolutely nothing more about what the customer is *** ! about. Further interaction is unlikely to be very revealing, but will probably lead to more stress, further misunderstanding, and loss of custom. The member of staff will feel angry, or irritated, and will be apt to bark at anyone, whether colleague or further customers. Not good!

Non reflective approach 2
Customer. Look here, I'm absolutely *** !
Staff. I'm very sorry about that, and if it's something we can change, I would hope we can put it right.
Customer. Well, it jolly well is, and I want it put right straight away!
Staff. Right, sir, nothing is too much trouble. What was the problem exactly?
Customer. Well, it all started . . .

Much better. The staff member has been on a customer care course, and remembered some of the basic rules. The customer is probably mollified, but what has the organisation been signed up to? We do not yet know, but it could be expensive.

Reflective approach
Customer. Look here, I'm absolutely *** !
Staff. Hello, sir, I'm sorry to hear that you're obviously *** ! [*Pause.*]
Customer. Yes, I really am ***. It made me absolutely livid, to be put into that situation.
Staff. I really do appreciate that the situation made you feel absolutely livid. [*Another slight pause.*]
Customer. When I ordered the wine, I mistakenly ordered a bottle of *x*, and the waiter brought a bottle of *y*. I suppose I was eager to impress my guests or something. It was a delicious bottle of wine, but the bill was far more than I was willing to pay. And now I just don't have enough money to pay . . .

The example may be a little off the wall. But even reflecting back on one or two phrases takes some of the steam out of the situation. The customer begins to tell the story, and when you hear it through, it may not be the one you were expecting, or which the staff member assumed in the second situation.

> Encourage team members to record all they remember of undesirable encounters, and then to analyse them to show where a more reflective approach might have revealed more about what the real problem was. Only when the real problem is identified can the solution be found.

Ideas for further study or work

1   Choose a community of any size (close to where you work, study or live) and draw up plans as to how you would work with local people to find how they would wish to develop more opportunities for leisure and recreation. Consider particularly how you would include the young and older people.

2   For the same community, sketch out ideas which you could develop (as above) to establish a festival. Identify the benefits for the community, noting also any disadvantages and displacement effects.

3   Choose a successful organisation involved in some aspect of leisure and recreation, and study its history, and its current state of health. Consider what critical management issues it will face, and in particular consider what plans it has (or should have) to ensure continued enthusiasm in management (succession).

4   Consider the passage of a visitor through some facility, say a theatre, or visitor attraction, and consider the changing position in respect of safety and liability. Record who has what balance of responsibility at the various stages of the visit.

5   Carry out a simple customer care audit at two different locations where the same form of leisure or recreation is provided for. Compare the results and send the anonymous findings back to the operators as thanks!

Reading

People

The more enlightened books on management provide much of value, and notable among them are books by Handy, who also provides help with understanding the motivation behind volunteering , and Peters and Waterman.

For techniques such as Participatory Appraisal, or for working in close partnership, read Chambers and Wilcox. More valuable (but more costly, initially!) than reading might be attending brief courses. In Britain the Environment Council has taken a leading role in introducing mediation techniques to this area, and Stewart provides a useful introductory text on conflict resolution. For an insight into how rangers work, websites are useful, and books such as Frost *et al.*

Funds

In North America many more organisations support themselves or find corporate funding, and this becomes a useful hunting ground for sources of wisdom, such as Crompton (whether collected case studies or advice on pricing) and, if you can still find it, the story of how the Children's Museum of Denver earned $600,000 a year (Simons *et al*. 1984). Each country is also likely to have the equivalent guides to those produced in Britain by the Directory of Social Change: Brown and Smyth, Clarke and Norton and FitzHerbert *et al*. These are invaluable, although, in my experience, in most organisations there are some individuals who are just able to develop the appropriate networks to make the difference when it comes to fund raising (for which they probably have a genetic predisposition). General marketing texts such as Kotler are also of help.

Managing visits

In the past the focus has been on managing sites and facilities, rather than settings or experiences, as can be seen looking at standard texts like Torkildsen. The best sources in this area focus on elements of the visit. Read Bee and Bee for an insight into the really important aspects of customer care and Ball for approaches to risk and visitor safety management. For an inclusive approach in respect of people with disabilities, most of the texts focus on specifications, the BT *Countryside for All* Guidelines have some useful specifications concerning access to the countryside for people with disabilities on physical access and programme access.

Chapter 9

Managing future

environments

Overview

Our ambitions continue to grow. Although longevity is increasing, there is no hint that people are content to spread out a lifetime's experience over the new extended lifetime. Quite the reverse, our high-speed communications and technology allow us to achieve more, and encourage us to live in the fast lane, if not perpetually overtaking. This may be the preferred route for some, but by no means all. Others wish to live out their lives at a more natural pace, in step with nature. Preferences look set to continue to fragment. The family, as the normal unit for a household, is in decline at present in Britain. There is every reason to suppose that more flexible patterns will persist (at least in the short to mid term). However, many trends seem to oscillate, like the return of religious fundamentalism. Perhaps the family, or other similar substitute, will again take its place as the dominant unit. Given the ageing population in a number of developed countries, pressures are likely to encourage extended families or community groups to take greater responsibility for caring for older people, rather than rely on the state. Whatever the favoured unit is to be, fragmentation in patterns of living will echo the development of different lifestyles and activities. Increasing population pressures, traffic congestion and the desire to achieve more in less time will increase stress. Rising stress, and the associated cost of mental health, will encourage the adoption of leisure as an effective preventive cure for society. The benefits-based approach (Parks and Recreation Federation of Ontario 1992; Canada Parks/Recreation Association 1999; Sport England 1999a, b) will encourage policy makers to put a greater value on leisure provision.

We are already beginning to ascribe a higher value to our environmental well-being and the Environmental Management Ethic will continue to grow. As we manage our land ever more productively, we will be able to divert resources to rehabilitating damaged land. We will be able to reclothe our land with trees, in smaller-scale woodlands networked together. We will take more care of our coast and the marine environment, and strive to stem the tide of species lost. We will look after the fundamental resources – water, soil, air and energy – with renewed vigour. West and East, and North and South, will trade more. The developed countries will deploy technology to overcome the increasing pressures, and hopefully learn how to curb excessive material desires, and learn to slow down. Within the built environment, we will explore re-use of buildings rather than demolition and rebuilding. We will take care to protect our culture and all records of our way of life, guarding against their loss (and considerable effort will be expended on protecting electronic information from interference and decay).

The boundaries between work and leisure will become blurred and fuzzy, and core leisure and micro-leisure within work will be increasingly valued. People will pay much more attention to finding a satisfactory balance between their obligations, to others and to themselves. Access to leisure will be seen to be as important (even to Britons) as the pursuit of happiness was to the Americans in drawing up their constitution. New activities will emerge to allow us to go higher, faster and farther, and to store and utilise energy more efficiently. Many of the motivations and sources of pleasure are fundamental and will not change, but we will always be searching for new expressions of recreation and leisure. Lifelong learning will become an inseparable part of leisure.

Our rising concern for sustainable issues will encourage a continuing search for sustainable leisure, within a more sustainable way of life. For packaged components or entire holidays, certification will continue to develop and with it various green guides. Increasing the leisure benefit–cost ratio will be the aim, in economic, environmental and social terms. Our introduction to the subject closes by looking at possible futures. We look at some of the conflicting trends and see a resolution in the way that patterns of work and leisure evolve. Ironically, people will take leisure more seriously. We consider some of the ways forward, including the matching of reality to expectations; changing awareness, understanding and behaviour; developing as communities and as individuals. For the good of the environment, we owe it to ourselves (and to everyone else) to enjoy life to the full.

Writing the future that we want

Forecasting is a tricky business. Weather forecasting has advanced enormously since the advent of ever more powerful computers to model weather systems. Despite the application of Cray computers, forecasting (in any detail) beyond a few days is very much harder (Gleick 1988). We should not be surprised at how much more difficult it is to forecast what we should take account of over the next twenty years, in respect of managing for leisure and recreation. Look back to any texts which suggest what approaches for managing leisure will be relevant in the coming decades, and you will be either disappointed or amused (Epperson 1977). Texts that point to some of the major thrusts to bear in mind are probably more helpful (Kelly and Godbey 1992).

How far ahead can we reasonably foretell what goals we wish to pursue? What will our goals be? Do we, as individuals, have a clear idea? If we do, we may be able to write our own futures. In Chapter 4 we recounted some of the medium-term trends. People will be living in more flexible arrangements. There will be fewer families of the conventional nuclear type. Many people will live singly or in loose association with friends and colleagues, but within large settlements. With the increase in freedoms, and respect for the individual, the influence of institutions seems set to continue to diminish. In many developed countries the changing demographics will put a greater strain on the state. Extended families or other groups may be expected to play a larger role in the care of the elderly. People will generally live longer, and arguably with greater physical health ('arguably', on the one hand because allergies and other minor complaints are on the increase, and on the other because the evidence suggests that with increasing longevity there may also be a longer period at the end of life when

BOX 9.1 The future nobody wants – war, famine and disease

Our planet is rich in natural and human capital, and we need to look after what we have. If we shared out the resources we could avoid war, famine and disease. The problems are well catalogued, e.g.:

Water. The world's population has increased since 1900 by 100 per cent, but water consumption has increased by 600 per cent. In that great country, China, half of its 600 or more cities suffer water shortage. We each require 20–40 litres a day, whilst a washing machine may use 70 litres in one wash. Bhoutros-Ghali, former Egyptian Foreign Minister, is quoted as having said, 'The next war in the Middle East will not be over oil, but water' (Porter 1999).

Waste. Recognised as a key product of tourism (Tribe 1995), the developed countries are continuing to produce more than their share of waste, through excessive use of packaging and the shorter life span of what were previously known as consumer durables. Much waste is buried in landfill sites or burnt in incineration plants.

Energy. Ever since we discovered that wood burns to produce a comforting heat, and one which will tenderise the toughest meat, we have been exploiting different sources of energy: coal, oil, wind power, wave power, hydro power, solar power, biomass and even nuclear power. Because many of our sources are dependent on fossil fuels, we have begun to realise that we should focus more effort on ensuring that we are efficient in our use of energy, making full use of heat insulation, for example. Tourism is very energy-hungry, chiefly through transport.

Transport. Our cities are congested, and suffering increased pollution. It is not clear which is the bigger problem (Everitt and Higman 2000), but in developed countries we are profligate in our use of energy. In producing energy for transport from fossil fuels (oil) we contribute massively to the greenhouse gases which are contributing to climate change. On the other hand the forest fires of 1998 in Indonesia contributed more emissions than all the industries of Western economies (Porrit 2000)

Quite aside from our management of natural resources, we seem to be inept in a number of departments. Undoubtedly, we have developed great skills and technologies. Yet, despite the great history of human development, and the great intellectual heights scaled, as a species we still have some very basic characteristics. We seem unable to share the earth's resources so that they adequately support the world's growing population. Many different solutions have been offered, but instead of taking a reasoned approach we resort to violence to impose views on others, or to secure resources. If just a quarter of military expenditure was diverted, it could fund world programmes to provide:

- ◆ sufficient food (US$19 billion);
- ◆ health care (US$15 billion);
- ◆ shelter (US$21 billion);

- ◆ safe, clean water (US$50 billion);
- ◆ literacy (US$5 billion);
- ◆ erase developing nations' debt (US$30 billion);
- ◆ prevent acid rain (US$8 billion);
- ◆ prevent global warming (US$8 billion);
- ◆ stop ozone depletion (US$5 billion);
- ◆ stop deforestation (US$7 billion);
- ◆ prevent soil erosion (US$24 billion);
- ◆ provide safe, clean energy, with efficiency measures (US$33 billion); and renewable energy (US$17 billion);
- ◆ stabilise the population (US$10 billion).

(Henderson, in Barrow 1999: 268)

With what was left over we could have quite a good time as well! Leisure, recreation and tourism contribute to the profligate use of resources (mostly at the margins), but there is great scope for schemes to encourage wiser use of these resources.

health is poor). Stress seems set to increase, for those who find it hard to change and adapt, as people will strive to satisfy their different wants. Some will wish to experience more within their allotted time, covering a greater distance, or experiencing a wider range of events, seeking extensive novelty. Others will wish to gain a more intense or in-depth experience, seeking intensive novelty. A greater proportion of people seem set to suffer from problems of mental health.

We seek faster (real-time) communications, faster transport, just-in-time, more experience in less time, and a more hectic pace of life. All this makes forecasting more difficult. At the same time, we will be continuing to develop our model of the way the environment operates. Our understanding of some systems will inevitably falter as new and competing theories develop to explain what we see unfolding, as with global climate change. Our more extensive use of computers will allow us to integrate different disciplines. We will use a more holistic approach, no longer categorising effects solely as social, economic and environmental or political, economic, sociological and technological – or will we?

We should hope to deliver as much leisure as we can, for individuals and society to achieve whatever they individually and corporately wish. Leisure demands a balance (a constructive tension) between obligation to others and self-expression, and we need shared values and shared beliefs, at least at some scale of resolution, to deliver the optimum mix. A useful area for research might be in deriving indices that could be used to gauge the extent to which society is making use of leisure and recreation to its greatest advantage. A number of indices could contribute to a more fundamental measure of societal well-being.

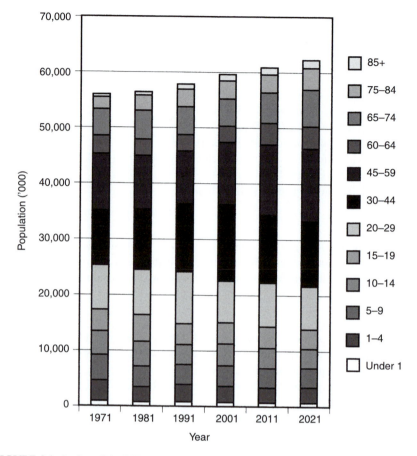

FIGURE 9.1 Ageing of the UK population
Source: Office of National Statistics, *Annual Abstract of Statistics* (1999)

Forces for change

The world population continues to expand at an alarming rate. At present rates of growth, the population of China will double by 2075 or so. That would result in the same overall population density across China as is currently the position in Britain. In those parts of the world where the population is more or less static, the distribution of age classes is changing. In Britain the population is ageing. There are concerns about the decline in the number of people who will be working to support older people who have retired, and who will have increased longevity. Our settlement patterns change with technological change (Bracey 1970), particularly in relation to transport and communications.

Particularly important influences have been the invention of the internal combustion engine and the development of telecommunications. Overall patterns are sometimes hard to discern, as they are often masked by intervening variables. In England people are moving from towns back into the countryside, and there is a steady drift south. Just across the English Channel, or La Manche, lies France, where the concerns are all about people moving away from the country into the towns. During the summer in Europe there is mass transhumance between centres of industry in the north and the holiday resorts of the south, around the Mediterranean. Terms like 'countryside' and 'town' are relative. Remoteness, to a resident of Britain, will be rather different from what it is to a resident of Tibet. The distinction people draw between countryside and town is in part dependent on the extent of development and intrusion, for example of roads. There is evidence that some roads are seen as extensions of urban development, and only when roads have been left behind do visitors perceive themselves as arriving in the countryside proper.

Two major forces for change are working in opposition: standardisation or globalisation to take account of economies of scale and differentiation, to celebrate local character and distinctiveness. Standardisation could be regarded as a phase, a stage in the cycle that necessarily gives way to differentiation, and the development of niches. Once niches become too small to survive, they collapse into larger units, and systems coalesce and tend to become standardised. Successful managers recognise the dynamic, and administer alternate doses of integration and differentiation to supply what people wish. It seems entirely probable that, if we carried out some massive factorial analysis of the influences, which will direct how we develop our recreation, we might find two families of drivers: the technological and the sociological.

The *technological* influence will chase development and economic goals. It will encourage the development of technology to help us advance as quickly as possible, because of the shortage of time – the need for efficiency, and economies of scale, leading to a further push for globalisation and standardisation. Volume and the cost of recreation will be important. Competition and the rights of the individual will be paramount. This could be described as a largely Western style of thinking.

The *sociological* influence will chase societal and environmental goals. It will encourage us to identify the complexities, and cherish the diversity and distinctiveness – a more Eastern approach, in which it will be easier to recognise that not every problem needs to have a simple solution. It recognises that many processes will have to operate across space and time, in quite different scales from those normally used by humans. Co-operation and the rights of communities, and of different species, will be important. This approach will keep pressing for understanding, and a search for meaning. Quality of life and the benefits of recreation will be important.

People who live in more pleasant surroundings, who engage in healthy activities and take healthy nourishment are living longer and longer. There is a large group of people who now have the time, money and health to engage in a broader range of activities than ever before. The trend is continuing. Our technological skills have far outstripped our ability to resolve the philosophical and ethical problems being posed by our ability to tinker with life. We still need to develop criteria to help us allocate scarce resources. One of the criteria to date has often been quality of life. In the United Kingdom economists have been employed to develop a methodology to help in choosing priorities in the allocation of funds within the health service. Perhaps it would be more cost effective still to spend a little more on preparing people for healthy lifestyles, and ensuring the enjoyment of healthy leisure (Sport England 1999b). Ultimately, we need to make choices about how we choose to live. Do we wish to have a still more extensive volume and range of experiences, or do we wish to have fewer and more intense experiences?

The growth in travel has been very marked in Western countries. It is impossible to conceive that the growth will continue at the current rate, yet no other rational model provides alternative scenarios. There are still plenty of countries and parts of the world where the population appears to be heading along a similar curve. The relative cost of travel continues to fall (and will, so long as externalities continue to be ignored, and energy is priced so low). Exceptions to this general rule seem to be caused by the interventions of governments (but each with a different view, as exemplified by the varying tax rates on fuel across Europe). We may though find more planet-friendly forms of energy, to help us continue to feed our energy-expensive habit of travelling. Technological developments will help improve fuel efficiency, and our use of scarce resources could be further reduced (Wright and Wilsdon 1999). A more likely result may be that the reduced cost will encourage still more travel. Each unit of travel may consume fewer of our scarce resources, but the volume of travel will be so great that our scarce resources will be under greater pressure. If we were able to find a source of fuel or a means of energy storage that was sustainable, there is every indication that almost everything would be done on the move. There are people who think that atomic or nuclear power is that very fuel, and others for whom the attached risks are too great.

Personal mobility is one of the most highly valued of freedoms, but travel may have to be rationed. Under pressure, people may cut out journeys that only serve to bring information or items together. Post, courier service, facsimile, e-mail and video-conferencing are increasingly preferred as a means of delivering factual information and goods. Where colour and tone are to be added to messages, or where slight nuances are to be detected, personal face-to-face meetings are at present the only real solution. Video-conferences are still a little clumsy and can be expensive. However, the technology is developing fast. With video-walls, virtual meetings could be considerably improved, and the technology will allow extra sensory suits which would record, say, a sequence of smells from one end of the video link and transfer them to the other end. The same could be arranged for touch, etc., although it may take some time to simplify the technology to prevent participants from being weighed down, tethered (or even strangled) by the wiring. Pearson (2001) gives many examples of technological developments in store. In this scenario, leisure travel would be seen as core travel, enabling people to reach and meet each other, for real. One of the chief motivations for travel is the search for novelty and this will continue to be much in evidence. If travel continues to become easier, the trend towards increasing the number of holidays taken each year seems set to continue (Department for Culture, Media and Sport 1999).

By whatever means, individuals are likely to encounter a greater number of new ideas, new people and new places in a shorter time than has been possible in the past.

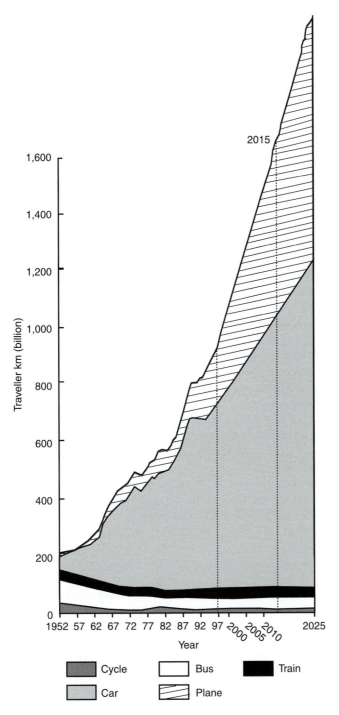

FIGURE 9.2 Travel by Britons: cycle, bus, train, car and plane, 1952–2025
Source: Adams (2000)

The amount of time available in any one encounter will therefore be very much smaller. Stress will result. There is insufficient time available in these high-speed encounters to decode meanings adequately. Too many ambiguities remain, and dissonance results. While we are able to increase our apparent physical health through our clever technologies, we seem not to have been so successful in relation to the mental health of our societies (Greenfield 1997). Increasing the stream of information available may increase levels of excitement for some, but it is less likely to do very much for their mental health. The scope for information overload is all too readily seen, within just a few minutes on the Internet.

New environments

As we manage to enhance the productivity of our agriculture and our manufacturing industries we should be able to restore some of our landscapes to their former glory. Woodlands, and other extensive forms of land cover (with more edges, ecotones, and rich complexity), will be developed to provide new multipurpose environments close to where people live. We should not be constrained by the models of the past, and there are plenty of examples of new, or rediscovered, approaches (e.g. greenways, and community orchards and forests). More and more, communities will be keen to see integrated development locally, with organic vegetables grown in allotments and gardens, aquaculture in a pond which provides a pleasing environment (Wright *et al.* 2000), reedbeds purifying waste water, and (why not?) sewage processing within parks (Smale 1999). Perhaps commuting will be contained, with more working from home, through the use of information and communication technology. The skills and techniques have now been developed to allow us to plant and nurture woodlands on the most poisoned of soils, where former mineral extraction has raised heavy metals and other noxious elements and compounds to the surface.

Areas formerly abandoned can be brought into the recreation portfolio. For areas rich in aggregates and yet to be quarried we can plan future lakes for recreation, conservation or both. In the future we may regard our landscape as a temporary blend of land uses. The objective will not always be to create new or alternative landscapes. We may wish to share existing landscapes with future generations to provide a glimpse of how things were. Not every mine could be kept open for interpretation; many have to be very carefully sealed to prevent mishap or accident. Yet there are also others, which are described as 'underground cathedrals' with great cavernous spaces, whose future we might review.

The coast provides the dynamic interface between the sea and the land, and will continue to fascinate mankind for all time. We will continue to be attracted to, and to value, water (Butler 1999: 107), to enjoy the opportunities it provides for bathing when (and where) the temperature makes this an enticing option, and elsewhere for diving, sailing, windsurfing and messing about in boats of all kinds. Some will value it also for the opportunity to fish for, or study, marine life. With the implementation (and observance) of tighter regulations to prevent pollution of the seas, this environment should become an even more attractive setting for our leisure. Unfortunately, in

---

BOX 9.2 Greenspace targets

English Nature recommends that people in towns and cities should have:

- an accessible natural green space less than 300 m from home;
- at least one accessible 20 ha site within 2 km of home;
- one accessible 100 ha site within 5 km of home;
- one accessible 500 ha site within 10 km of home.

Sources: Radford (1999), Harrison *et al.* (1995)

---

some parts of the world population growth and economic pressures have caused marine pollution, overfishing and the spread of diseases through poor practice. Resorts of the future will have greater respect for natural processes, of which the shore is a manifestation, and will aim to work with these processes rather than attempt to block them.

A good deal of leisure time will be spent indoors. Given the number of people who pass through particular built indoor spaces, and the time that is cumulatively spent within them, we should pay much greater attention to our interaction with these spaces. Some buildings are sufficiently flexible to allow managers of these settings to respond to changing demand. Others require a flash of inspiration to demonstrate what is possible. Flexibility will be the key. Fashions move from the ornate and Renaissance to the crisp and Bauhaus and back again. New buildings will have to be flexible to allow for recladding to match evolving tastes. In any building, especially where public funds are used, the appearance will be judged more important than has been the case in the last fifty years. Artist craftsmen will be used more often.

Our desire for quality and distinctiveness will probably determine minimum levels of expenditure on embellishment by artists. More countries may adopt the approach (first taken by Sweden over sixty years ago) of 'per cent art' schemes (where a fixed percentage of the cost of the project must be invested in public art connected with it). Such schemes have been adopted in the Netherlands and elsewhere in Europe. They have also been applied in the private sector with good results (Waters 1987). The formula may be adjusted to recognise the amount of time spent by people in the building. There may be an increased requirement to involve the people concerned in choosing what should adorn the space. Some of the great buildings of the past will have new leisure purposes reassigned. Urban nature reserves will continue to be important for preserving some hint of nature as a contribution to constant natural assets. Unless the resources are critical, these urban reserves are unlikely to be fixed. Instead, policies will ensure a matrix of spaces at hand.

More tricky will be the conservation of culture, the protection and preservation of sufficient buildings, and more important the culture and less tangible aspects of human interrelationships. Recording techniques allow us to compress more and more data into smaller spaces, and we will have to ensure that these are secure, and not to be wiped out by some electromagnetic pulse, or storm. New recordings may provide for smell, the most evocative of senses. It should soon be feasible to specify the make-up of smells, so that in time the playback of experiences will take into account not only sight and sound, but also smell and movement. These recordings will be miniaturised so that they can be played in the home. The home will develop as the most controlled of environments. We will be able to select experiences, almost from the shelf, and to be downloaded from the Internet. People suffocated by the indoors, will seek excitement in new encounters, with new people, new challenges and new places, outdoors.

There is sure to be greater pressure on the land as populations increase. Speeding up the pace at which we live will increase the pressure. People will wish to have more space, and this desire could manifest itself in the continued pursuit of exclusivity. Membership of clubs will guarantee a certain quality of space or experience. Politics will determine the extent to which populations will have a right to access leisure. We will be more attentive to the detailed structure and linkages within our landscapes and townscapes (Tibbalds 1992), spaces for pedestrians, and life at a human scale. More consideration may be given to the social structures, down to the level of households (Boyle 2000). Rights will be reassigned – for example, more museums and galleries will hand back treasures to their rightful owners, and debate will continue to rage as to who the rightful owners are.

New leisure

It is often assumed that work and leisure are at opposite ends of the spectrum. Freedom of choice, of association, of expression, of movement; and freedom from a sense of obligation are the major characteristics respectively of recreation and leisure. Much hard work, whether physical or mental, demands the kind of focus and concentration which gives the sense of fulfilment, or peak experience, best described by Csikzentmihalyi's (1992) concept of flow. Hard work and intense leisure and recreation activity have much in common.

The freedoms sought in leisure are at the heart of satisfaction in work, and indeed in life as a whole. Leisure is not peripheral, it is central to our lives. Some of us make a clear separation between what we think of as work and what we think of as leisure, but even so, the work will be interspersed with moments of leisure if not recreation. Micro-leisure (the moments of leisure scattered through work) will be increasingly recognised as making a major contribution to health, well-being, productivity and enjoyment. Social interaction, humour and laughter all play a part in work, just as they do in leisure. The divisions are unclear, very fuzzy indeed. After all, for the average academic, leisure may well be the opportunity to study, but (crucially) with no strings attached.

What seems central is our sense of obligation. It is easy to identify clearly individuals to whom we feel affiliation or obligation. It becomes more complex when we describe obligations to groups, and to society. To whom will we feel obliged in the future? Personal mobility has undoubtedly loosened the bonds or ties which kept families so prominent as units of society, in Britain. Will something take the place of the family? One might expect that more attention would be paid to establishing real communities of neighbourhood, which will at least exist during leisure hours. Communities of interest brought together for work will increasingly exist as virtual communities alongside virtual companies, and virtual associations of many kinds, thanks to the development of high-speed communications and e-mail (Pearson 2001). This may allow us to refocus our efforts on establishing communities of neighbourhood. Recently revealed trends of increased dependence of youngsters on the home, and on electronic entertainment, pose a challenge which some of the other forms of recreation and leisure may help to overcome.

BOX 9.3 Green guide

Each of us could make up a twelve-point guide, thinking carefully about what we could personally contribute. An example follows:

◆   Move less.
◆   Relate more.
◆   Hurt less.
◆   Laugh more.
◆   Smile more.
◆   Take part more.
◆   Focus on senses.
◆   Raw material conscious – refuse, reduce, reuse, recycle.
◆   Add on (pleasant aspects to unpleasant tasks).
◆   Share more.
◆   Enjoy more.
◆   Think more.

New recreation

Will we change fundamentally the nature of the activities we enjoy? There would seem to exist fundamental activities that lead to the pleasure and the fulfilment we seek. Reviewing activities, and studying clusters, are likely to give clues as to the source of new activities. Some activities seem so fundamental it is difficult to foresee their decline. Many focus on physical activity, and provide the opportunity for social interaction. The motive for taking part, though, will be different for different participants. We may have greater control over the results of recreational activity. Technology will continue to be applied to almost all sports and recreational activity to help ensure a better match between intention and outcome. While many of us remain risk-averse, there are those who will go to great lengths to ensure adequate doses of risk and excitement remain. They will prefer a measured application of technology, and a maximum of human effort.

Social settings for recreation and leisure will reflect continuing changes in society. Increasingly younger people are maturing early (probably closer to the situation in prehistoric times!) physically and mentally. In some countries there is a gap in the provision for teenagers, just as they are beginning to explore farther away from the family nest. Peer groups determine what is acceptable, and there is every reason why this will continue to be at odds with what is enshrined in the laws made by adults. For example, in Britain there is an increasing problem with under-age drinking of alcohol. As this is the natural age for rebellion and for developing peer pressure (Caldwell and Darling 1999; Coleman and Hendry 1998), such problems seem likely to persist.

People will also continue to seek out pleasure, taking substances and indulging in activities that change their moods. What the popular substances or activities may be in the future is not easy to foresee. Cannabis and some of the drugs that are socially acceptable now seem on the verge of fresh censure. Despite concerted campaigns over the past forty years, cannabis is still an illegal drug in the United Kingdom, although many commentators would suggest that its effects are less serious than those of smoking (tobacco) and drinking. Tobacco smoking is hitting the headlines in respect of legal battles in the United States, and markets in the West appear to be shrinking. Trade with countries in the East is expanding, and it seems new markets are opening up in India and China. (This is rather ironic, given that it is in part the return of Hong Kong and Macau to China, wrested from her in the time of the Opium Wars, that has been responsible for the opening up of trade with the West.) The tobacco companies will be hoping for an easier time with the law, no doubt. Damage from tobacco smoking tends to be focused on what happens to the end user. However, attention is increasingly being drawn to the harm being done in developing countries, which produce some 80 per cent of the world's tobacco crop, and suffer deforestation, soil depletion and the build-up of residues from pesticides and fertilisers, among other ills (Denker 2000). The World Health Organisation is exhorting governments to encourage farmers to diversify into other crops.

Kenco and Friends of the Earth have teamed up to address some of the social and environmental issues which arise from growing coffee bushes across Kenya, Costa Rica and Colombia. Such large ventures, in tea, coffee, tobacco and alcohol, are often run by multinational companies whose prime concern is profit, and not environmental gain for the host communities (although in the long term these issues may be linked). The poorer the country the harder it is for governments to have any degree of control. There is some scope for developed countries to support more environmental approaches, using debt swap schemes. To feed the desire for cocaine in fashionable (and not so fashionable) circles, large parts of countries in Central America have been forced into the grip of organised crime.

Meanwhile alcohol seems certain to keep its place. Findings that there are medicinal properties in beer (which may fight cancer) and in red wine (which in certain doses may be good for the heart) come as little surprise. Yet there are also concerns about the misuse of alcohol, and the amount of suffering caused as a result. There will be greater understanding of the effects of different activities and the chemical transmitters the activities stimulate. Growing interest in folk history, culture and alternative medicines is likely to reveal (or rediscover) secrets about other plants and substances.

Technology will continue to be deployed to allow individuals to re-use the skills of others, perhaps to compose music in the style of a contemporary musician, whether rap, heavy metal, or classical (e.g. Philip Glass or Mozart); or to paint a picture in the style of an Impressionist or an Old Master. While creative activity will be paramount to some, the more mundane activities like shopping will continue to be necessary, and for some a recreational activity. It seems likely that at least two trends will shape our aproach to shopping: the continued development of shopping over the internet, via electronic markets and auctions, and the recasting of leisure as a social activity, pursued in groups for pleasure.

Car designers are hard at work producing vehicles that will better accommodate modern transport needs, for fewer people to be transported at less cost (economic, social, and environmental), and in greater comfort. For some, driving on the roads is recreation: with the sense of speed, and the skills required for keeping the car on the trajectory sought. For most, travelling on roads is merely a means to an end. Some people take their vehicles off road in search of excitement and pleasure.

The ultimate in personal freedom must be to develop further single-person motorised transport. The motor cycle, the scooter and the motorised bicycle used in some Mediterranean countries provide popular forms of transport, and there are developments under way (http://www.thinkmobility.com/neighbor.html). Motor cycles have been developed for all types of terrain. Vehicles originally developed for work, such as the quad bikes devised for shepherds, are now extensively used for motor recreation. Personal water craft may be re-engineered to reduce the noise but the real revolution is likely to be in personal air travel. The strap-on helicopter, SoloTrek, is a personal flying machine, in prototype form, which could soon be commercially developed (Kleiner 2000) and would take things further than the auto-gyro and micro-lite. The early stages of congestion or conflict would stimulate the use of such machines for personal mobility, but increasing congestion would render the pleasure null and void.

Greater effort will be expended in developing quiet engines, although the problems associated with bystanders being unaware of oncoming (noise-free) vehicles still requires solution. With increasing wealth, it is likely that people will have their own personal library or garage of different gadgets (or access to them). People will have cycles for different occasions or weathers.

There has been something of a reawakening to the idea that we spend our whole lives learning. The same is true of leisure. Leisure is not reserved for a particular age or social group. People will recognise an increasing range of activities and forms of recreation and leisure in which they can take part, at different stages of their lives, and will seek a balanced diet of leisure, just as we need a balanced diet of food. Inequities in the distribution of leisure will be challenged repeatedly, and more innovative techniques used to ensure that people have adequate leisure, not so much for their own good as for that of society. Leisure will become more highly valued and prized. People may make choices aided by decision support systems, which point out the relative merits and pleasures to be gained from different types of recreation.

New tourism

As social animals we have built into our psyche the need to compete, as well as to co-operate. We continually explore boundaries, testing the limits in every area of endeavour, and tourism is no different. As technology develops, so possibilities expand. A premium is attached to travelling faster. Supersonic flights are still marketed as a novelty. Space flight is becoming more common, and this coupled with the greater reliance on private finance and on innovative schemes of co-financing endeavour can only encourage more space tourism (c.f. Tito in 2001). For those who seek speed thrills, there are dragsters, and production motor cycles to try. There are options to try out motor racing at some of the Grand Prix circuits, in comparative safety under the supervision of instructors. Speed has always been an attraction, and the sensation is available now in simulators for those who wish to have the sense of speed without the physical risk. Journeys that used to take months and even years can now be accomplished in minutes and hours. Some people now travel across the Atlantic for weekend shopping trips, in preparation for Christmas.

There may be few parts of the earth's surface which have not felt the touch of man's hand, but there are still places where we can feel the full force of the elements and the grandeur of creation. Most of the world's mountains have probably been climbed, but there are still plenty of areas of the sea bed yet to be explored. Tourists with enough money can already take up options to travel on miniature submarines and bathyscopes to descend to the ocean floor to view parts of the wreck of the RMS *Titanic*. There will also be those who wish to travel unencumbered by technology. People can now sail across some of the world's oceans in company (ARC rallies, and similar) and more opportunities for such travel will open up for people who wish to take a step or two beyond conventional travel (Ogilvie 1996).

People will continue to follow the sun (if ordinarily they see too little of it) and to cooler regions (if they see too much of it). Precisely because some comfort can be manufactured, with heating and air conditioning, the authenticity of real sun or real snow will increase in value. People will value their own culture more and more, and resist the trampling of their values. Treks and trips to the Himalayas and other remote areas of the world will require an approach that takes much more notice of the needs of the host communities (Nepal 2000), as well as the need to safeguard the natural environment. The approach adopted by the Himalayan country of Bhutan, where the number entering is regulated, will become more widespread as cultural as well as natural resources are defended. This will become an issue in developed parts of the world just as much as, if not more than, elsewhere. Great effort will be diverted into demonstrating how benign tourism is, not least because some nations will continue to use tourism as an economic tool to transfer wealth until such time as we devise systems to take into account the externalities, the real costs to the global environment.

Sustainable leisure

Recreation cannot be impact-free. Any activity must have consequences, just as it must have antecedent causes. What kind of impact, upon whom, or upon what, in what manner, over what time, to what effect? These are all questions that will be asked. No longer is it politically correct to engage in activity to great personal benefit at someone else's great personal cost, unless that someone else is adequately recompensed. It was never morally correct, but somehow exchanges that took place were rationalised afterwards as fair. Some of the impacts are naturally outside the usual rules of exchange, because the costs or the benefits have yet to be experienced, and may not accumulate or mature until generations to come. More effort will be made to take into account all (known) costs and benefits. People will adopt an environmental management ethic that will help to shape patterns of leisure.

BOX 9.4 Sustrans

An active core of individuals (with engineering expertise, grit and persistence) has been behind a most audacious plan, to create a National Cycle Network in the United Kingdom. The first 5,000 miles opened on midsummer's day 2000, and the target is to increase the extent to 10,000 miles by 2005. Sustrans has been the champion of sustainable transport for some while, particularly in the development of shared-use paths along routes that often make use of abandoned railway lines. The group strongly believes that paths made fit for cycles would also open up scope for those pushing prams, wheeling buggies, or merely wanting a gentle walk on level ground. The technical skills involved in the process of developing these paths have been translated into advice, so that the word has spread.

Sustrans's ambitions were fired by a combination of changing attitudes to sustainability, following Rio, and the possibility of securing significant funding from the UK lottery. The Millennium Commission has supported the venture with more than £40 million, and about 400 local authorities, public bodies and organisations from the private sector have helped to make it happen. There are now projects to link routes with schools, and to transport exchanges, with public art along the way. See the website for details (http://www.sustrans.org.uk).

Leisure is more than just time left over after work. Real leisure gives a sense of resolved balance among obligations – that supreme sense of well-being, or interrelationship and interdependence with other people and our environment. This feeling of being at one – with ourselves, our colleagues and our environment – is surely everybody's goal. This does not reside solely in the realm of any one discipline. It involves philosophy, ethics, religion and economics. It involves body-centred, mind-centred and other activity integrated with the activity of others.

Ways to move forward

People's expectations may differ from what is actually available. Surveys seem to indicate that people are very happy to watch hours and hours of television, absorbing life vicariously, rather than experience the world at first hand. People consume ever-increasing volumes of alcoholic drink rather than engage in strenuous sports. Access to personal mobility will determine many of the boundaries of what is possible. Many people will live within their own city all their lives, and a high proportion will never leave even for a weekend. Education is patently not just for the classroom (real or virtual), and will ensure society relearns many of the skills previously taught within

---

BOX 9.5 Actions for the manager

Improvements in the physical environment outdoors:
- Remove litter.
- Remove unwanted sound.
- Consider planting more trees, shrubs and herbs or otherwise enhancing the three-dimensional nature of the space.
- Soften the impact of buildings and infrastructure.
- Use water features.

Improvements in the physical environment indoors:
- Remove litter and dirt.
- Make use of light.
- Use colour.
- Match mood to function.
- Build in flexibility.

Improvements to encourage in the social setting:
- Smile, and let others smile.
- Include humour.
- Encourage settings where people have the opportunity to speak and listen to each other.
- Provide some settings where the clock is absent.
- Provide personal spaces and refuges.

Improvements to make in the economic environment:
- Ensure exchanges are fair.
- Be prepared to use many currencies, not just pounds and dollars.
- Ensure that all costs and benefits are taken into account.
- Encourage economies of scale, but include reference to what would have been externalities.
- Ensure future generations have similar choices.

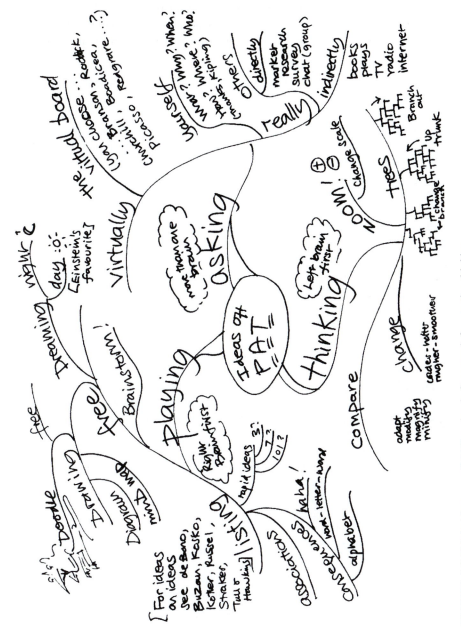

FIGURE 9.3 Bright ideas . . . off PAT: a simple framework for creative thinking

families and neighbourhood communities. Pleasure and happiness are not the preserve of the wealthy. There is no rationing of laughter and smiling, although it may be a little easier to appreciate this if the basic physical and physiological needs are met.

If the reality fails to match expectations, we can either reduce the expectation by informing about the reality, or change the reality so that it better matches the expectations. There is often a good case for doing both. Here are some improvements, managers can put into effect.

BOX 9.6 Leave No Trace

In the Britain of the 1960s, increasing use of the motor car made many planners extremely nervous about what was in store (Dower 1965). In that same decade the use of US National Forest primitive areas and wilderness areas tripled, and the use of public land continues to increase. By 1992 combined visits to federal lands had reached 570 million.

The federal agencies reacted to the pressures of the 1960s and 1970s by printing all kinds of codes and explanatory leaflets. By 1987 the US Forest Service Bureau of Land Management and the National Park Service had cooperated to produce a joint leaflet, promoting 'leave no trace' ethics, a land ethic. The US Forest Service developed a partnership with the National Outdoor Leadership School to factor in the school's considerable skills in education programmes. The other agencies (including by this stage the US Fish and Wildlife Service) later joined forces by means of a memorandum of understanding.

Leave No Trace Inc. is supported financially largely by the private sector, with support in kind from the public agencies, and continues to grow. The aim of the educational programme is 'to avoid or minimise impacts to protected area resources and help ensure a positive recreational experience for all visitors'. Recently the principles have been reviewed and boiled down to seven actions:

◆ Plan ahead and prepare.
◆ Travel and camp on durable surfaces.
◆ Dispose of waste properly.
◆ Leave what you find.
◆ Minimise camp fire impacts.
◆ Respect wildlife.
◆ Be considerate of other visitors.

Key to the education programme are the five-day Master Educator (and other shorter) courses, and the development of (some sixteen to date) Outdoor Skills and Ethics booklets. These are based on science, for which recreation ecology has been fundamental.

Recreation ecologists like Jeff Marion have been able, through observations taken all over the country, to identify that location is absolutely crucial to

---

the recreation impact. For example, campers usually search for a level area close to the trail on which to pitch their tent. These areas tend to grow, as they are often not so resilient, in part because they are not so well drained. By creating pitches on platforms along the slope, in areas just off the trail, the experiences of all can be improved. Concentrating recreation impacts on slopes will automatically improve drainage. The importance of location is borne out in the development of the Outdoor Skills and Ethics booklets, which deal with particular regions or activities.

The programme is now reaching into 'front country' situations like 'car camp grounds', day visit areas and urban parks. Managers in Britain, and in other densely used countries, should be very grateful for this move, because they can learn a great deal from it (http://www.lnt.org).

Source: Marion and Reid (in press)

Each of us, as an individual, can develop awareness, understanding and experience; doing so will help develop knowledge and skills. This will help us to enjoy our leisure more, and to contribute to the leisure of others. So long as people pursue this course because of the added pleasure it will bring them (and not out of a sense of obligation), it should be possible to achieve much more with less. On the other hand, any sense of added obligation to others could reduce the quality of leisure and take the form of 'serious leisure' (Stebbins 2000).

Without perfect knowledge, we are bound to make mistakes and occasionally harm the environment (physical or social) without intending to do so. We are super-adaptive beings, continually learning. We may, like amoebae, move toward pleasant things and away from unpleasant ones (Campbell 1973), but we can spend plenty of time anticipating, appreciating and reflecting on that pleasant experience. Because we are super-adaptive, we do not have to carry out our first intentions. We can be flexible, and take into account all the latest information, much of which is literally at our fingertips, and think before we act. We should encourage managers to deploy lateral (and every other kind of) thinking so that we have as many different responses available as we could wish.

As a species we are part of this most wonderful natural world, and have a fascinating range of habits, skills and potentialities. Some of the habits are destructive and antisocial. The litany of war-torn areas and countries with famine would seem bizarre to a Martian. Nevertheless, our ability to organise is legendary. Leisure, recreation and tourism are not going to put to rights all the wrongs that exist, but they can be used to spread understanding of different natural and cultural environments, and crucially of different peoples. Travel broadens the mind. We have built up fantastic cultures. A few individuals have given pleasure to millions, through their poetry, their writing, their painting and their example. As managers or practitioners of recreation and leisure we owe it to ourselves, our friends, our families, to our communities and to society to take very good care to enjoy ourselves to the full. We should also strive to make it easier for others to enjoy themselves and gain the full range of benefits from leisure and recreation, for themselves, their friends and families, and for society.

Ideas for further study or work

1   Look at five old (say 1950s–1970s) textbooks, and collate their prognoses about what will be happening in the future. Compare them with what actually happened. (This will help to keep a sense of perspective! Forecasting the future was not easy then. It is more difficult now.)

2   Choosing one aspect of leisure and recreation, consider what will be the most important changes, and drivers of change, which are likely to affect the way in which managers should operate, in the short term (say the next two years) and in the longer term (say five to ten years). (Keep a note of these, and resume after the appropriate time has elapsed, for amusement.)

3   Explore different futures for our approach to work and leisure; across different cultures (choose three countries) and for different groups (at least three). Are there ways we could combine the best of all worlds?

4   With the exponential growth in use of the Internet, identify some of the advantages and disadvantages for leisure and recreation at three levels: individual, family or other social group, and society.

5   Has the West paid more attention to achievement in the realm of technology, and the East focused more on the spiritual side of life? Work through some examples in relation to leisure and recreation from history or the present day. Can we learn from this as managers of future environments for leisure and recreation? What are the main lessons, and for whom?

Reading

Future management

Management relies on the communication of information and on responding to the content and nature of that information. There is a good deal of information about. For views on what the future may hold, read Toffler, or science fiction. To understand how sensitive this may be to any one step, read Gleick. For the future of management, read Handy.

Future environments

Patterns of settlement in the past have depended greatly on the need for transport and communication. One way of looking at the future is to adopt the anthropological approach, and look at cultures which seem to be ahead (in whatever direction you wish to choose). Looking at these environments can be instructive. The magazine

Green Futures is full of articles that show some potential future environments, and tackle some of the more awkward questions about sustainability. We may yet conquer the motor car, and read Tibbalds, to look back into the past, and into the future to see how we might care better for people on foot.

Future leisure and recreation

As we have seen, many forms of leisure are old indeed, but, for an insight into some of the possible futures of leisure and recreation, read journals, newspapers and websites to augment study of the texts and references mentioned in this book. There is a useful gateway site at UK Foresight and Pearson BT's Futurologist has an interesting site. The websites are listed in the appendix. Alternatively, shut your eyes and imagine what could be. Take a sheet of paper and use some creative thinking techniques to help envisage the future. Meeting and talking to colleagues will continue to be an important catalyst for action, and is unlikely to be surpassed in terms of enjoyment.

Diagrams for
different tasks

As mentioned in the preface, many people find it important to try to visualise solutions. All sorts of different ways of making images can help in our quest to see the pattern that explains a relationship, or which helps us suggest what might happen. Here are some favourites, which help both analytical (left-brain) and creative (right-brain) approaches.

'Back of an envelope' calculations

Actually it does not have to be an envelope, but the fact is you can use any scrap of paper, at any time. Spontaneity may just help.

Brainstorming

Unleash a stream of consciousness, working alone or with others. Set down the problem, and then let the mind freewheel. Humour helps spark ideas and make crazy ideas possible, and encourages people to produce off-the-wall ideas, bringing ideas together in novel combinations or solutions. Save evaluation for another day (Buzan, Russell).

Charts

Using simple packages like Microsoft Excel, you can try out any number of different charts to try to reveal a pattern, or portray the meaning you intend. Keep it simple.

Clustering

Working on your own or with others, using stickies, gather together ideas which share something in common. Whilst there are formal procedures you could employ, it can be as simple and as complex as intuitive factor analysis. With stickies you can return after a break for further iterations until you are satisfied with the result. For a full range of techniques making use of stickies see Straker.

Flow charts

For the process problems, setting out each stage and looking at the particular process at each stage can help focus on what is important.

Fuzzy maps

Using this notation the relationship between different actors or variables can be described and given directions and values (crude or refined, see Buzan, Kosko).

Graphs

Using graphs can eloquently focus on what really matters, and so can be very helpful and (on occasion) equally dangerous (Huff).

H diagrams

This is from the armoury of those who practise participatory appraisal techniques. Use H diagrams to gather views about whether an idea is good or bad, collecting scores, and stickies with reasons as to why it is good or bad. Then, under the crossbar, ask for stickies about what would improve the situation (McPhillimy).

Illustrations

Sketches or doodles, allow the subject to relax and think at the edges of problems where the solution may sometime arise, and can help to convey the essence of an idea swiftly.

Maps

Those produced by map makers can give an objective view of the distribution of attributes. Those produced by individuals can give an idea of the personal significance of particular features of the environment (Moore).

Matrices

Two-way tables can be used to make arrays of numbers, or words in cells, and to help make simple relationships explicit.

Mind maps

These can be used to bring many ideas to bear on the one piece of paper (or other surface). Use words, lines, images of any kind, diagrams, drawings, symbols, colours, arrows, humour, anything you like, to represent the situation under analysis. Very effective for joint working, where you can pass the pen to others for them to make their contribution (Buzan, Russell).

Pie charts

A special kind of chart, which is really good at showing proportions in a way which most people can readily assimilate.

Photographs

Can be potent, either as snapshots (points in time) or as a series of moving images, video or cine.

SWOT analysis

A simple, swift and effective tool for preliminary analysis: of Strengths (internal factors on which to build); Weaknesses (internal factors to remedy, minimise or over-come); Opportunities (external, to seize); and Threats (external, but to be aware of, and to avoid or counter). Use it to look at the situation you are in, or to plan what a competitor might do.

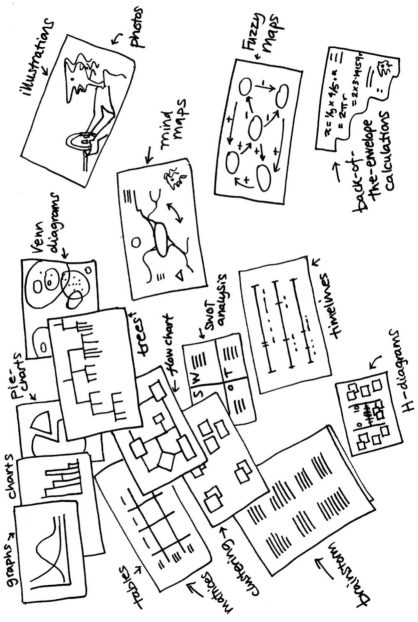

FIGURE A1 Diagrams for different tasks

Tables

Arranging ideas, whether words or figures, of a similar kind in rows and columns, to reveal a pattern.

Timelines

A simple way of putting events in order, and comparing them, e.g. events for different people, different streams or themes of events (McPhillimy).

Trees

For classifiying and exploring options this is an invaluable tool. Often used to break down a problem or issue into separate components, it can equally be used to collapse components into a larger entity, or concept. Used in combination it allows the exploration of many branches in many trees (Straker).

Venn diagrams

Simply charting out sets, to see what relationships may exist.

Web sites

This is a list of some useful, and some interesting sites. The list is necessarily selective. Using the search engines will often find you something you need, and often a good deal besides. No weighting is given to these entries, other than to give a special mention to gateway sites.

Gateways

Anything connected with government in the UK, by topic or organisation http://www.ukonline.gov.uk/
Art sites and projects in Scotland http://www.sac.org.uk/
British Library http://www.bl.uk/
Culture, media and sport in the UK http://www.culture.gov.uk/
Newspapers of all kinds http://www.newspapersoc.org.uk/
Publishers http://www.thebookseller.com/links/links.html
Universities of the UK http://www.hesa.ac.uk/links/he_inst.htm
USDA Forest Service http://www.fs.fed.us/
US National Parks Service http://www.nps.gov/

Access for all

http://www.mobility-international.org/2_2.htm
http://www.radar.org.uk/
http://www.mobility-international.org/tourisme.htm
http://www.disabilitynet.co.uk/groups/tourism/

http://www.cae.org.uk
http://www.freespace.virgin.net/hol.care
http://www.thinkmobility.com/neighbor.html

Art and cultural heritage

http://www.artscouncil.org.uk/
http://www.welfare-state.org/
http://www.artcircus.co.uk
http://www.classicfm.com/
http://www.design-council.org.uk/
http://www.culture.gov.uk/
http://www.lottery.culture.gov.uk/
http://www.arts.org.uk/
http://www.sac.org.uk/
http://www.dds.nl/~afk/beelden/english.html
http://www.acw-ccc.org.uk/
http://www.wai.org.uk/home.html
http://www.scottishmuseums.org.uk/
http://www.nms.ac.uk/
http://www.historic-scotland.gov.uk/
http://www.scran.ac.uk/
http://www.scottishscreen.com/static.taf
http://www.welfare-state.org/

Children and education

http://www.cmdenver.org/recipes4fun/index.html
http://www.spaceship-earth.com/hht.html
http://www.the-childrens-society.org.uk/

Countryside (including coast, water, forests, parks)

http://www.cf.ac.uk/crn/about/
http://rbge-sun1.rbge.org.uk/research/celtica/scot2000/
http://www.nationaltrust.org.uk/
http://www.ramblers.org.uk/
http://www.countryside.gov.uk/
http://www.maff.gov.uk/
http://www.thebigsheep.co.uk/
http://www.chycor.co.uk/tourism/cata/index.htm
http://www.wildlifetrust.org.uk/london/

http://www.entsweb.co.uk/waterways/uk/index.html
http://www.british-waterways.org/
http://www.british-waterways.co.uk/
http://www.snh.org.uk/
http://www.english-nature.org.uk/
http://enviroweb.org/greenaction/projects/c21/index.html
http://www.ontheline.org.uk
http://www.wwf-uk.org
http://www.sustainable.scotland.gov.uk
http://www.abdn.ac.uk/fef/res/mcs.htm
http://www.goodbeachguide.co.uk/
http://www.fs.fed.us/
http://www.fs.fed.us/oldpage.html
http://www.srs.fs.fed.us/trends/unit.html
http://www.srs.fs.fed.us/trends/nsre.html
http://www.fs.fed.us/intro/speech/20000627.htm
http://roadless.fs.fed.us/
http://mf.ncr.forestry.ca/
http://www.modelforest.net/DOCS/eomf_measuringe.pdf
http://www.for.gov.bc.ca/
http://www.freeourforests.org/
http://www.futureforests.com/
http://www.forestry.gov.uk
http://www.nufu.org.uk/
http://frcc.forestry.gov.uk/
http://www.nmw.ac.uk/ectf/
http://www.smy.fi/koulut/carbon/

Edinburgh festivals

http://www.edinburghfestivals.co.uk/main_set.html
http://www.eif.co.uk
http://www.edfringe.com
http://www.edbookfest.co.uk
http://www.edinburghshogmanay.org

Environmental management and recycling

EMAS http://europa.eu.int/comm/environment/emas/
Details ref to 450 laws and deteriorating environment http://europa.eu.int/comm/
   environment/emas/intro_en.htm
http://www.tc207.org/articles/index.html

http://www.hmso.gov.uk/cgi-bin/htm_hl3?URL = http://www.hmso.gov.uk/
legislation/scotland/ssi1999/99900102.htm&STEMMER = en&WORDS =
scotland + environment + impact + 9711ec + eea + town + annex + assess +
85337eec + project + &COLOUR = Red&STYLE = s#muscat_highlighter_
first_match for EIA Scotland

EAI England and Wales http://www.hmso.gov.uk/si/si1999/99029305.htm#sch1

http://www.rcep.org.uk/epissues.html

http://www.montana.com/sem

http://www.democrats.org.au/issue/enhinchinbrook.html

http://www.ieem.org.uk/

http://www.prototypecarbonfund.org/

http://www.bre.co.uk/welcomer/index.html

http://www.doingyourbit.org.uk/

http://www.wastewatch.org.uk/

http://www.emeraldawards.com/1997.html

http://www.leep.org.uk

http://www.ollierecycles.com/uk/

http://www.naturalstep.org.uk/frameset4.htm

http://www.cat.org.uk/shopping/

Funding

http://www.d-s-c.demon.co.uk/

http://sbbc.co.uk/grants/

Golf and the environment

http://members.aol.com/CBOMexico/golf.html

http://web2.airmail.net/ntgcsa1/effluent.html

http://www.asiangolf.com/designer/society1.html

Government and agencies

http://www.unep.org/

http://www.environment.detr.gov.uk/greening/ems/ce/c&epart5.pdf

http://firewall.unesco.org/

http://www.unesco.org/whc/nwhc/pages/home/pages/homepage.htm

http://firewall.unesco.org/opi/eng/unescopress/upanglo.htm

http://www.statcan.ca/

http://www.ec.gc.ca/eco/education/index_e.htm

http://www.ec.gc.ca/

http://www.agf.gov.bc.ca/

http://www.doc.govt.nz/

http://www.statistics.gov.uk/
http://www.ons.gov.uk/
http://www.mimas.ac.uk/surveys/ghs/ghs_info.html
http://www.audit-commission.gov.uk/
http://www.ukonline.gov.uk/
http://www.scottish.parliament.uk/
http://www.scottish.parliament.uk/official_report/cttee99-00/rural.htm
http://www.scotland.gov.uk/
http://www.doh.gov.uk/
http://www.culture.gov.uk/
http://www.lottery.culture.gov.uk/
http://www.parliament.the-stationery-office.co.uk/pa/cmselect/cmscotaf/711/
    9072101.htm.
http://www.roads.detr.gov.uk/roadnetwork/nrpd/heta2/nrtf97/index.htm
http://www.scotent.co.uk/
http://www.sepa.org.uk/index.htm
http://www.ccw.gov.uk/
http://www.northernireland.gov.uk/

Health

http://healthpromis.hea.org.uk/
http://www.hpe.org.uk/
http://www.haznet.org.uk/
http://www.ohn.gov.uk/ohn/ohn.htm
http://www.doh.gov.uk/
http://www.nhsdirect.nhs.uk/main.jhtml
Teacher promotional campaign on health and exercise http://www.mhie.ac.uk/HERO/
    healthed/
http://www.hebs.scot.nhs.uk/index.htm
http://somis.ais.dundee.ac.uk/physed/wellbg.htm

Inspiration

http://www.kipling.org.uk

Leave No Trace

http://www.lnt.org

Libraries

http://www.lib.ed.ac.uk/
http://copac.ac.uk/copac

Miscellaneous

http://www.commonground.org.uk/
http://www.think.no/
http://www.serm.co.uk/
http://www.bsi.org.uk/
http://plato.stanford.edu/entries/prisoner-dilemma/
http://netec.mcc.ac.uk/BibEc/data/Papers/fthbiluec98-07.html
http://www.ilam.co.uk
http://www.blra.co.uk/
http://www.runet.edu/~gsantopi/ee.html
http://www.wildsong.demon.co.uk/BA/bioacoustica.html
http://www.wsa.org.uk/wsa_report_toc_1997.htm
http://www.eea.dk/frames/main.html
http://www.bsi.org.uk/
http://www.royalmail.co.uk/paf
http://www.businesszone.co.uk/keynote/hore9701.html

Motor manufacturers

http://www.ford.com
http://www.generalmotors.com
http://www.daimlerchrysler.com

News

http://www.newspapersoc.org.uk/
http://www.independent.co.uk/
http://www.the-times.co.uk/news/pages/changing-times.html?999
http://www.ft.com/
http://www.guardian.co.uk/
http://www.scotsman.com/
http://www.met-office.gov.uk/
http://www.newscientist.co.uk/
http://www.bbc.co.uk/home/today/
http://news.bbc.co.uk/
http://www.merseyworld.com/ash/

Parks

http://www.nps.gov/
http://www.nps.gov/deto/
http://www.pps.org/urbanparks/resbib10.html
Seattle Adopt-a-park programmes http://seeh.spl.org/homepages/Adopt-a-Park.htm
Adopt-a-street programmes http://www.ci.san-jose.ca.us/prns/parkst.htm
http://www.bpfp.org/AdoptAParkGroups/Sixty-thirdStreet/
http://www.peakdistrict.org/
http://www.env.gov.bc.ca/bcparks/
http://parkscanada.pch.gc.ca/
Australia's National Parks http://www.atn.com.au/index.htm
Details of national parks across the world http://www.world-national-parks.net/
    index.html
Africa http://hyperion.advanced.org/16645/national_parks/national_parks.shtml
Trends articles http://www.prr.msu.edu/trends2000/trendsarticles.htm
http://www.doc.govt.nz/rec/recgen/issues/value.htm
http://parkscanada.pch.gc.ca/library/PC_Guiding_Principles/Park37_e.htm
http://www.worldweb.com/parkscanada-banff/mp_texte.html
http://www.worldweb.com/parkscanada-banff/mp_map_e.html

Pets

American Pets Association http://www.apapets.com/
http://www.vetweb.co.uk/scas.htm
http://netvet.wustl.edu/
Details of UK Dogs' World (7 April 2000) http://www.k9netuk.com/

Policy studies and the future

http://www.psi.org.uk/
http://www.peabody.org.uk/
http://www.forumforthefuture.org.uk/ns/nsindex.htm
http://www.labs.bt.com/people/pearsonid
http://aleph0.clarku.edu/~djoyce/julia/julia.html
http://www.newscientist.co.uk/
http://www.foresight.gov.uk/

Sport

http://www.english.sports.gov.uk/
http://www.npfa.co.uk/nsframes.htm

http://www.highwaymen.demon.co.uk/prices/new.htm
http://www.uksport.gov.uk
http://www.english.sports.gov.uk
http://www.pch.gc.ca/sportcanada/SC_E/EscF.htm
http://sportscotland.org.uk
http://www.sportsdirectory.co.uk
http://www.sports.com
http://capoeira.com
http://www.rya.org.uk

Theme parks

http://www.informatik.tumuenchen.de/~schaffnr/etc/disney
http://www.oitc.com/Disney/Paris/English/secrets.html
http://www.disneylandparis.com
http://www.seekfuture.co.uk/dlp/

Tourism

http://www.englishtourism.org.uk/who/default1.htm
http://www.tourismtrade.org.uk/pg6.htm
http://www.wttc.org
http://www.wto.org/
http://www.orbitex.ch/uluru/place/ayersrock.html
http://www.northernterritory.com/3–2.html
http://www.holiday.scotland.net/
http://www.thinkquest.org/tqfans.html
http://www.thenepaltrust.demon.co.uk
http://www.tq.com.au/research/trends/issue17/stateoverview.htm
http://www.britishairways.com/21stcentury/
http://www.aboutvenice.com/
http://www.techreview.com/articles/oct95/TrendTyson.html
http://www.mi.cnr.it/WOI/deagosti/regions/veneto2.html#Venice
http://www.staruk.org.uk/
http://www.b-mercer.demon.co.uk/links.htm
http://www.scotland-inverness.co.uk/ruareidh.htm
http://www.scottishborders.com/
http://www.holiday.scotland.net/
http://www.thisislondon.co.uk/dynamic/visitors/review.html?in_review_id = 15729
http://www.city2000.com/tl/thorpe-park.html
http://www.thorpepark.co.uk/press.html
http://musky.oitc.com/Disney/Paris/English/EuroDisneyLinks.html
http://www.traveltax.msu.edu/barometer/geneva.htm

Green tourism

http://www.greenglobe21.com
http://www.crctourism.com.au
http://www.cha-cast.com
http://www.couran-cove.com.au/
http://www.greentourism.org.uk/

Voluntary

http://www.atconf.org/
http://members.aol.com/franklogue/atmap.html
http://www.lnt.org/
http://www.vois.org.uk/
http://www.cat.org.uk/
http://www.rnli.org.uk
http://www.rspb.org.uk/
http://www.sustrans.org.uk
http://www.wwf-uk.org

Yoga

http://www.yoga.co.uk

Government departments and agencies in the UK

A selection, involved in various forms of leisure, recreation and tourism:

Arts

Arts Council of England
14 Great Peter Street
London SW1P 3NQ

Tel.: 020 7333 0100
General enquiries 020 7973 6517
Fax: 020 7973 6590
E-mail: enquiries@artscouncil.org.uk
Web: http://www.artscouncil.org.uk

Scottish Arts Council
12 Manor Place
Edinburgh EH3 7DD

Tel.: 0131 226 6051
Fax: 0131 225 9833
Help desk: 0131 240 2433
E-mail: help.desk.SAC@artsfb.org.uk
Web: http://www.sac.org.uk/

Arts Council of Wales
9 Museum Place
Cardiff
South Glamorgan CF10 3NX

Tel.: 029 2037 6500
Fax: 029 2022 1447
E-mail: information@ccc-acw.org.uk
Web: http://www.ccc-acw.org.uk

Department of Culture, Media and Sport
2–4 Cockspur Street
London SW1Y 5DH

Tel.: 020 7211 6000
E-mail: arts@culture.gov.uk
General e-mail:
enquiries@culture.gov.uk
Web: http://www.culture.gov.uk/

Heritage

English Heritage
Customer Services
NMRC
Kemble Drive
Swindon
Wiltshire SN2 2YP

Tel.: 01793 414910
E-mail: customers@ english-
    heritage.org.uk
Web: http://www.english-
    heritage.org.uk/

Historic Scotland
Longmore House
Salisbury Place
Edinburgh EH9 1SH

Tel.: 0131 668 8887
Fax: 0131 668 8789
Web: http://www.historic-
    scotland.gov.uk/

Cadw Welsh Historic Monuments
Crown Building
Cathays Park
Cardiff
South Glamorgan CF10 8NQ

Tel.: 029 2050 0200
Fax: 029 2082 6375
E-mail: cadw@wales.gsi.gov.uk
Web: http://www.cadw.wales.gov.uk/

Countryside

British Waterways
Willow Grange
Church Road
Watford WD1 3QA

Tel.: 01923 226422
E-mail: info@canalshq.demon.uk
Web: http://www.British-Waterways.org/

Countryside Agency
John Dower House
Crescent Place
Cheltenham
Gloucestershire GL50 3RA

Tel.: 01242 521381
Fax: 01242 584270
E-mail: info@countryside.gov.uk
Web: http://www.countryside.gov.uk

Countryside Council for Wales
Plas Penrhos
Fford Penrhos
Bangor
Gwynedd LL57 2LQ

Tel.: 01248 370444
Fax: 01248 355782
E-mail: n.sanpher@ccw.gov.uk
Web: http://www.ccw.gov.uk/

English Nature
Northminster House
Peterborough
Cambridgeshire PE1 1UA

Tel.: 01733 455000
Fax: 01733 568834
E-mail: enquires@english-nature.org.uk
Web: http://www.english-nature.gov.uk

Environment Agency
Rio House
Waterside Drive
Aztec West
Almondsbury
Bristol BS32 4UD

Tel.: 01454 624400
Fax: 01454 624409
E-mail: enquiries@environment-
    agency.gov.uk
Web: http://www.environment-
    agency.gov.uk/

Information and Education
Environment and Heritage Service
Commonwealth House
Castle Street
Belfast BT1 1GU

Tel.: 028 9025 1477
Fax: 028 9054 6660
E-mail: mbr@doeni.gov.uk
Web: http://www.ehsni.gov.uk/

Forestry Commission
231 Corstorphine Road
Edinburgh EH12 7AT

Tel.: 0131 334 0303
Fax: 0131 334 4473
E-mail: enquiries@forestry.gsi.gov.uk
Web: http://www.forestry.gov.uk

Scottish Natural Heritage
12 Hope Terrace
Edinburgh EH9 2AS

Tel.: 0131 447 4784
Fax: 0131 446 2277
E-mail: enquiries@snh.gov.uk
Web: http://www.snh.gov.uk/

Sport

Sport England
16 Upper Woburn Place
London WC1H 0QP

Tel.: 020 7273 1500
Fax: 020 7383 5740
E-mail: info@english.sports.gov.uk
Web: http://www.english.sports.gov.uk/

sportscotland
Caledonia House
South Gyle
Edinburgh EH12 9DQ

Tel.: 0131 317 7200
Fax: 0131 339 5361
E-mail: gen.info@sportscotland.org.uk
Web: http://www.sportscotland.org.uk/

Sports Council for Wales
Welsh Institute for Sport
Sophia Gardens
Cardiff
South Glamorgan CF1 9SW

Tel.: 02920 300500
Fax: 02920 300600
E-mail: scw@scw.co.uk
Web: http://www.sports-council-
   wales.co.uk/

Tourism

British Tourism Authority
Thames Tower
Black's Road
Hammersmith
London W6 9EL

Tel.: 020 8563 3186
Fax: 020 8563 3352
Web: http://www.visitbritain.com/
Trade enquiries:
   tradehelpdesk@bta.org.uk
Trade web site: http://www.
tourismtrade.org.uk/scripts/
   index.asp/

English Tourism Council
Thames Tower
Black's Road
Hammersmith
London W6 9EL

Tel.: 020 8563 3000
Fax: 020 8563 0302
Web: http://www.
   englishtourism.org.uk

visitscotland (Scottish Tourist Board)
23 Ravelston Terrace
Edinburgh EH4 3EU

Tel.: 0131 332 2433
Fax: 0131 343 1513
E-mail: info@stb.gov.uk
Web: http://www.stb.gov.uk

Wales Tourism Board
Brunel House
2 Fitzalan Road
Cardiff
South Glamorgan CF24 0UY

Tel.: 029 2049 9909
Fax: 029 2048 5031
E-mail: info@tourism.wales.gov.uk
Web: http://www.visitwales.com/

Bibliography

Abraham, G. (1979) *The Concise Oxford History of Music*, London: Oxford University Press.

Access Forum (1996) *Scotland's Hills and Mountains: a concordat on access*, Edinburgh: Scottish Natural Heritage.

Adams, J. (1994) 'Putting alchemy to work . . . for open space?' in Wood, R. (ed.) *Environmental Economics, Sustainable Management and the Countryside*, Cardiff: Countryside Recreation Network.

—— (2000) 'Hypermobility: too much of a good thing', *Countryside Recreation*, 8, 1: 5–7.

Adaptive Environments Center and Barrier Free Environments (1997) 'The Americans with Disabilities Act checklist for readily achievable barrier removal', in Bullock, C.C. and Mahon, M.J. (eds) *Introduction to Recreation Services for People with Disabilities: a person centered approach*, Champaign, IL: Sagamore, 439–70.

Agee, J.K. and Johnson, D.K. (eds) (1988) *Ecosystem Management for Parks and Wilderness*, London: University of Washington Press.

Agyeman, J. (ed.) (1996) *Involving Communities in Forestry . . . through Community Participation*, Forestry Practice Guide 10, Edinburgh: Forestry Commission.

Aitken, R. (1984) *Scottish Mountain Footpaths: a reconnaissance review of their condition*, Perth: Countryside Commission for Scotland.

Ajzen, I. (1991) 'Benefits of leisure: a social psychological perspective' in Driver, B.L., Brown, P.J. and Peterson, G.L. (eds) *Benefits of Leisure*, State College, PA: Venture Publishing, 411–17.

Akama, J.S. (1998) 'The evolution of tourism in Kenya', *Journal of Sustainable Tourism*, 7, 1: 6–25.

Alder, J. (1996) 'Costs and effectiveness of education and enforcement, Cairns section of the Great Barrier Reef Marine Park', *Environmental Management*, 20, 4: 541–51.

Aldridge, D. (1975) *Guide to Countryside Interpretation* I, *Principles of Interpretation and Interpretative Planning*, Edinburgh: HMSO.

Alexander, K. (ed.) (1996) *Facilities Management: theory and practice*, London: Spon.

Allen, N. (1993) 'Caring for guests at a major leisure venue in the UK', in Wood, R. (ed.) *Customer Care in the Countryside*, proceedings of the 1993 Countryside Recreation Conference, Cardiff: Countryside Recreation Network, 8–14.

Altman, I. and Wohlwill, J.F. (eds) (1978) *Children and the Environment* (Human Behaviour and Environment III), New York: Plenum.

Altman, Y. (1995) 'A theme park in a cultural straitjacket: the case of Disneyland Paris', *Managing Leisure*, 1: 43–56.

Anderson, J.R. (2000) *Cognitive Psychology and its Implications*, fifth edition. New York: Worth.

Angus Rural Partnership and Angus Council Community Education Service (1999) *Sustainable Newtyle: a working paper describing the progress of a collaborative project to develop indicators of community sustainability in an Angus village*, Newtyle: Angus Rural Partnership and Angus Council Community Education Service.

Anton, A.E. (1991) *Rights of Way: a guide to the law in Scotland*, revised edition, Edinburgh: Scottish Rights of Way Society.

Appleton, J. (1975) *The Experience of Landscape*, London: Wiley.

Argyle, M. (1996) *The Social Psychology of Leisure*, London: Penguin.

—— (1987) *The Psychology of Happiness*, London: Methuen.

Armstrong, D.M. (1962) *Bodily Sensations*, London: Routledge.

Arnheim, R. (ed.) (1967) *Towards a Psychology of Art*, London: Faber.

Arnold, N.D., (1976) *The Interrelated Arts in Leisure: perceiving and creating*, St Louis, MO: Mosby.

ASH Environmental Design Partnership (1986) *Environmental Design and Management of Ski Areas in Scotland: a practical handbook*, Perth: Countryside Commission for Scotland.

Ashley, C.W. (1947) *The Ashley Book of Knots*, London: Faber and Faber.

Audit Commission (1999) *The Price is Right?*, London: Audit Commission. Online http:// www.audit-commission.gov.uk/ (24 September 2000).

Augustyn, M. (1998) 'National strategies for rural tourism development and sustainability', *Journal of Sustainable Tourism*, 6, 3: 191–209.

Badman, T. (1996) 'Regional strategies for coastal recreation', in Goodhead, T. and Johnson, D. (eds) *Coastal Recreation Management: the sustainable development of maritime leisure*, London: Spon, 252.

Bailey, J. (1997) 'Environmental impact assessment and management: an underexplored relationship', *Environmental Management*, 21, 3: 317–27.

Bailey, R. (ed.) (1994) *France*, London: Dorling Kindersley.

Bailey, S.J. (1995) *Public Sector Economics: theory, policy and practice*, Basingstoke and London: Macmillan.

Baines, C. (1985) *How to Make a Wildlife Garden*, London: Elm Tree Books.

Baines, J. (1995) *Beyond Compromise: building consensus in environmental planning and decision making*, London: Environment Council.

Baker, D.A. and Crompton, J.L. (2000) 'Quality, satisfaction and behavioural intentions', *Annals of Tourism Research*, 27, 3: 785–804.

Ball, D. (1995) 'In search of a balanced approach to visitor safety', *Countryside Recreation Network News*, 3, 3: 4–6.

—— (1998) 'Leisure walking and health', *Countryside Recreation*, 6, 2: 4–6.

Ballinger, R. (1996) 'Looking ashore: a comparison with inland waters and waterways', in Goodhead, T. and Johnson, D. (eds) *Coastal Recreation Management: the sustainable development of maritime leisure*, London: Spon, 225–45.

Bammel, G. and Burrus-Bammel, L.L. (1996) *Leisure and Human Behaviour*, third edition, Dubuque, IA: Brown and Benchmark.

Bannock, G., Baxter, R. and Rees, R. (1984) *The Penguin Dictionary of Economics*, Harmonds-worth: Penguin.

Barrow, C.J. (1999) *Environmental Management: principles and practice*, London: Routledge.

Bateman, I. (1992) 'The United Kingdom', in Wibe, S. and Jones, T. (eds) *Forests: Market and Intervention Failures*, London: Earthscan, 10–57.

Bateman, I. and Bryan, F. (1994) 'Recent advances in monetary evaluation of environmental preferences', in Wood, R. (ed.) *Environmental Economics: sustainable management and the countryside*, Cardiff: Countryside Recreation Network.

Bateman, I., Lovett, A., and Brainard, J. (1999) 'Developing a methodology for benefit transfers using geographical information systems: modelling demand for woodland recreation', *Regional Studies*, 33, 3: 191–205.

Bateman, I., Garrod, G., Brainard, J. and Lovett, A. (1996) 'Measurement, valuation and estimation issues in travel cost method: a geographical information systems approach', *Journal of Agricultural Economics*, 47, 2: 191–205.

Bauby, J-D. (1998) *The Diving-bell and the Butterfly*, London: Fourth Estate.

Bayfield, N.G. and Aitken, R. (1992) *Managing the Impacts of Recreation on Vegetation and Soils: a review of techniques*, report to the Countryside Commission, Countryside Com-mission for Scotland, English Nature and Countryside Council for Wales, Banchory: Institute of Terrestrial Ecology.

Bayfield, N.G. and McGowan, G.C. (1995) 'Monitoring and managing the impacts of ski development: a case study of Aonach Mor resort 1989–1995', in Griffiths, G.H. (ed.) *Landscape Ecology Theory and Applications*, proceedings of fourth IALE conference at Reading, Aberdeen: IALE.

Baynes, K. (1975) *Art in Society*, London: Lund Humphries.

Beard, J.G. and Ragheb, M.G. (1983) 'Measuring leisure motivation', *Journal of Leisure Research*, 15, 2: 19–28.

Bee, F. and Bee, R. (1999) *Customer Care*, London: Institute of Personnel and Development.

Beeho, A. and Prentice, R. (1996) 'Understanding visitor experiences as a basis for product development: ASEB grid analysis and the Black Country Museum in the West Midlands of England', in Harrison, L.C. and Husbands, W. (eds) *Practising Responsible Tourism: international case studies in tourism planning, policy and development*, Chichester: Wiley, 472–94.

Bell, S. (1997) *Design for Outdoor Recreation*, London: Spon.

Bell, S. and Morse, S. (1999) *Sustainability Indicators: measuring the immeasurable?* London: Earthscan.

Belnap, J. (1998) 'Choosing indicators of natural resource condition: a case study in Arches National Park, Utah, USA', *Environmental Management*, 22, 4: 635–42.

Benington, J. and White, J. (eds) (1988) *The Future of Leisure Services*, Harlow: Longman.

Bennett, R. and Tranter, R. (1997) 'Assessing the benefits of public access to the countryside', *Planning Practice and Research*, 12, 3: 213–22.

Benson, J.F. and Willis, K.G. (1992) *Valuing Informal Recreation on the Forestry Commission Estate*, Forestry Commission Bulletin 104, London: HMSO.

Bergler, E. (1974) 'The psychology of gambling', in Haliday, J. and Fuller, P. (eds) *The Psychology of Gambling*, London: Allen Lane.

Beskine, D. (ed.) (1993) *Public Access to Woodlands sold off by the Forestry Commission*, London: Ramblers' Association.

Binks, G., Dyke, J. and Dagnall, P. (1988) *Visitors Welcome: a manual on the presentation and interpretation of archaeological excavations*, London: HMSO.

Bird, S., Smith, A. and James, K. (1998) *Exercise: Benefits and Prescriptions*, Cheltenham: Thornes.

Bishop, J. and Rose, J. (1994) 'Community action – an overview', in Etchell, C. (ed.) *Communities in their Countryside*, proceedings of a conference held at the University of York, 1994, by Countryside Recreation Network, Cardiff: Countryside Recreation Network, 14–29.

Bishop, K. (1992a) 'Assessing the benefits of community forests: an evaluation of the recreational use benefits of two urban fringe woodlands', *Journal of Environmental Planning and Management*, 35, 1: 63–77.

—— (ed.) (1992b) *Off the Beaten Track: access to open land in the UK*, proceedings of 1992 Countryside Recreation Conference Cardiff: Countryside Recreation Research Advisory Group.

Black, J. (1996) *Mindstore for Personal Development*, London: Thorson.

Blamey, E. (ed.) (1998) *GIS and Countryside Management: theory and application*, proceedings of a workshop held at Fairbairn House, University of Leeds, Cardiff: Countryside Recreation Network.

Blanshard, B. (1961) 'The case for determinism', in Hook, S. (ed.) *Determinism and Freedom in the Age of Modern Science*, London: Collier Macmillan.

Blunden, J. and Turner, G. (1985) *Critical Countryside*, London: BBC.

Bond, M. (1996) 'Plagued by noise', *New Scientist*, 16 November. Online http://www.newscientist.co.uk/ns/970510/noise2.html (24 September 2000).

Boni, M.B. (1947), *Fireside Book of Folk Songs*, New York: Simon & Schuster.

Booth J. (1998) 'Filthy lucre', *EdiT*, 14, 8–11, Edinburgh: University of Edinburgh.

Borland, E.M. (1999) 'Laughter can be the best medicine', *Scotsman*, 25 September, Weekend, p. 6, Edinburgh.

Bovaird, T. (1994) 'Monitoring and review of the effectiveness of programmes', in Etchell, C. (ed.) (1994) *Communities in their Countryside*, proceedings of a conference held at the University of York, 1994, by Countryside Recreation Network, Cardiff: Countryside Recreation Network, 71–88.

Boyle, D. (2000) 'Why can't we live together?' *Green Futures*, 22: 32–4.

Bracey, H.E. (1970) *People and the Countryside*, London: Routledge & Kegan Paul.

Bradshaw, J.W.S. and Brown, S.L. (1990) 'Behavioural adaptation of dogs to domestication', in Burger, I.H. (ed.) *Pets: Benefits and practice*, Waltham Symposium 20, London: British Veterinary Association. 18–24.

Brainard, J., Bateman, I. and Lovett, A. (1997) 'Using isochrone surfaces in travel cost models', *Journal of Transport Geography*, 5, 5: 117–26.

Brainard, J., Lovett, A., and Bateman, I. (1999) 'Integrating geographical information systems into travel costs analysis and benefit transfer', *International Journal of Geographical Information Systems*, 13, 3: 227–46.

Bramham P., Henry I., Momaas, H. and van der Poel, H. (eds) (1989) *Leisure and Urban Processes: critical studies of leisure policy in West European cities*, London: Routledge.

Bramwell, B. (1997) 'Strategic planning before and after a mega-event', *Tourism Management*, 18, 3: 167–76.

Bramwell, B. and Sharman, A. (1999) 'Collaboration in local tourism policymaking', *Annals of Tourism Research*, 26, 2: 392–415.

Brent Ritchie, J.R. (1998) 'Managing the human presence in ecologically sensitive tourist destinations: insights from the Banff-Bow Valley study', *Journal of Sustainable Tourism*, 6, 4: 293–313.

Brewers' & Licensed Retailers Association (1999) Press release. Online http://www.blra.co.uk/newpages/sec6newp/fpress57.html (5 March 1999).

British Balloon and Airship Club (1996) *Code of Conduct for Pilots and Farmers*, second edition, issued annually in April for inclusion in the BBAC Handbook.

British Helicopter Advisory Board (1999) *Guidelines*, Online http://www.bhab.demon.co.uk/contact.htm (24 September 2000).

British Mountaineering Council (1996) *Bird Nesting Restrictions 1995/6: information for climbers 1996 update*, Manchester: British Mountaineering Council in co-operation with the Royal Society for the Protection of Birds.

British Museum (1968) *Flint Implements: an account of stone-age techniques and cultures*, third edition, London: British Museum, 82–103.

Broadhurst, R. (1987) 'Forest recreation: orienteering in woodlands and forests', in Talbot-Ponsonby, H. (ed.) *Recreation and Wildlife: working in partnership*, proceedings of the Annual Countryside Recreation Conference, Bristol: Countryside Recreation Research Advisory Group, 80–3.

—— (1989) 'The search for new funds', in Uzzell, D. (ed.) *Heritage Interpretation* II, *The Visitor Experience*, London: Belhaven Press, 29–43.

—— (1997) 'People, trees and woods across Europe', in ILO, *People, Forests and Sustainability: social elements of sustainable forest management in Europe*, Geneva: International Labour Organisation, 53–66.

Broadhurst, R.T. and Harrop, P. (2000) 'Forest tourism: putting policy into practice in the Forestry Commission', in Font, X. and Tribe, J. (eds) *Forest Tourism and Recreation: case studies in environmental management*, Wallingford: CABI Publishing, 183–99.

Bromley, P. (1990) *Countryside Management*, London: Spon.

—— (1994) *Countryside Recreation: a handbook for managers*, London: Spon.

Broom, G. (1992) 'The context', in Talbot, H. (ed.) *Our Priceless Countryside: should it be priced?* Bristol: Countryside Recreation Research Advisory Group, 21–33.

Brown, D. (1970) *Bury my Heart at Wounded Knee: an Indian history of the American west*, London: Vintage.

Brown, D.O. (1998) 'Debt funded environmental swaps in Africa: vehicles for tourism development', *Journal of Sustainable Tourism*, 6,1: 60–79.

Brown, P. and Smyth, J. (1988) *The Guide to UK Company Giving*, London: Directory of Social Change.

Brunt, P. and Courtney, P. (1999) 'Host perceptions of sociocultural impacts', *Annals of Tourism Research*, 26, 3: 493–515.

Buchanan C. (1964) *Traffic in Towns: the specially shortened version of the Buchanan Report*, Harmondsworth: Penguin.

Buckley, R. (1999) 'Tools and indicators for managing tourism in parks', *Annals of Tourism Research*, 26,1: 207–10.

Bull, S.J., Albinson, J.G. and Shambrook, C.J. (1996) *The Mental Game Plan: getting psyched for sport*, Eastbourne: Sports Dynamics.

Bullock, C.C. and Mahon, M.J. (1997) *Introduction to Recreation Services for People with Disabilities: a person centered approach*, Champaign, IL: Sagamore.

Burch, Jr, W.R. and Hamilton-Smith, E. (1991) 'Mapping a new frontier: identifying, measuring, and valuing social cohesion benefits related to nonwork opportunities and activities', in Driver, B.L., Brown, P.J. and Peterson, G.L. (eds) *Benefits of Leisure*, State College, PA: Venture Publishing, 369–82.

Burdge, R., Fricke, P., Finsterbusch, K., Freudenburg, W.R., Gramling, R., Llewellyn, L., Petterson, J.S., Thompson, J. and Williams, G. (1993) *The Interorganizational Committee (1994) Guidelines and Principles for Social Impact Assessment*, NOAA Technical Memorandum NMFS-F/SPO-16, Washington, DC: US Department of Commerce.

Burger, I.H. (1990) 'Concluding remarks', in Burger, I.H. (ed.) *Pets: Benefits and Practice*, Waltham Symposium 20, London: British Veterinary Association.

Burger, J. (1998) 'Attitudes about recreation, environmental problems, and estuarine health along the New Jersey shore, USA', *Environmental Management*, 22, 6: 869–76.

Burgess, J. (1995) *Growing with Confidence: understanding people's perceptions of urban fringe woodlands*, Manchester: Countryside Commission.

Burke, V. and Collins, D. (1998) 'The great outdoors and management development: a framework for analysing the learning and transfer of management skills', *Managing Leisure*, 3: 136–48.

Burstyn, V. (1999) *The Rites of Men: manhood, politics, and the culture of sport*, Toronto: University of Toronto Press.

Butler, R.W. (1999) 'Understanding tourism', in Jackson, E.L. and Burton, T.L. (eds) *Leisure Studies: prospects for the twenty-first century*, State College PA: Venture Publishing, 97–116.

Buttle, F. (1993) 'Quality management: theories and themes', in Wood, R. (ed.) *Customer Care in the Countryside: a practical review of the techniques to meet customer needs and expectations in the countryside*, proceedings of the 1993 Countryside Recreation Conference, Cardiff: Countryside Recreation Network, 39–57.

Buzan, T. (1974) *Use your Head*, London: BBC.

—— (1989a) *Use your Memory*, revised edition, London: BBC.

—— (1989b) *Master your Memory*, revised edition, Newton Abbot: David & Charles.

Buzan, T. and Buzan, B. (1993) *The Mind Map Book: radiant thinking – the major evolution in human thought*, London: BBC.

Bygren, L.O., Konlaan, B.B. and Johansson, S-E, (1996) 'Unequal in death: attendance at cultural events, reading books or periodicals, and making music or singing in a choir as determinants for survival: Swedish interview survey of living conditions', *British*

Medical Journal, 313, 7072, London: BMJ Publishing. Online http://www.bmj.com/archive/7072ud2.htm (24 September 2000).

Caalders, J. (1997) 'Managing the transition from agriculture to tourism: analysis of networks in Auvergne', *Managing Leisure*, 2: 127–42.

Caldwell, L.L. and Darling, N. (1999) 'Leisure context, parental control, and resistance to peer pressure as predictors of adolescent partying and substance use: an ecological perspective', *Journal of Leisure Research*, 31, 1: 57–77.

Campbell, H.J. (1973) *The Pleasure Areas*, London: Eyre Methuen.

Campbell, J. (2000) 'Campaign against back pain', *BBC News*, 2 February. Online http://bbc.news.co.uk (2 February 2000).

Canada Parks/Recreation Association (1999) *The Benefits Catalogue: summarizing why recreation, sport, fitness, arts, culture, and parks are essential to personal, social, economic and environmental well-being*, Gloucester, Ont.: Canada Parks/Recreation Association.

Cardonnery, L. (1999) 'Implementing the Protocol on Environmental Protection to the Antarctic Treaty: future application of geographic information systems within the Committee for Environmental Protection', *Journal of Environmental Management*, 56: 285–98.

Carley, M. (1990) *Housing and Neighbourhood Renewal: Britain's new urban challenge*, London: Policy Studies Institute.

Carlson, R. (1999) *The Don't Sweat the Small Stuff Workbook*, London: Hodder & Stoughton.

Carlson, R. and Bailey, J. (1998) *Slowing Down to the Speed of Life*, London: Hodder & Stoughton.

Carmichael, A. (1972) *Carmina Gadelica: hymns and incantations with illustrative notes on words, rites, and customs, dying and obsolete/orally collected in the Highlands and islands of Scotland and translated into English by Alexander Carmichael*, second edition, two volumes, Edinburgh: Scottish Academic Press.

Carr, D.S., Selin, S.W. and Schuette, M.A. (1998) 'Managing public forests: understanding the role of collaborative planning', *Environmental Management*, 22, 5: 767–76.

Carson, R.T., Flores, N.E., Martin, K.M. and Wright, J.L. (1996) ' Contingent valuation and revealed preference methodologies: comparing the estimates for quasi-public goods', *Land Economics*, 72, 1: 80–99.

Carter, J. and Masters, D. (1998) *Arts and the Natural Heritage*, Scottish Natural Heritage Review 109, Perth: Scottish Natural Heritage.

Cashmore, E. (1990) *Making Sense of Sport*, London: Routledge.

Chairmen's Policy Group (1983) *Leisure Policy for the Future*, London: Sports Council.

Chambers, R. (1983) *Rural Development: putting the last first*, Harlow: Longman.

—— (1992) 'Rural appraisal: rapid, relaxed and participatory', Discussion Paper 311, Brighton: Institute of Development Studies, University of Sussex.

Chambers, T.W.M. and Price, C. (1986) 'Recreation congestion: some hypotheses tested in the Forest of Dean', *Journal of Rural Studies*, 2, 1: 41–52.

Chapman, A.J. and Foot, H.C. (eds) (1996) *Humor and Laughter: theory, research and application*, New Brunswick, NJ: Transaction.

Chebakova, I.V. (1997) *National Parks of Russia: a guidebook*, Moscow: Biodiversity Conservation Centre.

Cheers, A. and Sampson, A. (1991) *The Leisure Environment*, London: Macmillan.

Children of the world, in association with the United Nations (1994) *Rescue Mission Planet Earth: a children's edition of Agenda 21*, London: Kingfisher Books.

Chippendale, C. and Taçon, P.S.C. (1998) *The Archaeology of Rock Art*, Cambridge: Cambridge University Press.

Christiansen, M.L. (1983) *Vandalism Control Management for Parks and Recreation Areas*, London: Spon.

Chubb, M. and Chubb, H.R. (1981) *One-third of our Time? An introduction to recreation behavior and resources*, Chichester: Wiley.

Clark, G., Darrall, J., Grove-White, R., Macnaghten, P. and Urry, J. (1994) *Leisure Landscapes, Leisure, Culture and the English Countryside: challenges and conflicts*, London: Council for the Protection of Rural England.

Clark, R.N. and Stankey, G.H. (1979) *The Recreation Opportunity Spectrum: a framework for planning, management and research*, Washington, DC: General Technical Report PNW-98, US Department of Agriculture, Forest Service.

Clarke, S. and Norton, M. (1997) *The Complete Fundraising Handbook*, third edition, London: Directory of Social Change in association with ICFM.

Clawson, M. and Knetsch, J.L. (1971) *Economics of Outdoor Recreation*, paperback edition, Baltimore, MD: Johns Hopkins University Press.

Cloke, P. (1993) 'The countryside as commodity: new spaces for rural leisure', in Glyptis, S. (ed.), *Leisure and the Environment: essays in honour of Professor J.A. Patmore*, London: Belhaven Press, 53–67.

Coalter, F. (1988) *Sport and Anti-social Behaviour: a literature review*, Scottish Sports Council Research Report 2, Edinburgh: Scottish Sports Council.

—— (1999) 'Sport and recreation in the United Kingdom: flow with the flow, or buck the trends?', *Managing Leisure*, 4: 24–39.

Coalter, F., Dowers, S. and Baxter, M. (1995) 'The impact of social class and education on sports participation: some evidence from the General Household Survey', in Roberts, K. (ed.) *Leisure and Social Stratification*, Brighton: Leisure Studies Association, 59–73.

Coalter, F., MacGregor, C. and Denman, R. (1996) *Visitors to National Parks: summary report of the 1994 survey carried out by the Centre for Leisure Research and JMP Consultants, carried out for the Countryside Commission, Countryside Council for Wales and the 21 other organisations who funded the project*, Cheltenham: Countryside Commission.

Codling, R.J. (1995) 'The precursors of tourism in the Antarctic', in Hall, C.M. and Johnston, M.E. (eds) *Polar Tourism: tourism in the Arctic and Antarctic regions*, Chichester: Wiley, 167–77.

Cohen, E. (1991) 'Leisure: the last resort: a comment', in Driver, B.L., Brown, P.J. and Peterson, G.L. (eds) *Benefits of Leisure*, State College, PA: Venture Publishing, 439–44.

—— (1992) 'Tourist arts', in Cooper, C.P. and Lockwood, A. (eds) *Progress in Tourism, Recreation and Hospitality Management* IV, London: Belhaven Press, 3–32.

Cohen, J.M. and Cohen, M.J. (1960) *The Penguin Dictionary of Quotations*, Harmondsworth: Penguin.

Cole, D.N. (1987) 'Research on soil and vegetation in wilderness: a state-of-knowledge review', in Lucas, R.C. (comp.) *Proceedings – National Wilderness Research Conference: Issues, State-of-knowledge, Future Directions; Fort Collins, CO*, General technical report INT-220, Ogden, UT: USDA Forest Service, Intermountain Research Station, 135–77.

Cole, S. (1965) *Races of Man*, London: British Museum (Natural History), 82–4.

Coleman, J. and Hendry, L. (1998) *The Nature of Adolescence*, third edition, London: Routledge.

Coles, J. (1973) *Archaeology by Experiment*, London: Hutchinson, 158–67.

Collins, M.F. and Kennett, C. (1999) 'Leisure, poverty and social inclusion: the growing role of passports to leisure in Britain', *European Journal of Sport Management*, 6,1: 19–29.

Cooke, A. (1994) *The Economics of Leisure and Sport*, London: Routledge.

Coopers & Lybrand Associates (1981) *Service Provision and Pricing in Local Government*, London: HMSO.

Cope, A., Doxford, D.E. and Millar, G. (1999), 'Counting users of informal recreation facilities', *Managing Leisure*, 4: 229–44.

Cope, D.R. and Sisman, K. (1994) *A Life Cycle Analysis of a Holiday Destination: Seychelles*, British Airways Environment Report 41/94, Cambridge: UK CEED.

Coppock, J.T. and Duffield, B.S. (1975) *Recreation in the Countryside: a spatial analysis*, London: Macmillan.

Cornwall Association of Tourist Attractions (2001) Online http://www.cata.co.uk/cata-guest/index.htm.

Cotte, J. (1997) 'Chances, trances and lots of slots: gambling motives and consumption experiences', *Journal of Leisure Research*, 29, 4: 380–406.

Cotton, N. and Grimshaw, J. (2001) *The Official Guide to the National Cycle Network*, Bristol: Sustrans.

Country Landowners' Association, National Farmers' Union, and Countryside Commission (1994) *Managing Public Access: a guide for farmers and landowners*, Cheltenham: Countryside Commission.

Countryside Commission (1992) *Out in the Country: where you can go and what you can do*, Cheltenham: Countryside Commission.

—— (1994) *Informal Countryside Recreation for Disabled People: a practical guide for countryside managers*, Cheltenham: Countryside Commission.

—— (1995a) 'The Market for Recreational Cycling in the Countryside', unpublished report, Countryside Commission.

—— (1995b) *Sustainable Rural Tourism: opportunities for local action*, Cheltenham: Countryside Commission.

—— (1995c) *Principles of Sustainable Rural Tourism: opportunities for local action*, Cheltenham: Countryside Commission.

—— (1998) *Greenways*, Research Notes Issue CCRN4, Cheltenham: The Countryside Agency.

Countryside Commission for Scotland (1974) *A Park System for Scotland*, Perth: Countryside Commission for Scotland.

Countryside Council for Wales (1995) *Access to the Welsh Countryside: a draft document for consultation*, Bangor: Countryside Council for Wales.

Countryside Group (1991) *Final Report to the Tourism and Environment Task Force*, London: Department of Employment.

Countryside Recreation Network (2000) *Countryside Recreation*, Cardiff: Countryside Recreation Network.

Cousins, N. (1979) *Anatomy of an Illness as Perceived by the Patient*, London: Norton.

Cramb, A. (1996) *Who owns Scotland now?* Edinburgh: Mainstream.

Crandall, R. (1980) 'Motivations for leisure', *Journal of Leisure Research*, 12: 45–54.

Cranz, G. (1998) *The Chair: rethinking culture, body and design*, London: Norton.

Critcher, C., Bramham, P. and Tomlinson, A. (eds) (1994) *Sociology of Leisure: a reader*, London: Spon.

Crompton, J.L. (1982) 'Psychological dimensions of pricing leisure services', *Recreation Research Review*, October: 12–20.

—— (1987) *Doing More with Less in the Delivery of Parks and Recreation Services: a book of case studies*, State College, PA: Venture Publishing.

—— (1999) 'Beyond grants: other roles for foundations in facilitating delivery of park and recreation services in the US', *Managing Leisure*, 4: 1–23.

Crompton, J.L. and McKay, S.L. (1997) 'Motives of visitors attending festival events', *Annals of Tourism Research*, 24, 2: 425–39.

Cross, G. (1990) *A Social History of Leisure since 1660*, State College, PA: Venture Publishing.

Crystal, D. (1995) *The Cambridge Encyclopaedia of the English Language*, Cambridge and London: Cambridge University Press.

Csikszentmihalyi, M. (1975) *Beyond Boredom and Anxiety: the experience of play in work and games*, London: Jossey-Bass.

—— (1992) *Flow: the Psychology of Happiness*, London: Rider.

Cummings, L.E. (1997) 'Waste minimisation supporting urban tourism sustainability: a mega-resort case study', *Journal of Sustainable Tourism*, 5, 2: 93–108.

Curry, N. (1994) *Countryside Recreation: access and land use planning*, London: Spon.

—— (2000) 'Community participation in outdoor recreation and the development of Millennium Greens in England', *Leisure Studies*, 19: 17–35.

Curry, N. and Pack, C. (1993) 'Planning on presumption: strategic planning for countryside recreation in England and Wales', *Land Use Policy*, 10: 140–50.

Curtis, J., McTeer, W. and White, P. (1999) 'Exploring effects of school sport experiences on sports participation in later life', *Sociology of Sport Journal*, 16, 4: 348–65.

Cushman, G., Veal, A.J. and Zuzanek, J. (1996) *World Leisure Participation: free time in the global village*, Wallingford: CAB International.

Dahles, H. and Bras, K. (1999) 'Entrepreneurs in romance: tourism in Indonesia', *Annals of Tourism Research*, 26, 2: 267–93.

HH Dalai Lama and Cutler, H. (1998) *The Art of Happiness*, London: Hodder & Stoughton.

Daley, H.G. and Cobb, J.B. (1989) *For the Common Good: redirecting the economy toward the community, the environment and a sustainable future*, Boston: Beacon Press.

Davies, P. and Knipe, T. (1984) *A Sense of Place: sculpture in the landscape*, Sunderland: Ceolfrith Press.

Davis, K.A. (1994) *Sport Management: successful private sector business strategies*, Dubuque, IA: Brown & Benchmark.

Deadman, P.J. (1999) 'Modelling individual behaviour and group performance in intelligent agent-based simulation of the tragedy of the commons', *Journal of Environmental Management*, 56: 159–72.

de Bono, E. (1983) *Atlas of Management Thinking*, Harmondsworth: Penguin.

—— (1990a) *Six Thinking Hats*, Harmondsworth: Penguin.

—— (1990b) *The Happiness Purpose*, Harmondsworth: Penguin.

—— (1994) *Water Logic*, Harmondsworth: Penguin.

de Lange, P. (1998) *The Games Cities Play*, Monument Park, South Africa: de Lange.

de Santillana, G. (1956) *The Age of Adventure: the Renaissance philosophers*, London: New English Library.

Deem, R. (1995) 'Time for a change? Engendered work and leisure in the 1990s', in McFee, G., Murphy, W. and Whannel, G. (eds) *Leisure Cultures: values, genders, lifestyles*, LSA Publication 56, Eastbourne: Leisure Studies Association, 3–22.

Denker, K. (2000) 'Smoke clearing', *Green Futures*, 20: 22–4.

Denman, R. (1994) *Delivering Countryside Information: a good practice guide for promoting enjoyment of the countryside*, Cheltenham: Countryside Commission.

—— (n.d.) *Tourism in National Parks: a guide to good practice*, Countryside Commission, Countryside Council for Wales, English Tourist Board, Rural Development Commission and Wales Tourist Board, Cheltenham: Countryside Commission.

Denman, R. and Ashcroft, P. (1997) *Visitor Payback: encouraging tourists to give money voluntarily to conserve the places they visit*, Ledbury: Tourism Company.

Department for Culture, Media and Sport (1999), *Tomorrow's Tourism: a growth industry for the new millennium*, London: Department of Culture, Media and Sport.

Department of Employment and English Tourist Board (1991) *Tourism and the Environment: maintaining the balance*, London: Department of Employment.

Department of Environment, Transport and the Regions, Scottish Executive, National Assembly for Wales and Department of the Environment (in Northern Ireland) (2000) *Climate Change: draft UK programme*, London: Department of the Environment, Transport and the Regions.

Department of National Heritage and Scottish Office (1997) *Guide to Safety at Sports Grounds*, London: Department of National Heritage and Scottish Office.

Department of Trade and Industry (1992), *Best Practice Benchmarking*, London: Department of Trade and Industry.

—— (1998), *Opportunities in Sustainable Tourism*, London: Department of Trade and Industry.

Derks, P. (1996) 'Introduction to the Transaction edition: twenty years of research on humor: a view from the edge', in Chapman, A.J. and Foot, H.C. (eds) *Humor and Laughter: theory, research and application*, New Brunswick, NJ: Transaction Publishers.

Dewar, J. (1998) at a seminar, Snowdonia, Forestry Commission.

Diaz, N. and Apostol, D. (1993) *Forest Landscape Analysis and Design: a process for developing and implementing land management objectives for landscape patterns*, Washington DC: USDA Forest Service.

Dimitrius, J-E. and Mazzarella, M. (1999) *Reading People*, London: Vermilion.

Dower, M. (1965) *Fourth Wave: the challenge of leisure*, a Civic Trust survey, reprinted from the *Architects' Journal*, London: Civic Trust.

—— (1993) 'Fourth Wave revisited', in Glyptis, S. (ed.), *Leisure and the Environment: essays in honour of Professor J.A. Patmore*, London: Belhaven Press, 15–21.

Dower, M., Buller, H. and Asamer-Handler, M. (1998) *The Socio-economic Benefits of national parks: a review prepared for Scottish Natural Heritage*, Scottish Natural Heritage review 104, Edinburgh: Scottish Natural Heritage.

Dower, M. and Downing, P. (1975) 'Attitudes to man, land and leisure', in Haworth, J.T. and Smith, M.A. (eds) *Work and Leisure*, London: Lepus Books, 53–60.

Dowling, R.K. (1992) 'Tourism and environmental integration: the journey from idealism to realism', in Cooper, C.P. and Lockwood, A. (eds), *Progress in Tourism, Recreation and Hospitality Management*, IV, London: Belhaven Press, 33–46.

Downes, D.M., Davies, B.P., David, M.E. and Stone, P. (1976) *Gambling, Work and Leisure: a study across three areas*, London: Routledge & Kegan Paul.

Driscoll, J. (1995) 'The legacy of Lyme Bay', *Countryside Recreation Network News*, 3, 3: 12–13.

Driver, B.L. (ed.) (1974) *Elements of Outdoor Recreation Planning*, Ann Arbor, MI: University of Michigan Press.

Driver, B.L., Brown, P.J. and Peterson, G.L. (eds) (1991a) *Benefits of Leisure*, State College, PA: Venture Publishing.

—— (1991b) 'Research on leisure benefits: an introduction', in Driver, B.L., Brown, P.J. and Peterson, G.L. (eds) *Benefits of Leisure*, State College, PA: Venture Publishing, 3–11.

Driver, B.L., Tinsley, H.E.A. and Manfredo, M.J. (1991) 'The paragraphs about leisure and recreation preference scales: results from two inventories designed to assess the breadth of the perceived psychological benefits of leisure', in Driver, B.L., Brown, P.J. and Peterson, G.L. (eds) *Benefits of Leisure*, State College, PA: Venture Publishing, 263–86.

Driver, B.L., Dustin, D., Baltic, T., Elsner, G. and Peterson, G. (eds) (1997) *Nature and the Human Spirit: toward an expanded land management ethic*, State College, PA: Venture Publishing Inc.

Duff, D., (ed.) (1983) *Queen Victoria's Highland Journals*, Exeter: Webb & Bower.

Dumazedier, J. (1967) *Toward a Society of Leisure*, London: Collier Macmillan.

—— (1974) *Sociology of Leisure*, Amsterdam and New York: Elsevier.

Eardley, J. and Mawby, J. (eds) (1999) *Managing Personal Watercraft: a guide for local and harbour authorities*, British Marine Industries Federation, Canterbury City Council, Cyngor Gwynedd Council, Poole Harbour Commissioners, Royal Yachting Association and Wirral Jet Ski Club.

Edington, J.M. and Edington, M.A. (1986) *Ecology, Recreation and Tourism*, Cambridge: Cambridge University Press.

Elson, M.J., Heaney, D. and Reynolds, G. in association with Sidaway, R. (1995) *Good Practice in the Planning and Management of Sport and Active Recreation in the Countryside*, London: Countryside Commission and Sports Council.

EMAS (2001) Eco-management Audit System. Online http://europa.eu.int/comm/environment/emas/intro_en.htm.

Emmet, J. (1999) 'Doctor's orders', *Leisure Manager*, June: 13–15.

Engel, J.F. and Blackwell, R.D. (1982) *Consumer Behaviour*, fourth edition, London: Dryden Press.

English Nature (1994) *Access and Nature Conservation: Working Together*, report of an English Nature seminar held at the Royal Society of Arts, London, December 1994, Peterborough: English Nature.

English Tourist Board (1989) *Tourism for All: providing accessible accommodation*, London: English Tourist Board and Holiday Care Service.

English Tourist Board and Department of Employment (1991) *Tourism and the Environment: maintaining the balance*, London: Department of Employment.

Environmental Resources Management (2000), *Potential UK Adaptation Strategies for Climate Change*, Wetherby: Department of the Environment, Transport and the Regions.

Enzenbacher, D.J. (1995) 'The regulation of Antarctic tourism', in Hall, C.M. and Johnston, M.E. (eds) *Polar Tourism: Tourism in the Arctic and Antarctic regions*, Chichester: Wiley, 179–215.

Epperson, A.F. (1977) *Private and Commercial Recreation: a text and reference*, New York: Wiley.

Erhmann, M. (1986 edition) *The Desiderata of Happiness*, London: Souvenir Press.

Esteve, R., Martín, S. and Lopéz, D. (1999) 'Grasping the meaning of leisure: developing a self-report measurement tool', *Leisure Studies*, 18: 79–91.

Etchell, C. (ed.) (1996a) *Consensus in the Countryside*, proceedings of a workshop jointly arranged by the Countryside Recreation Network and the Environment Council, held in Exeter in February 1996, Cardiff: Countryside Recreation Network.

—— (ed.) (1996b) *A Brush with the Land: art in the countryside*, proceedings of a workshop held in Grizedale Forest Park, Cumbria, on 16 and 17 May 1995, Cardiff: Countryside Recreation Network.

—— (ed.) (1996c) *Playing Safe: managing visitor safety in the countryside*, proceedings of a workshop held at the Royal York Hotel, York, on 16 June 1995, Cardiff: Countryside Recreation Network.

Evans, G.L. (1999) 'The economics of the national performing arts: exploiting consumer surplus and willingness-to-pay; a case of cultural policy failure', *Leisure Studies* 18: 97–118.

Everitt, P. and Higman, R. (2000) 'So what's the problem, congestion or pollution?' *Green Futures*, 21: 46–8.

Farrell, P. and Lundegren, H.M. (1991) *The Process of Recreation Programming: theory and technique*, third edition, State College, PA: Venture Publishing.

Faulkner, B. and Tideswell, C. (1997) 'A framework for monitoring community impacts of tourism', *Journal of Sustainable Tourism*, 5,1: 3–28.

Fausold, C.J. and Lilieholm, R.J. (1999) 'The economic value of open space: a review and synthesis', *Environmental Management*, 23, 3: 307–320.

Feist, A. (1998) 'Comparing the performing arts in Britain, the US and Germany: making the most of secondary data', *Cultural Trends*, 31: 29–47.

Fentem, P.H., Turnbull, N.B. and Bassey, E.J. (1990) *Benefits of Exercise: the evidence*, Manchester: Manchester University Press.

Ferber, P. (ed.) (1974) *Mountaineering: the Freedom of the Hills*, third edition, Seattle, WA: The Mountaineers.

Field, D.R., (1976) 'The social organization of recreation places', in Cheek, N., Field, D. and Burdge, R. (eds) *Leisure and Recreation Places*, MI: Ann Arbor, University of Michigan Press.

Fieldfare Trust (1997) *BT Countryside for All Standards and Guidelines: a good practice guide to disabled people's access into the countryside*, Sheffield: BT Countryside for All and Fieldfare Trust.

Fishbein, M. and Ajzen, I. (1975) *Belief, Attitude, Intention, and Behaviour: an introduction to theory and research*, Reading, MA: Addison-Wesley.

FitzHerbert, L., Addison, D. and Rahman, F. (1999) *A Guide to the Major Trusts I 1999/2000*, seventh edition, London: Directory of Social Change.

FitzHerbert, L., Rahman, L. and Harvey, S. (1999) *The National Lottery Yearbook*, third edition, London: Directory of Social Change.

Floyd, M.F. (1998) 'Getting beyond marginality and ethnicity: the challenge of race and ethnic studies in leisure research', *Journal of Leisure Research*, 30, 1: 3–22.

Forestry Commission (1992) *Forest Recreation Guidelines*, London: HMSO.

—— (1996) *Guidelines for Continued Public Access*, Edinburgh: Forestry Commission.

—— (1998) *The United Kingdom Forestry Standard*, Edinburgh: Forestry Commission

Forest Enterprise (2000a) *Working with Communities in Britain: how to get involved*, second edition, Edinburgh: Forestry Commission.

—— (2000b) *Working with Communities in Scotland: our commitment*, second edition, Edinburgh: Forestry Commission.

Fowles, J. (1963) *The Collector*, London: Jonathan Cape.

Fraser Allander Institute of Strathclyde University (1990) *The Economic Impact of Sporting Shooting in Scotland: summary report*, Dunkeld: British Association for Shooting and Conservation and Scottish Development Agency.

Fredline, E. and Faulkner, B. (2000) 'Host community reactions: a cluster analysis', *Annals of Tourism Research*, 27,3: 763–84.

French, P.W. (1997) *Coastal and Estuarine Management*, London: Routledge.

Friedmann, E. (1990) 'The value of pets for health and recovery', in Burger, I.H. (ed.) *Pets: Benefits and Practice*, Waltham Symposium 20, London: British Veterinary Association, 8–17.

Friends of the Earth (1999) Atmosphere and Transport Campaign. London: Friends of the Earth. Online http://www.foe.org.uk/ (29 July 1999)

—— (2000) 'Which causes more pollution damage? One trip to the US or one year's motoring in the UK?' Press release, online http://www.foe.org.uk/ (3 March 2001). Also available in *Aviation and Global Climate Change*, Online http://www.foe.org.uk/campaigns/atmosphere_and_transport/pdf/aviation_and_global_climate_change.pdf (12 March 2001).

Frosdick, S. and Walley, L. (eds) (1997) *Sport and Safety Management*, Oxford: Butterworth Heinemann.

Frost, G., Care, N., Wharton, A., Henderson, J. and Daley, I. (eds) (1998) *Rangers: the Eyes and Ears of the Peak National Park*, Castleford: Yorkshire Art Circus.

Fry, W.F. and Allen, M. (1996) 'Humour as a creative experience: the development of a Hollywood humorist', in Chapman, A.J. and Foot H.C. (eds), *Humor and Laughter: theory, research and application*, New Brunswick NJ: Transaction Publishers, 245–50.

Future Foundation, BT and First Direct (1998) *Twenty-four Hour Society*, London: Future Foundation, BT and First Direct.

Galbally, R. (1996) 'Leisure for health: developing leisure settings for health promotion', *Managing Leisure*, 1: 115–20.

Gauntlett, D. and Hill, A. (1999) *TV Living: television, culture and everyday life*, London: Routledge, in association with the British Film Institute.

Geoff Broom Associates and Macaulay Land Use Research Institute (1999) *Socio-economic Benefits from Natura 2000*, Edinburgh: Scottish Office Central Research Unit.

GEOprojects (1997) *Inland Waterways of Britain* (map), Reading: GEOprojects.

Gershuny, J. and Fisher, K. (1999) 'Leisure in the UK across the twentieth century', Institute of Social and Economic Research Working Paper 99–3, published as a chapter in Halsey, A.H. (ed.) (2000) *British Social Trends: the twentieth century*, London: Macmillan.

Online http://www.iser.essex.ac.uk/pubs/workpaps/wp99–03.php (24 September 2000).

Gill, S. and Fox, J. (1996) *The Dead Good Funerals Book*, Ulverston: Engineers of the Imagination.

Gillespie, J. and Shepherd, P. (1995) *Establishing Criteria for Identifying Critical Natural Capital in the Terrestrial Environment: A Discussion Paper*, English Nature Research Reports 141, Peterborough: English Nature.

Gillmeister, H. (1997) *Tennis: A cultural History*, London: Leicester University Press.

Glasgow University Veterinary School (1995/6) *BVMS1 Notes on Welfare, Legislation and Framework/transport*, Glasgow: Glasgow University Veterinary School.

Glass, N. (ed.) (1997) *'The Green Book': appraisal and evaluation in central government: Treasury guidance*, London: Stationery Office.

Gleeson, J. (1998) *The Arcanum*, London: Bantam Press.

Gleick, J. (1988) *Chaos*, London: Sphere.

Glyptis, S. (1989) *Leisure and Unemployment*, Milton Keynes: Open University Press.

—— (1991) *Countryside Recreation*, ILAM Leisure Management Series, Harlow: Longman.

—— (1993) 'Perspectives on leisure and the environment', in Glyptis, S. (ed.), *Leisure and the Environment: essays in honour of Professor J. A. Patmore*, London: Belhaven Press.

Glyptis, S., McInnes, H. and Patmore, J.A. (1987) *Leisure and the Home*, London: Sports Council and Economic and Social Research Council.

Gnoth, J. (1997) 'Tourism motivation and expectation formation', *Annals of Tourism Research*, 24, 2: 283–304.

Godbey, G. (1982) *The Future of Leisure: implications for leisure services*, Salford: Centre for Work and Leisure Studies.

—— (1994) *Leisure in Your Life: an exploration*, State College, Pennsylvania: Venture Publishing Inc.

Goldsmith, F. and Munton, R. (1971) 'The ecological effects of recreation', in Lavery, P. (ed.) *Recreational Geography*, Newton Abbot: David & Charles

Goldsworthy, A. (1996) *Wood*, London: Viking.

Goleman, D. (1999) *Emotional Intelligence: why it can matter more than IQ*, London: Bloomsbury.

Gombrich, E. (1995) *The Story of Art*, sixteenth edition, London: Phaidon.

Goodale, T.L. and Cooper, W. (1991) 'Philosophical perspectives on leisure in English-speaking countries', in Driver, B.L., Brown, P.J. and Peterson, G.L. (eds) *Benefits of Leisure*, State College, PA: Venture Publishing, 25–35.

Goodale, T.L. and Godbey, G.C. (1988) *The Evolution of Leisure: historical and philosphical perspectives*, State College, PA: Venture Publishing, 185–203.

Goodall, B. (1990) 'Opportunity sets as analytical marketing instruments: a destination area view', in Ashworth, G. and Goodall, B. (eds) *Marketing Tourism Places*, London: Routledge, 63–84.

—— (1992) 'Environmental auditing for tourism', in Cooper, C.P. and Lockwood, A. (eds) *Progress in Tourism, Recreation and Hospitality Management*, IV, London: Belhaven Press, 60–74.

Goode, J., Callender, C. and Lister, R.(1998) *Purse or wallet? Gender inequalities and income distribution within families on benefit*, PSI report 853, London: Policy Studies Institute.

Goodhead, T., Kasic, N. and Wheeler, C. (1996) 'Marinas and yachting', in Goodhead, T. and Johnson, D. (eds) *Coastal Recreation Management: the sustainable development of maritime leisure*, London: Spon.

Gordon, C. (1988) 'Major tourism attractions', in Talbot-Ponsonby, H. (ed.) *Changing Land Use and Recreation*, proceedings of the 1988 Countryside Recreation Conference, Bristol: Countryside Recreation Research Advisory Group, 56–7.

Gordon, J., and Grant, G. (eds) (1997) *How we Feel: an emotional insight into the emotional world of the teenager*, London: Jessica Kingsley.

Grant, B. and Harris, P. (eds) (1991) *The Grizedale Experience; sculpture, arts and theatre in a Lakeland forest*, Edinburgh: Canongate Press.

Gratton, C. and Taylor, P. (1988) *Economics of Leisure Services Management*, Harlow: Longman.

Gratton, C., Dobson, N. and Shibli, S. (2000) 'The economic importance of major sports events: a case study of six events', *Managing Leisure*, 5: 17–28.

Greene, L. (1996) 'How to make it happen: constructing an arts strategy', in Etchell, C. (ed.) *A Brush with the Land: art in the countryside*, proceedings of a workshop held in Grizedale Forest Park, Cumbria, Cardiff: Countryside Recreation Network, 34–8.

Greenfield, S. (1997) *The Human Brain: a guided tour*, London: Weidenfeld & Nicolson.

Greenhalgh, L. and Worpole, K. (1995) *Park Life: urban parks and social renewal*, Stroud and London: Comedia and Demos.

Greenland, D. (1983) *Guidelines for Modern Resource Management: soil, land, water, air*, Columbus, OH: Merrill.

Gregory, R.L. (1967) *Eye and Brain: the psychology of seeing*, London: Weidenfeld & Nicolson.

—— (1974) *Concepts and Mechanisms of Perception*, London: Duckworth.

Grenfell, R. (1975) *Your Kitchen Garden*, London: Mitchell Beazley.

Grimal, P. (ed.) (1965) *Larousse World Mythology*, London: Hamlyn.

Gross, R. (1996) *Psychology: the Science of Mind and Behaviour*, third edition, London: Hodder & Stoughton, 95–118.

Grove-White, R. (1997) 'Environment, risk and democracy', in Jacobs, M. (ed.) *Greening the Environment*, Oxford: Blackwell.

Gülez, S. (1996) 'Relationship between recreation demand and some natural landscape elements in Turkey: a case study', *Environmental Management*, 20, 1: 113–22.

Gump, P. (1978) 'School environments', in Altman, I. and Wohlwill, J.F. (eds) *Children and the Environment* (Human Behaviour and Environment III), New York: Plenum, 131–74.

Gunse, B. and McAleer, J. (1997) *Children and Television*, second edition, London: Routledge.

Guyer, C. and Pollard, J. (1996) 'Tourism, environment and the Shannon-Erne Waterway', in Etchell, C. (ed.) *Today's Thinking for Tomorrow's Countryside: recent advances in countryside recreation management*, Cardiff: Countryside Recreation Network, 123–34.

Haggard, L.M. and Williams, D.R. (1991) 'Self-identity benefits of leisure activities', in Driver, B.L., Brown, P.J. and Peterson, G.L. (eds) *Benefits of Leisure*, State College, PA: Venture Publishing, 103–19.

Haldane, A.R.B. (1968) *The Drove Roads of Scotland*, Edinburgh: Edinburgh University Press.

Hall, C.M. and Johnston, M.E. (1995) 'Visitor management and the future of tourism in polar regions', in Hall, C.M. and Johnston, M.E. (eds) *Polar tourism: Tourism in the Arctic and Antarctic regions*, Chichester: Wiley, 329 pp.

Hamilton, A., Hunt, J. (ed.), Jones, B. and Thomas, M. (1999) *Upland Pathwork: construction standards for Scotland*, Perth: Scottish Natural Heritage.

Hamilton-Smith, E. (1991) 'The construction of leisure', in Driver, B.L., Brown, P.J. and Peterson, G.L. (eds) *Benefits of Leisure*, State College, PA: Venture Publishing, 445–50.

Hammit, W. (1980) 'Outdoor recreation: is it a multi-phase experience', *Journal of Leisure Research*, 19: 115–30.

Handy, C. (1984) *The Future of Work: a guide to a changing society*, Oxford: Blackwell.

—— (1988) *Understanding Voluntary Organisations*, Harmondsworth: Penguin.

—— (1999) *Thoughts for the Day*, London: Arrow.

Hanley, N. and Spash, C.L. (1994) *Cost–Benefit Analysis and the Environment*, Aldershot: Edward Elgar.

Hardin, G. (1968) 'The tragedy of the commons', *Science*, 162: 1243–8.

Hargrave, C.P. (1966) *A History of Playing Cards*, London: Constable.

Harper, W. (1997) 'The future of leisure: making leisure work', *Leisure Studies*, 16: 189–98.

Harries, R., Bishop of Oxford (2000) 'Thought for the day', BBC Radio 4 *The Today Programme* (Friday 31 March), London: BBC.

Harris, I., Hill, Z., Hill, R., Pourzanjani, M. and Savill, T. (1996) 'Legislation or self-regulation', in Goodhead, T. and Johnson, D. (eds) *Coastal Recreation Management: the sustainable development of maritime leisure*, London: Spon.

Harrison, C. (1991), *Countryside Recreation in a Changing Society*, London: TMS Partnership.

Harrison, C., Burgess, J., Millward, A. and Dawe, G. (1995) *Accessible Natural Greenspace in Towns and Cities: a review of the appropriate size and distance criteria*, English Nature Research Report 153, Peterborough: English Nature.

Harrison, D. (1992) 'Tourism to less developed countries: the social consequences', in Harrison, D. (ed.) *Tourism and the Less Developed Countries*, Chichester: Wiley, 19–34.

Harrop, D.O. and Nixon, J.A. (1999) *Environmental Assessment in Practice*, London: Routledge.

Harvey, P. (ed.) (1967) *The Oxford Companion to English Literature*, fourth edition, rev. D. Eagle, London: Oxford University Press.

Hauser, A. (1982) *The Sociology of Art*, London: Routledge & Kegan Paul.

Haworth, J.T. (1997) *Work, Leisure and Well-being*, London: Routledge.

Haworth, J.T. and Parker, S.R. (eds) (1973) *Forecasting Leisure Futures*, LSA Conference Paper 3, Leeds: Leisure Studies Association.

Haworth, J.T. and Smith, M.A. (eds) (1975) *Work and Leisure: an interdisciplinary study in theory, education and planning*, London: Lepus Books.

Haywood, L. (ed.) (1994) *Community Leisure and Recreation*, Oxford: Butterworth Heinemann.

Heathcote, K. (1988) *The Gym Business*, Newton Abbot: David & Charles.

Henderson, K.A., Bialeschki, M.D., Shaw, S.M. and Freysinger, V.J. (1989) *A Leisure of One's Own: a feminist perspective on women's leisure*, State College, PA: Venture Publishing.

Hendricks, B. (1995) 'Politically correct play', in Lawrence, L., Murdoch, E. and Parker, S. (eds) *Professional and Development Issues in Leisure, Sport and Education*, LSA Conference Paper 56, Brighton: Leisure Studies Association, 89–98.

Hendry, L.B., Shucksmith, S. and Cross, J. (1991) 'Healthy minds and healthy bodies: a view of the relationship between leisure activity patterns and feelings of wellbeing amongst adolescents', in Long, J. (ed.) *Leisure, Health and Wellbeing*, LSA Conference Paper 44, Eastbourne: Leisure Studies Association, 129–44.

Hendry, L.B., Shucksmith, J., Love, J.G. and Glendinning, A. (1993) *Young People's Leisure and Life-styles*, London: Routledge.

Henry, I. (1988) 'Alternative futures for the public leisure service', in Benington, J. and White, J. (eds) *The Future of Leisure Services*, Harlow: Longman, 201–43.

Hey, V. (1986) *Patriarchy and Pub Culture*, London: Tavistock.

Hirst, J. (1997) 'Peak Park leisure walks: a model for increasing physical activity levels in "low participation" groups through regular social walking', *International Journal of Health Education*, 35, 2: 1–6.

Hockings, M., Carter, B. and Leverington, F. (1998) 'An integrated model of public contact planning for conservation management', *Environmental Management*, 22, 5: 643–54.

Hodgkinson, L. (1987) *Smile Therapy: how smiling and laughter can change your life*, London: Optima, Macdonald.

Holden, A. (1999) 'High impact tourism: a suitable component of sustainable policy? The case of Downhill Skiing Development of Cairngorm, Scotland', *Journal of Sustainable Tourism*, 7, 2: 97–107.

Holmes, T., Zinkhan, C., Alger, T. and Mercer, E. (1996) *Conjoint Analysis of Nature Tourism Values in Bahia, Brazil*, FPEI Working Paper 57, 19 pp. Online http://www.srs.fs.fed.us/pubs/viewpub.jsp?index = 563 (24 September 2000).

Hook, S. (1967) *The Paradoxes of Freedom*, Berkeley and Los Angeles: University of California Press.

Hoskins, W.G. (1967) *Fieldwork in Local History*, London: Faber.

—— (1991 edition) *The Making of the English Landscape*, Harmondsworth: Penguin.

House of Commons Culture, Media and Sport Committee (1999) Fourth Report, *Staging International Sporting Events*, London: Stationery Office.

House of Commons Environment Committee (1995a) *The Environmental Impact of Leisure Activities*, London: HMSO.

House of Commons Environment Committee, (1995b) *The Environmental Impact of Leisure Activities: Government Response to the Fourth Report of the Committee in Session 1994–95*, London: HMSO.

Howard, D.R. and Crompton, J.L. (1980) *Financing, Managing and Marketing Recreation and Park Resources*, Dubuque, IA: Brown.

Howitt, S., (1997) Summary Online http://www.businesszone.co.uk/keynote/hore9701.html (24 September 2000).

HPI Research Group (1997) *Public Attitudes to the Countryside*, Cheltenham: Countryside Commission.

Huff, D. (1973) *How to Lie with Statistics*, Harmondsworth: Penguin.

Hughes, G. (1995) 'Authenticity in tourism', *Annals of Tourism Research*, 22, 4: 781–803.

Hughes, G.B. (1963) *The* Country Life *Collector's Pocket Book*, London: Hamlyn.

Hughes, T. (ed.) (1997) *By Heart: 101 poems to remember*, London: Faber.

Huizinga, J. (1949) *Homo Ludens*, London: Routledge.

Hultsman, J. (1995) 'Just tourism: an ethical framework', *Annals of Tourism Research*, 22, 3: 553–67.

Hunter, J. (1995) *On the Other Side of Sorrow: nature and people in the Scottish Highlands*, Edinburgh: Mainstream.

Husbands, W. and Harrison, L.C. (1996) 'Practicing responsible tourism: understanding tourism today to prepare for tomorrow', in Harrison, L.C. and Husbands, W. (eds) *Practicing Responsible Tourism: international case studies in tourism planning, policy and development*, Chichester: Wiley, 1–15.

Inglis, G.J., Johnson, V.I. and Ponte, F. (1999) 'Crowding norms in marine settings: a case study of snorkeling on the Great Barrier Reef', *Environmental Management*, 24, 3: 369–81.

Inhaber, H. (1976) *Environmental Indices*, New York: Wiley.

Institute of Leisure and Amenity Management (1997) 'The Green Flag Park Award: achieving best value', Fact Sheet 97/11, Basildon: ILAM.

Institute of Leisure and Amenity Management (1999), *Leisure Bulletin*, 9 September, Reading: Institute of Leisure and Amenity Management.

International Hotels Environment Initiative (1996) *Environmental Management for Hotels: the industry guide to best practice*, second edition, Oxford: Butterworth Heinemann.

Irving, J.A. (1985) *The Public in Your Woods*, Chichester: Packard for the Land Decade Educational Council.

Iso-Ahola, S.E. (1980) *The Social Psychology of Leisure and Recreation*, Dubuque, IA: Brown.

Itten, J. (1987) *The Art of Colour: the subjective experience and objective rationale of colour*, London: Van Nostrand Reinhold.

Jackson, C. (1995) 'Play strategies: the vision and the reality?' in Lawrence, L., Murdoch, E. and Parker, S. (eds) *Professional and Development Issues in Leisure, Sport and Education*, LSA Conference Paper 56, Brighton: Leisure Studies Association, 99–119.

Jackson, T. and Marks, N. (1992) *Measuring Sustainable Economic Welfare: a pilot index, 1950–90*, London and Stockholm: New Economic Foundation and Stockholm Environment Institute.

Jamal, T. and Getz, D. (1999) 'Community roundtables for tourism – related conflicts: the dialectics of consensus and process structures', *Journal of Sustainable Tourism*, 7, 3–4: 290–330.

Jamieson, A. D. (1984) 'Inland water and its leisure use in Scotland', *Scottish Field Studies* 11–18.

Jansen-Verbeke, M. (1990) 'Leisure + Shopping = Tourism product mix', in Ashworth, G. and Goodall, B. (eds) *Marketing Tourism Places*, London: Routledge, 128–37.

Jenkins, C. and Sherman, B. (1981) *The Leisure Shock*, London: Eyre Methuen.

Jenkins, G. (1999) personal communication.

Jerome, J.K. (1957 edition) *Three Men in a Boat*, Harmondsworth: Penguin.

Jim, C.Y. (1998) 'Soil characteristics and management in an urban park in Hong Kong', *Environmental Management*, 22,5: 683–95.

John, G. and Campbell, K. (eds) (1993) *Handbook of Sports and Recreational Building Design* I, *Outdoor Sports*, second edition, London: Architectural Press.

John, G. and Campbell, K. (eds) (1995) *Handbook of Sports and Recreational Building Design* II, *Indoor Sports*, second edition, London: Architectural Press.

John, G. and Campbell, K. (eds) (1996) *Handbook of Sports and Recreational Building Design* III, *Ice Rinks and Swimming Pools*, second edition, London: Architectural Press.

John, G. and Heard, H. (eds) (1981) *Handbook of Sports and Recreational Building Design*, London: Architectural Press.

Johnson, H. (1971) *The World Atlas of Wine*, London: Mitchell Beazley.

Jones, G. (1968) *A History of the Vikings*, London: Oxford University Press.

Jones, T. (1987) *The Improbable Voyage*, London: Grafton.

Jong, H-C., Lee, B., Park, M. and Stokowski, P. (2000) 'Measuring underlying meanings of gambling from the perspective of enduring involvement', *Journal of Travel Research*, 38, 3: 230–38.

Jubenville, A. and Twight, B.W. (1993) *Outdoor Recreation Management: theory and application*, third edition, State College, PA: Venture Publishing.

Kay, T. (1996) 'Women's leisure and the family in contemporary Britain', in Samuel, N. (ed.) *Women, Leisure and the Family in Contemporary Society: a multinational perspective*, Wallingford: CAB International, 143–59.

Kelly, A. and Whitaker, R. (2000) 'Not welcome at Chelsea: the Japanese 'triffid' ravaging Britain's gardens', *Independent on Sunday*, 21 May.

Kelly, J.R. (1983) 'Social benefits of outdoor recreation: an introduction', in Lieber, S.R. and Fesenmaier, D.R. (eds) *Recreation Planning and Management*, London: Spon, 3–25.

—— (1987) *Freedom to Be: a new sociology of leisure*, London: Macmillan.

—— (1993) *Leisure Identities and Actions*, London: Allen and Unwin.

—— (1999) 'Leisure behaviors and styles: social, economic, and cultural factors', in Jackson, E.L. and Burton, T.L. (eds) *Leisure Studies: prospects for the twenty-first century*, State College, PA: Venture Publishing, 135–50.

Kelly, J.R. and Godbey, G. (1992) *The Sociology of Leisure*, State College, PA: Venture Publishing.

Kempe, P. (ed.) (1993) *Access to the Countryside in Europe for Walkers*, proceedings of a Ramblers' Association seminar, London: Ramblers' Association.

Kent, P. (1990) 'People, places, and priorities: opportunity sets and consumers' holiday choice', in Ashworth, G. and Goodall, B. (eds) *Marketing Tourism Places*, London: Routledge, 42–62.

Khan, N. (1976) *The Arts Britain Ignores: the arts of ethnic minorities in Britain: a report to the Arts Council of Great Britain, Calouste Gulbenkian Foundation, and Community Relations Commission*, London: CRC.

Kipling, R. (1908) *Just So Stories for Little Children*, London: Macmillan, p. 75.

Kleiner, K. (2000) 'Up and away', *New Scientist*, 26 August, p. 17. Online http://www.newscientist.com.

Klemm, M.S. and Martin-Quiros, M.A. (1996) 'Changing the balance of power: tour operators and tourism suppliers in the Spanish tourism industry', in Harrison, L.C. and Husbands, W. (eds) *Practicing Responsible Tourism: international case studies in tourism planning, policy and development*, Chichester: Wiley, 126–44.

Kosko, B. (1994) *Fuzzy Thinking: the new art of fuzzy logic*, London: Flamingo (Harper Collins).

Kotler, P. (1984) *Marketing Management: analysis, planning and control*, fifth edition, London: Prentice Hall.

Kotler, P., Bowen, J. and Mackens, J. (1998) *Marketing for Hospitality and Tourism*, Upper Saddle River, NJ: Prentice Hall.

KPMG Peat Marwick and the Tourism Company (1994) *A Study of Activity Holidays in Scotland*, for Scottish Enterprise, Scottish Tourist Board, Highlands and Islands Enterprise, Dumfries and Galloway Enterprise and Forth Valley Enterprise.

Krane, V. and Kane, M.J. (1998) 'Psychosocial aspects of sport and exercise', in Parks, J.B.,

Zanger, B.R.K. and Quarterman, J. (eds) *Contemporary Sport Management*, Champaign, IL: Human Kinetics, 33–47.

Kremer, J. and Scully, D.M. (1994) *Psychology in Sport*, London: Taylor & Francis.

Krippendorf, J. (1987 edition) *The Holiday Makers: understanding the impact of leisure and travel*, London: Heinemann.

Kuhn, S.T. (2000) 'Prisoners' dilemma', in *Stanford Encyclopedia of Philosophy*. Online http://plato.stanford.edu/entries/prisoner-dilemma/ (29 April 2000).

Kutiel, P. (1999) 'Environmental auditing: tendencies in the development of tracks in open areas', *Environmental Management*, 23, 3: 401–8.

Lane, B. (1994) 'The role of community action in rural recreation and tourism', in Etchell, C. (ed.) *Communities and their Countryside*, proceedings of a conference held at the University of York, 1994, by Countryside Recreation Network, Cardiff: Countryside Recreation Network, 3–13.

Lankford, S.V., Knowles-Lankford, J. and Povey, D.C. (1996) 'Instilling community confidence and commitment in tourism: community planning and public participation in Government Camp, Oregon, USA', in Harrison, L.C. and Husbands, W. (eds) *Practicing Responsible Tourism: international case studies in tourism planning, policy and development*, Chichester: Wiley, 330–49.

Lavery, P. (ed.) (1971) *Recreational Geography*, Newton Abbot: David & Charles.

Lawrence, D.P. (1997) 'Integrating sustainability and environmental impact assessment', *Environmental Management*, 21, 1: 23–42.

Lee, T.R. (1981) 'Perception of risk: the public's perception of risk and the question of irrationality', *Proceedings of the Royal Society*, 376: 5–16.

—— (1998) 'Evaluating the effectiveness of heritage interpretation', in Uzzell, D. and Ballantyne, R. (eds) *Contemporary Issues in Heritage and Environmental Interpretation: problems and prospects*, London: Stationery Office, 203–31.

Leisure Industries Research Centre (1997) *Economic Impact of Major Sports Events: six events in England and Scotland*, London: UK Sport.

Lekakis, J.N. (2000) 'Environment and development in a southern European country: which environmental Kuznets curves?' *Journal of Environmental Planning and Management*, 43, 1: 139–53.

Lentell, R. (2000) 'Untangling the tangibles: physical evidence and customer satisfaction in local authority leisure centres', *Managing Leisure*, 5: 1–16.

Leung, Y-F. and Marion, J.L. (1999) 'Characterizing backcountry camping impacts in Great Smoky Mountains National Park, USA', *Journal of Environmental Management*, 57: 193–203.

—— (2000) 'Recreation impacts and management in wilderness: a state-of-knowledge review', in Cole, D., McCool, S., Borrie, W. and O'Loughlin, J. (comps) *Wilderness Science in a Time of Change V, Proceedings RMRS-P-15-Vol-5*, Ogeden, UT: USDA Forest Service, Rocky Mountain Research Station.

Lewey, S. (1996) 'Water quality and pollution' in Goodhead, T. and Johnson, D. (eds) *Coastal Recreation Management: the sustainable development of maritime leisure*, London: Spon.

Liddle, M. (1997) *Recreation Ecology: the ecological impact of outdoor recreation and ecotourism*, London: Chapman and Hall.

Lieber, S. and Fesenmaier, D. (eds) (1983) *Recreation Planning and Management*, London: Spon.

Lieberman, J.N. (1977) *Playfulness: its Relationship to Imagination and Creativity*, Academic Press, London.

Lipscombe, N. (1999) 'The relevance of the Peak experience to continued skydiving participation: a qualitative approach to assessing motivations', *Leisure Studies*, 18: 267–88.

Lipsey, R.G. (1983) *An Introduction to Positive Economics*, sixth edition, London: Weidenfeld & Nicolson.

Liston-Heyes, C. and Heyes, A. (1998) 'Recreational benefits from the Dartmoor National Park', *Journal of Environmental Management*, 55: 69–80.

Lock, H.C. (1972) *Cricket: Take care of your Square*, Wakefield: EP Group.

Loikkanen, T., Simojoki, T. and Wallenius, P. (1999) *Participatory Approach to Natural Resource Management: a guide book*, Vantaa, Finland: Metsähallitus Forest and Park Service.

London Wildlife Trust (2001) Online http://www.wildlifetrust.org.uk/london/

Love, A. (1996) *Market Research for Countryside Recreation: a practical guide to market research for the providers of recreation in the countryside*, Cheltenham: Countryside Commission.

Lovett, A., Brainard, J., and Bateman, I. (1997) 'Improving benefit transfer demand functions: a GIS approach', *Journal of Environmental Management*, 51: 373–89.

Lovich, J.E. and Bainbridge, D. (1999) 'Anthropogenic degradation of the southern Californian desert ecosystem and prospects for natural recovery and restoration', *Environmental Management*, 24, 3: 309–26.

Lury, C. (1996) *Consumer Culture*, Cambridge: Polity Press.

Lynch, R. and Brown, P. (1999) 'Utility of large-scale leisure research agendas', *Managing Leisure*, 4: 63–77.

Mabey, R. (1972) *Food for Free: a guide to the edible wild plants of Britain*, London: Collins.

—— (1996) *Flora Britannica*, London: Sinclair-Stevenson.

McAllister, T.L., Overton, M.F. and Brill Jr, E.D. (1996) 'Cumulative impact of marinas on estuarine water quality', *Environmental Management*, 20, 3: 385–96.

MacArthur, B. (1996) *Historic Speeches*, Harmondsworth: Penguin.

McArthur, S. (2000) 'Beyond carrying capacity: introducing a model to monitor and manage visitor activity in forests', in Font, X. and Tribe, J. (eds) *Forest Tourism and Recreation: case studies in environmental management*, Wallingford: CABI Publishing, 259–78.

MacAuley, D. (ed.) (1999) *Benefits and Hazards of Exercise*, London: British Medical Journal.

McCold, L.N. and Saulsbury, J.W. (1996) 'Including past and present impacts in cumulative impact assessments', *Environmental Management*, 20, 5: 767–76.

McCormick E.J. and Ilgen D. (1981) *Industrial Psychology*, seventh edition, London: Allen & Unwin.

McDonald, B.L. and Schreyer, R. (1991) 'Spiritual benefits of leisure participation and leisure settings', in Driver, B.L., Brown, P.J. and Peterson, G.L. (eds) *Benefits of Leisure*, State College, PA: Venture Publishing, 179–94.

McEwen, J. (1977) *Who owns Scotland? A study in land ownership*, Edinburgh: Edinburgh University Student Publication Board.

—— (1981) *Who owns Scotland? A study in land ownership*, second edition, Edinburgh: Polygon Books.

McGrath, J. (1981) *The Cheviot, the Stag and the Black Black Oil*, second edition, London: Eyre Methuen.

McIntosh, A. and Edwards-Jones, G. (2000) 'DCF and weak sustainability analysis: discounting the children's future?' *Geophilus*, 1: 122–33. Online http://www.alastair-mcintosh.com/articles/2000_discounting.htm (10 March 2001).

McIntyre, N. and Roggenbuck, J.W. (1998) 'Nature/person transactions during an outdoor adventure experience: a multi-phasic analysis', *Journal of Leisure Research*, 30, 4: 401–22.

Mackay, E. (1999) personal communication.

Mackay Consultants (1989) *Economic Importance of Salmon Fishing and Netting in Scotland: a report for the Scottish Tourist Board and the Highlands and Islands Development Board*, Inverness: Mackay Consultants.

McKee, A. (1990) *A World too Vast: the four voyages of Columbus*, London: Souvenir Press.

McKercher, B. and Robbins, B. (1998) 'Business development issues affecting nature-based tourism operators in Australia', *Journal of Sustainable Tourism*, 6, 2: 173–88.

MacLeish, W.H. (1989) *The Gulf Stream: encounters with the blue god*, London: Hamish Hamilton.

McPhillimy, D. and Guy, S. (1998) *The Community Woodland Handbook: a guide for local groups setting up community woodlands and for organisations seeking to encourage participatory forestry*, Edinburgh: Reforesting Scotland.

Magee, B. (1985) *Popper*, third edition, London: Fontana.

Magnusson, M. and Palsson, H. (transl.) (1965) *The Vinland Sagas: the Norse discovery of America*, Harmondsworth: Penguin.

Mannell, R.C. and Kleiber, D.A. (1997) *A Social Psychology of Leisure*, State College, PA, Venture Publishing.

Margerison, C. and McCann, D. (1991a) *The Personal Development Manual*, second edition, York: TMS (UK).

—— (1991b) *The Team Development Manual*, York: TMS (UK).

Margerum, R.D. (1999) 'Integrated environmental management: the foundations for successful practice', *Environmental Management*, 24, 2: 151–66.

Marion, J.L. and Reid, S. (in press) 'Development of the US Leave No Trace program: a historical perspective', *Proceedings of the Scottish Natural Heritage Access Conference*, Edinburgh: Scottish Natural Heritage.

Mark, S.J., Broadmeadow, R. and Freer-Smith, P. (1996) *Urban Woodland and the Benefits for Local Air Quality: a report for Department of the Environment*, London: HMSO.

Martin, B. and Mason, S. (1993) 'Current trends in UK leisure: new views of countryside recreation', *Leisure Studies*, 12: 1–6.

Martin, W. and Mason, S. (1991) 'The Thatcher years', *Leisure Management*, 11, 4: 40–2.

Maslow, A.H. (1954) *Motivation and Personality*, New York: Harper & Row.

—— (1967) 'Lessons from the peak experience', *Journal of Humanistic Psychology*, 3, 1: 20–35.

Matthews, S. (1994) *With Animals*, Washington, Tyne and Wear: Pedalling Arts.

Mathieson, A. and Wall, G. (1982) *Tourism: Economic, Physical and Social Impacts*, Harlow: Longman.

Mayo, E. and MacGillivray, A. (n.d., *c.* 1992) *Growing pains? An index of sustainable economic welfare for the United Kingdom, 1950–1990*, London: New Economics Foundation.

Mayo, E., MacGillivray A. and McLaren, D. (n.d.) *More isn't always Better: a special briefing on growth and quality of life in the United Kingdom*, London: New Economics Foundation.

Meyer, L.B. (1956) *Emotions and Meaning in Music*, London: University of Chicago Press.

Middleton, V.T.C. with Hawkins, R. (1998) *Sustainable Tourism: a marketing perspective*, Oxford: Butterworth Heinemann.

Mieczkowski, Z. (1981) 'Some notes on the geography of tourism: a comment', *Canadian Geographer*, 215: 189.

Mignucci-Giannoni, A.A., Montoya-Ospina, R.A., Jiménez-Marrero, N.M., Rodríguez-López, M.A., Williams, Jr, E.H. and Bonde, R.K. (2000) 'Manatee mortality in Puerto Rico', *Environment Management*, 25, 2: 189–98.

Miles, C.W.N. and Seabrooke, W. (1977) *Recreational Land Management*, London: Spon.

Millar, G., Heap, J. and Henderson-Howat, D. (1992) 'Management in practice', in Talbot, H. (ed.) (1992) *Our Priceless Countryside: should it be priced?* Bristol: Countryside Recreation Research Advisory Group, 83–8.

Miller, L. (1997) *From the Heart: voices of the American Indian*, London: Pimlico Press.

Mitchell, G. (1999) 'Demand forecasting as a tool for sustainable water resource management', *International Journal of Sustainable Development and World Ecology*, 6: 231–41.

Moffat, I. (1996) *Sustainable Development: principles, analysis and policies*, Carnforth: Parthenon Publishing.

—— (1999) 'Is Scotland sustainable? A time series of indicators of sustainable development', *International Journal of Sustainable Development and World Ecology*, 6: 242–50.

Moore, C.A. (1992) *The Game Finder: a leader's guide to great activities*, State College, PA, Venture Publishing.

Moore, R.C. (1986) *Childhood's Domain: play and place in child development*, Beckenham: Croom Helm.

Moore, R. and Young, D. (1978) 'Childhood Outdoors: towards a social ecology of the landscape', in Altman, I. and Wohlwill, J.F. (eds) *Children and the Environment* (Human Behaviour and Environment, III), New York: Plenum, 83–130.

Moore, S.A. and Lee, R.G. (1999) 'Understanding dispute resolution processes for American and Australian public wildlands: towards a conceptual framework for managers', *Environmental Management*, 23, 4: 453–65.

Moorehead, A. (1968) *The Fatal Impact: an account of the invasion of the South Pacific*, Harmondsworth: Penguin.

Morant, H. (2000) 'BMA demands more responsible media attitude on body image', *British Medical Journal*, 320: 1495.

More, T.A. (1999) 'A functionalist approach to user fees', *Journal of Leisure Research*, 31, 3: 207–26.

Morgan, K.O. (1984) *The Oxford Illustrated History of Britain*, London: BCA and Oxford University Press.

Mortazari, R. (1997) 'The right of public access in Sweden', *Annals of Tourism Research*, 24, 3: 609–23.

Moscardo, G. (1996) 'Mindful visitors: heritage and tourism', *Annals of Tourism Research*, 23, 2: 376–97.

Moscardo, G., Pearce, P., Morrison, A., Green, D. and O'Leary, J.T. (2000) 'Developing a typology for understanding the visiting friends and relatives market', *Journal of Travel Research*, 38, 3: 251–9.

Mountaineering Council of Scotland (n.d., *c.* 1999) *Where to 'Go' in the Great Outdoors*, Perth: Mountaineering Council of Scotland (leaflet).

Muller, H. (1990) 'The case for developing tourism in harmony with man and nature', in Bramwell, B. (ed.) *Shades of Green: working towards green tourism in the countryside*, conference proceedings, Manchester: Countryside Commission, English Tourist Board, Rural Development Commission, 11–20.

Murray, C. and Nakajima, I. (1999) 'The leisure motivation of Japanese managers: a research note on scale development', *Leisure Studies*, 18: 57–65.

National Playing Fields Association (n.d., *c.* 1983) *The State of Play: the crisis in the provision of recreational playing space in the United Kingdom*, London: National Playing Fields Association.

National Playing Fields Association and Amateur Athletic Association (1980) *Facilities for athletics (track and field)*, second edition, London: National Playing Fields Association.

Nature Conservancy Council (1984) *Nature Conservation in Great Britain*, Shrewsbury: Nature Conservancy Council.

National Trust (1995) *Open Countryside: report of the Access Review Working Party*, London: National Trust.

Nelkin, D. (1982) 'Public participation in environmental planning in the USA', in Ahmad, J.Y. and Müller, F.G. (eds) *Integrated Physical Planning, Socio-economic and Environmental Planning*, Dublin: Tycooly, 84–9.

Nepal, S.K. (2000) 'Tourism in protected areas: the Nepalese Himalaya', *Annals of Tourism Research*, 27, 3: 661–81.

Nepal Trust (2000) 'Nepal Trust, working with health and community in the "hidden Himalayas", Forres: Nepal Trust (leaflet).

Neulinger, J. (1974) *The Psychology of Leisure: research approaches to the study of leisure*, Springfield, IL: Thomas.

—— (1981) *To Leisure, an Introduction: a psychological approach to leisure and its relationship to the quality of life*, London: Allyn & Bacon.

Newby, E. (1956) *The Last Grain Race*, London: Secker & Warburg.

Newlands, C. and Roworth, P. (2000) 'Managing the roughs', *Enact: managing land for wildlife*, 8, 1: 16–17.

Norton, M. and Eastwood, M. (1997) *Writing Better Fundraising Applications*, London: Directory of Social Change in association with ICFM.

Norton-Taylor, R. (1982) *Whose Land is it Anyway? Agriculture, planning and land use in the British countryside*, Wellingborough: Turnstone Press.

Ogilvy, D. (1983) *Ogilvy on Advertising*, London: Orbis.

Ogilvie, J. (1996) *Adventure Holidays Worldwide*, Brighton: In Print Publishing.

O'Hear, A. (1997) *NonSense about Nature*, Risk Controversies 9, London: Social Affairs Unit.

O'Neill, D. (ed.) (1990) *Motorised Sports in the Countryside: overcoming the challenges*, conference report, East Midlands: Regional Council for Sport and Recreation.

Onions, C.T. (ed.) (1966) *The Oxford Dictionary of English Etymology*, London: Oxford University Press.

Opie, I. and Opie, P. (1951) *The Oxford Dictionary of Nursery Rhymes*, London: Oxford University Press.

—— (1969) *Children's Games in Street and Playground*, London: Oxford University Press.

Opie, I. and Opie, P. (1997) *Children's Games with Things*, Oxford: Oxford University Press.

Oppermann, M. (1995) 'Travel life cycle', *Annals of Tourism Research*, 22,3: 535–52.

—— (1999) 'Sex tourism', *Annals of Tourism Research*, 26,2: 251–66.

Orams, M.B. (1995) 'Using interpretation to manage nature-based tourism', *Journal of Sustainable Tourism*, 4,2: 81–94.

O'Riordan, T. and Cameron, J. (eds) (1994) *Interpreting the Precautionary Principle*, London: Earthscan.

Orlick, T. (1990) *In Pursuit of Excellence: how to win in sport and life through mental training*, Champaign IL: Human Kinetics.

Ove Arup & Partners (1997) *The National Cycle Network: Guidlines and Practical Details*, issue 2. Bristol: Sustrans.

Owen, K. (1984) *Community Art and the State: storming the citadels*, London: Comedia.

Owens, S. (1992) 'The charging debate: the free countryside', in Talbot, H. (ed.) *Our Priceless Countryside: should it be priced?*, Bristol: Countryside Recreation Research Advisory Group.

The Oxford Dictionary of Quotations, (1953), second edition, London: Oxford University Press.

Packenham, T. (1996) *Meetings with Remarkable Trees*, London: Weidenfeld & Nicolson.

Palmer, H., Aldridge, J. and Measham, F. (1998) *Illegal Leisure: the normalization of adolescent recreational drug use*, London: Routledge, 132–47.

Parker, J. and Stimpson, J. (1999) *Raising Happy Children: what every child needs their parents to know – from 0 to 7 years*, London: Hodder and Stoughton, 275 ff.

Parker, S. (1976) *The Sociology of Leisure*, London: Allen & Unwin.

—— (1983) *Leisure and Work*, revised edition, London: Allen & Unwin.

Parks and Recreation Federation of Ontario (1992) *A Catalogue of the Benefits of Parks and Recreation*, Ontario: Parks and Recreation Federation of Ontario.

Patmore, J.A. (1970), *Land and Leisure*, Newton Abbot: David & Charles.

Patterson, M.E., Watson, A.E., Williams, D.R. and Roggenbuck, J.R. (1998) 'An hermeneutic approach to studying the nature of wilderness experience', *Journal of Leisure Research*, 30, 4: 423–52.

Peak District National Park (1999) Online http://www.peakdistrict.org./ (24 September 2000).

Pearce, D. (1994) 'The precautionary principle and economic analysis', in O'Riordan, T., and Cameron, J. (eds) *Interpreting the Precautionary Principle*, London: Earthscan, 132–51.

Pearce, D.G. (1999) 'Tourism in Paris: studies at the microscale', *Annals of Tourism Research*, 26,1: 77–97.

Pearce, D., Barbier, E., Markandya, A., Barrett, S., Turner, R.K., and Swanson, T. (1991) *Blueprint 2: Greening the World Economy*, London: Earthscan.

Pearce, D., Markandya, A. and Barbier, E.B. (1989) *Blueprint for a Green Economy*, London: Earthscan.

Pearce, D., Putz, F. and Vanclay, J.K. (1999) *A Sustainable Forest Future*, CSERGE Working Paper GEC 99–15, Norwich: CSERGE.

Pearson, A. (1988) *Arts for Everyone: Guidance on provision for the disabled*, London: Carnegie UK Trust and Centre on Environment for Handicapped.

Pearson, I. (2001) Online http://www.labs.bt.com/people/pearsonid/

Pelham, D. (1976) *The Penguin Book of Kites*, Harmondsworth: Penguin.

Peper, E., Ancoli, S. and Christon, M. (eds) (1979) *Mind/Body Integration: essential readings in biofeedback*, London: Plenum Press.

Peter Scott Planning Services (1991) *Countryside Access in Europe*, Review Series 23, Perth: Scottish Natural Heritage.

—— (1994) *Access to the Northern Ireland Countryside: summary report*, Environment Service, Northern Ireland Tourist Board and the Sports Council for Northern Ireland, Belfast: HMSO.

—— (1997) *Calmer Waters: guidelines for planning and managing watersports on inland waters in Scotland*, Edinburgh: Scottish Sports Council.

Peter Scott Planning Services in association with Howcroft, A., Bondo-Anderson, A., Stoll, P., Vistad, O.I. and Ståhl, S. (1998) *Access to the Countryside in selected European Countries: a review of access rights, legislation and associated arrangements in Denmark, Germany, Norway and Sweden*, Edinburgh: Scottish Natural Heritage.

Peters, T.J. and Waterman, R.H. (1982) *In Search of Excellence: lessons from America's best-run companies*, New York: Harper and Row.

Philpin, T. (1996) 'Access management for the Millennium', in Etchell, C. (ed.) (1996) *Today's Thinking for Tomorrow's Countryside: recent advances in countryside management*, proceedings of the 1995 Countryside Recreation Conference, Countryside Recreation Network, Cardiff: Countryside Recreation Network, 87–97.

Physical Activity Task Force (1995) *The Health of the Nation: more people, more active, more often*, London: Department of Health.

Pieper, J. (1965) *Leisure: the Basis of Culture*, London: Faber.

Plant, M. and Plant, M. (1992) *Risk Takers: alcohol, drugs, sex and youth*, London: Routledge.

Pledge, H.J. (1966) *Science since 1500*, London: HMSO.

Pomeroy, J. (1964) *Recreation for the Physically Handicapped*, London: Macmillan.

Porrit, J. (2000) 'High lights, low lights, green lights', *Green Futures*, 20: 26–7.

Porter, S. (1999) 'Sweet water solutions', *Green Futures*, July/August: 22–4.

Poucher, W.A. (1965) *The Scottish Peaks: a pictorial guide for walkers and climbers*, London: Constable.

Price, C. and Chambers, T.W.M. (2000) 'Hypotheses about recreational congestion: tests in the Forest of Dean (England) and wider management implications', in Font, X. and Tribe, J. (eds) *Forest Tourism and Recreation: case studies in environmental management*, Wallingford: CABI Publishing, 55–74.

Rackham, O. (1986) *The History of the Countryside*, London: Dent.

Radford, N. (1999) 'Recreation and the nature experience in our towns', in Hughes, J. (ed.) *Is the Honeypot Overflowing? How much recreation can we have?* Proceedings of the 1998 Annual Countryside Recreation Conference, Cardiff: Countryside Recreation Network.

Ragatz, R.L. and Crofts, J.C. (2000) 'US timeshare purchasers: who are they and why do they buy?', *Journal of Hospitality and Tourism Research*, 24,1: 49–66.

Ramblers' Association (1996) *Access to the Countryside: a draft bill*, Wandsworth: Ramblers' Association.

Rameau, J.P. (1971 edition), *Treatise on Harmony*, London: Dover

Ramsay, E. (in press, 2001) 'Managing visitor safety: principles developed by the Visitor Safety in the Countryside Group', *Countryside Recreation*, 9, 3.

Randall, G. (1980) *Church Furnishing and Decoration in England and Wales*, London: Batsford.

Range Rover of North America (1990) *Tread Lightly, please: a Range Rover guide to off-road responsibility*, Lanham, MD: Range Rover of North America.

Rapport, D.J., Gaudet, C., Karr, J.R., Baron, J.S., Bohlen, C., Jackson, W., Jones, B., Naiman, R.J., Norton, B. and Pollock, M.M. (1998) 'Evaluating landscape health: integrating social goals and biophysical process', *Journal of Environmental Management*, 53: 1–15.

Rapoport, R. and Rapoport, R.J. (1975) *Leisure and the Family Life Cycle*, London: Routledge & Kegan Paul.

Ratcliffe, D.A. (ed.) (1977) *A Nature Conservation Review* I, Cambridge: Cambridge University Press.

Ravenscroft, N. (1992), *Recreation Planning and Development*, London: Macmillan.

Rendel, S. (1996) 'A tranquil countryside?', *Countryside Recreation Network News*, 4, 1: 9–11, Cardiff: Countryside Recreation Network.

Rickwood, P. (1982) *The Story of Access in the Peak District*, Bakewell: Peak Park Joint Planning Board.

Rights of Way Law Review (1995) *Better Public Access within the Law: commons, recreational rights and land management*, a seminar held at Christ Church, Oxford, April.

Riley, R., Baker, D. and Van Doren, C.S. (1998) ' Movie induced tourism', *Annals of Tourism Research*, 25,4: 919–35.

Roadburg, A. (1983) 'Freedom and enjoyment: disentangling perceived leisure', *Journal of Leisure Research*, 15, 1: 15–26.

Roberts, A. (transl.) (1993) *Venice*, Everyman Guides, London: David Campbell, 50–4.

Roberts, K. (1970) *Leisure*, London: Longman.

—— (1978) *Contemporary Society and the Growth of Leisure*, London: Longman.

—— (1992) Lecture on a course for Forestry Commission staff at the Forest of Dean

—— (1997) 'Same activities, different meanings: British youth cultures in the 1990s', *Leisure Studies*, 16: 1–15.

Roberts, K. and Brodie, D.A. (1992) *Inner-city Sport: who plays, and what are the benefits?* Voorthuizen: Giordano Bruno Culemborg.

Robertson, P. (1994) *The Great Divide in the Mind of Music* (programme notes).

Robinson, I. (ed.) (1995) *The Waltham Book of Human–Animal Interaction: benefits and responsibilities of pet ownership*, Oxford: Pergamon.

Robinson, J.P. and Godbey, G. (1997) *Time for Life: the surprising ways Americans use their time*, University Park, PA: Pennsylvania State University Press.

Robinson, L. (1999) 'Following the quality strategy: the reasons for the use of quality management in UK public leisure facilities', *Managing Leisure*, 4: 201–17.

Rojek, C. (ed.) (1989) *Leisure for Leisure: critical essays*, London: Macmillan.

—— (1993) *Ways of Escape: modern transformations in leisure and travel*, London: Macmillan.

—— (2000) 'Leisure and the rich today: Veblen's theory after a century', *Leisure Studies*, 19: 1–15.

Rothbart, M.K. (1996) 'Incongruity, problem-solving and laughter', in Chapman, A.J. and Foot, H.C. (eds) *Humor and Laughter: theory, research and application*, New Brunswick, NJ: Transaction, 37–54.

Rowe, D. (1991) *Wanting Everything: the art of happiness*, London: HarperCollins.

Rowell, A. and Freerfusson, M. (1991) *Bikes not Fumes: the emission and health benefits of a modal shift from motor vehicles to cycling*, London: Cyclists' Touring Club.

The Rural Media Company (2000) *Not Seen, not Heard? Social exclusion in rural areas*, London: Countryside Agency.

Russel, R.V. (1996) *Pastimes: the Context of Contemporary Leisure*, London: Brown & Benchmark.

Russell, P. (1979) *The Brain Book: know your own mind and how to use it*, London: Routledge.

Ryan, C. (1991) *Recreational Tourism: a social science perspective*, London: Routledge.

—— (1998) 'The travel career ladder: an appraisal', *Annals of Tourism Research*, 25, 4: 936–57.

Sagoff, M. (1996) 'On the value of endangered and other species', *Environmental Management*, 20, 6: 897–911.

Saunders, E. and Sugden, J. (1997) 'Sport and community relations in Northern Ireland', *Managing Leisure* 2: 39–54.

Schofield, M. (1987) 'Golf courses: maximising ecological interest', in Talbot-Ponsonby, H. (ed.) *Recreation and Wildlife: working in partnership*, proceedings of the 1987 Countryside Recreation Conference, Bristol: Countryside Recreation Research Advisory Group, 84–6.

Schuetti, M.A. (1997) 'State park directors' perceptions of mountain biking', *Environmental Management*, 21, 2: 239–46.

Scottish Arts Council (1999) *Strategy: the Distribution of Arts National Lottery Funding in Scotland*, Edinburgh: Scottish Arts Council.

Scottish Canoe Association (n.d.) *Sea Kayaking: a guide to good environmental practice*, Edinburgh: Scottish Canoe Association (leaflet).

Scottish Executive (1999) *Land Reform: the proposals*, Edinburgh: Stationery Office.

—— (2001a) *Land Reform, The Draft Bill*, Edinburgh: Stationery Office. Online http://www.scotland.gov.uk/consultations/landreform/ Crdb-oo.asp (10 May 2001).

—— (2001b) *A Draft Scottish Outdoor Access Code*, Edinburgh: Stationery Office. Online http://www.scotland.gov.uk/library3/environment/soac-oo.asp (10 May 2001).

Scottish Health Service (1999) *Cancer Comparisons: cancer of the trachea, bronchus and lung*, Edinburgh: Scottish Health Service. Online http://www.show.scot.nhs.uk/hhb/publicat/cancercomparisons (3 March 2001).

Scottish Landowners' Federation, (1993) *Access: towards Access without Acrimony, Scottish Landowners' Federation – advice to members on the provision of access*, Edinburgh: Scottish Landowners' Federation.

Scottish Natural Heritage (1994) *Enjoying the Outdoors: a programme for action*, Perth: Scottish Natural Heritage.

—— (1997) *Rangers in Scotland: SNH policy statement*, Scottish Natural Heritage, Perth.

—— (1998) *Access to the Countryside for Open-air Recreation: Scottish Natural Heritage's Advice to Government*, Perth: Scottish Natural Heritage.

Scottish Natural Heritage (in association with the Convention of Scottish Local Authorities) (1993) *Public Access to the Countryside: a guide to the law: practice and procedure in Scotland*, Scottish National Heritage and Convention of Scottish Local Authorities.

Scottish Sports Council (1995a) *Technical Digest: Swimming Pools: small pools design (No. 300)*, Edinburgh: Scottish Sports Council.

—— (1995b) *Technical Digest: Swimming Pools: improvements and alterations (No. 301)*, Edinburgh: Scottish Sports Council.

—— (1995c) *Technical Digest: Swimming Pools: changing accommodation (No. 302)*, Edinburgh: Scottish Sports Council.

—— (1995d) *Statement of the Scottish Sports Council Policies for the Planning of Natural Resources for Sports and Physical Recreations*, Edinburgh: Scottish Sports Council.

Scottish Rights of Way Society (1994) *Rights of Way in Scotland: A Brief Overview of the Law*, Edinburgh: Scottish Rights of Way Society.

Scottish Tourist Board (1997) *Standardised Questions for Tourism Research*, revised edition, co-sponsored by English Tourist Board, Northern Ireland Tourist Board, Scottish Tourist Board, and Wales Tourist Board, Edinburgh: Scottish Tourist Board.

—— (1998) Leisure Day Visits Fact Sheets, Edinburgh: Scottish Tourist Board.

Scraton, S. and Watson, B. (1998), 'Gendered cities: women and public leisure space in the "postmodern city"', *Leisure Studies*, 17: 123–37.

Seabrook, J. (1989) *The Leisure Society*, Oxford: Blackwell.

Seaton, A.V. and Hay, B. (1998) 'The marketing of Scotland as a tourist destination,, 1985–96', in MacLellan, R. and Smith, R. (eds) *Tourism in Scotland*, London: International Thomson Business Press, 209–40.

Seckler, D.W. (1966) 'On the uses and abuses of economic science in evaluating public outdoor recreation', *Land Economics*, 42: 485–94.

Seeley, I.H. (1973) *Outdoor Recreation and the Urban Environment*, London: Macmillan.

Segrave, J.O. (2000), 'Sport as escape', *Journal of Sport and Social Issues*, 24, 1: 61–77.

Selwood, S. (1995) *The Benefits of Public Art: the polemics of permanent art in public places*, London: Policy Studies Institute.

Serpell, J.A. (1990) 'Evidence for long-term effects of pet ownership on human health', in Burger, I.H. (ed.) *Pets: Benefits and Practice*, Waltham Symposium 20, London: British Veterinary Association, 1–17.

Severin, T. (1978) *The Brendan Voyage*, London: Arrow.

Seymour, J. (1978) *The Complete Book of Self-sufficiency*, London: Corgi.

Shackleton, E. (1999) *South: the* Endurance *Expedition*, Harmondsworth: Penguin.

Shackley, M. (1998) '"Stingray City": Managing the impact of underwater tourism in the Cayman Islands', *Journal of Sustainable Tourism*, 6,4: 328–38.

Shafer, C.L. (1999) 'US National Park buffer zones: historical, scientific, social, and legal aspects', *Environmental Management*, 23, 1: 49–73.

Sharpley R. (1996) *Tourism and Leisure in the Countryside*, second edition, Huntingdon: Elm.

Sharpley, R. and Sharpley, J. (1997) *Rural Tourism: an introduction*, London: International Thomson Business Press,

Shaw, G. and Williams, A. (1992) 'Tourism, development and the environment: the eternal triangle', in Cooper, C.P. and Lockwood, A. (1992) *Progress in Tourism, Recreation and Hospitality Management* IV, London: Belhaven Press, 47–59.

Shaw, S.M. (1997) 'Controversies and contradictions in family leisure: an analysis of conflicting paradigms', *Journal of Leisure Research*, 29, 1: 98–112.

Shin, W.S. and Jaakson, R. (1997) 'Wilderness quality and visitors' wilderness attitudes: management implications', *Environmental Management*, 21, 2: 225–32.

Shipp, D. (1993) *Loving them to Death? Sustainable tourism in Europe's nature and National Parks*, Grafenau, Germany: Federation of Nature and National Parks of Europe.

Shivers, J.S. and de Lisle, L.J. (1997) *The Story of Leisure: context, concepts, and current controversies*, Champaign, IL: Human Kinetics.

Shivers, J.S. and Fait, H.F. (1975) *Therapeutic and Adapted Recreational Services*, Philadelphia, PA: Lear & Febiger.

Shoard, M. (1980) *The Theft of the Countryside*, London: Temple Smith.

—— (1987) *This Land is our Land: the struggle for Britain's countryside*, London: Grafton.

—— (1999) *A Right to Roam*, Oxford: Oxford University Press.

Sibly, E. (n.d., *c.* 1810) *A Genuine and Universal System of Natural History; comprising the Three Kingdoms of Animals, Vegetables, and Minerals, arranged under their respective classes, orders, genera, and species* (a translation of Linnaeus), fifteen volumes, London: Lewis.

Sidaway, R. (1988) *Sport, Recreation and Nature Conservation*, Study 32, London: Sports Council.

—— (1991) *Good Conservation Practice for Sport and Recreation*, a guide prepared for the Sports Council, Countryside Commission, Nature Conservancy Council and World Wide Fund for Nature, Study 37, London: Sports Council.

—— (1994a) *The Limits of Acceptable Change*, a report prepared for the Countryside Commission by Roger Sidaway, Cheltenham: Countryside Commission.

—— (1994b) 'Recreation and the natural heritage: a research review' in *Scottish Natural Heritage Review* 25, Perth: Scottish Natural Heritage.

Sidaway, R. (1998) with contributions from Judith Annett and David Rothe, *Conflict Resolution*, Good Practice in Rural Development 5, *Consensus building*, Edinburgh: Scottish Office, Central Research Unit.

Sidaway, R.M. and Thompson, D. (1991) 'Upland recreation: the limits of acceptable change', *ECOS* 12, 1: 31–9.

Sidaway, R.M., Coalter, J.A., Rennick, I.M. and Scott, P.G. (1986) *Access to the Countryside for Recreation and Sport*, Manchester: Countryside Commission and Sports Council.

Simmie, J.M. (1974) *Citizens in Conflict: the sociology of town planning*, London: Hutchinson.

Simons, R., Miller, L.F. and Lengsfelder, P. (1984) *Nonprofit Piggy goes to Market: how the Denver Children's Museum earns $600,000 annually*, Denver CO: Children's Museum of Denver.

Simpson, D. (2000) 'Links for Wildlife', *ENACT Managing Land for Wildlife*, 11–15.

Simpson, J. (1997) *Touching the Void*, London: Vintage.

Simpson, M. (1982) *Skisters: the Story of Scottish Skiing*, Carrbridge: Landmark Press.

Sims, S. and Hislop, M. (1998) personal communication.

Sirakaya, E. (1997) 'Attitudinal compliance with ecotourism guidelines', *Annals of Tourism Research*, 24, 4: 919–50.

Sjoberg, R. (2000) *RYA Cruising*, 46.

Skinner, B.F. (1972) *Beyond Freedom and Dignity*, London: Jonathan Cape.

Slocum, J., (1949) *Sailing alone around the World* and *Voyage of the* Liberdade, London: Reprint Society.

Smale, B.J.A. (1999) 'Spatial analysis of leisure and recreation', in Jackson, E.L. and Burton, T.L. (eds) *Leisure Studies: prospects for the twenty-first century*, State College, Pennsylvania: Venture Publishing, 177–97.

Smalley, M. (1998), 'WWW – a brave new world of marketing', *Business Review*, 5, 2.

Smith, D. (2000) 'The Benefits Catalogue and benefits-based approach – a Canadian perspective', *LSA Newsletter*, 55: 23–5.

Smith, M. (1995) 'A participant observer study of a "rough" working class pub', in Critcher, C., Bramham, P. and Tomlinson, A. (eds) *Sociology of Leisure: a reader*, London: Spon, 183–5.

Smith, R. (1988) 'The four R's', in Mountaineering Council of Scotland and Scottish Landowners' Federation (1988) *Heading for the Scottish Hills*, Edinburgh: Scottish Mountaineering Trust.

Smyth, J. (ed.) (1998) *Learning to Sustain*, Stirling: Scottish Environmental Education Council.

Social and Community Planning Research (1997) *UK Day Visits Survey: summary of the 1996 survey findings*, London: Social and Community Planning Research.

Social and Community Planning Research (1999) *Leisure Day Visits: summary of the 1998 UK day visits survey*, London: Social and Community Planning Research.

Sofield, T.H.B. and Li, F.M.S. (1998) 'Tourism development and cultural policies in China', *Annals of Tourism Research*, 25,2: 362–92.

South West Coast Path Project (1995) *Report of the South West Coast Path User Survey 1994*, Exeter: South West Coast Path Project.

Sparkes, J. (1995a) 'Visitor risk management: translating theory into practice', in Etchell, C. (ed.) *Countryside Recreation Network News*, 3, 3: 14–15.

—— (1995b) 'Visitor risk management: translating theory into practice', in Etchell, C. (ed.) *Playing Safe? Managing visitor safety*, proceedings of a workshop, Cardiff: Countryside Recreation Network, 38–41.

Spash, C.L. (1992) 'Techniques to value non-market goods', in Talbot, H. (ed.) *Our Priceless Countryside: should it be priced?* Bristol: Countryside Recreation Research Advisory Group, 72–3.

Sports Council (1993) *A Countryside for Sport: policy for sport and recreation*, London: Sports Council.

Sport England (1999a) *The Value of Sport*, London: Sport England.

—— (1999b) *The Value of Sport to the Health of the Nation*, London: Sport England.

—— (1999c) *The Value of Sport to Local Authorities*, London: Sport England.

—— (1999d) *The Value of Sport to Regional Development*, London: Sport England.

Stabler, M. (1990) 'The concept of opportunity sets as a methodological framework for the analysis of selling tourism places: the industry view', in Ashworth, G. and Goodall, B. (eds) *Marketing Tourism Places*, London: Routledge, 23–41.

—— (1997) *Tourism and Sustainability: principles to practice*, Oxford: CAB International.

—— (2000) 'Editorial introduction to the paper by D.R. Vaughan, H. Farr and R.W. Slee on estimating and interpreting the local economic benefits of visitor spending: an explanation', *Leisure Studies*, 19: 91–4.

Stadel, C. (1996) 'Divergence and conflict, or convergence and harmony? Nature conservation and tourism potential in Hohe Tauern National Park, Austria', in Harrison, L.C. and Husbands, W. (eds) *Practicing Responsible Tourism: international case studies in tourism planning, policy and development*, Chichester: Wiley, 445–71.

Stankey, G.H., Cole, D.N., Lucas, R.C., Peterson, M.E. and Frissell, S.J. (1985) *The Limits of Acceptable Change (LAC) of Wilderness Planning*, General Technical Report, INT– 176. USDA, Intermountain Research Station, Ogden, UT: Forest Service.

Stebbins, R. (2000) 'Obligation as an aspect of leisure experience', *Journal of Leisure Research*, 32: 152–5.

Stenton, F. (1971) *Anglo-Saxon England*, third edition, London: Oxford University Press.

Stevenson, R.L. (1896a) *An Inland Voyage*, undated edition, London: Nelson.

—— (1896b) *Travels with a Donkey in the Cevennes*, 1948 edition, London: Falcon Press.

Stewart, S. (1998) *Conflict Resolution: a foundation guide*, Winchester: Waterside Press.

Stewart, W.P. (1998) 'Leisure as multi-phase experiences: challenging traditions', *Journal of Leisure Research*, 30, 4: 390–400.

Stillman, F. (1966) *The Poet's Manual and Rhyming Dictionary*, London: Thames & Hudson.

Stoakes, R. (1982) 'Charging and pricing for countryside recreation', Countryside Recreation Research Advisory Group Conference Proceedings, Bristol: Countryside Recreation Research Advisory Group.

Stokowski, P.A. (1994) *Leisure in Society: a network structural perspective*, London: Mansell.

Storr, A. (1997) *Music and the Mind*, London: HarperCollins.

Storer, R. (1991) *Exploring Scottish Hill Tracks: for walkers and mountain bikers*, Newton Abbot: David and Charles.

Straker, D. (1997) *Rapid Problem Solving with Post-it Notes*, Aldershot: Gower.

Suetonius (1957) *The Twelve Caesars*, trans. Robert Graves, Harmondsworth: Penguin.

Sun, D. and Walsh, D. (1998) 'Review of studies on the environmental impact of recreation and tourism in Australia', *Journal of Environmental Management*, 53: 323–38.

Sustrans (1994) *Making Ways for the Bicycle: a guide to traffic-free path construction*, Bristol: Sustrans.

—— (1995) *The National Cycle Network*, Bristol: Sustrans.

Swinburne, R. (1974) *The Evolution of the Soul*, Oxford: Clarendon.

Talbot, H. (ed.) (1992) *Our Priceless Countryside: should it be priced?* Proceedings of the 1991 Annual Countryside Recreation conference, Bristol: Countryside Recreation Research Advisory Group.

Talbot, M. and Vickerman, R.W. (eds) (1978) *Social and Economic Costs and Benefits of Leisure*, LSA Conference Paper 8, Leeds: Leisure Studies Association.

Tasks, M., Renson, R. and Vanreusel, B. (1999) 'Consumer expenses in sport: a marketing tool for sports and sports facility providers', *European Journal for Sport Management*, 6, 1: 4–18.

Taylor, G. (1973) 'An approach to forecasting tourism futures', in Haworth, J.T. and Parker, S.R. (eds) *Forecasting Leisure Futures*, LSA Conference Paper 3, Leisure Studies Association, Leeds: 56–63.

Taylor, P. (1960) *The Notebooks of Leonardo da Vinci*, London: New English Library.

Taylor, R. and Stevens, T. (1995) 'An American adventure in Europe: an analysis of performance of Euro Disneyland, 1992–1994', *Managing Leisure*, 1: 28–42.

Taylor, R.P. (1992) *Football and its Fans: supporters and their relations with the game, 1885–1985*, Leicester: Leicester University Press.

Telfer, D.J. and Wall, G. (1996) 'Linkages between tourism and food production', *Annals of Tourism Research*, 23, 3: 635–53.

Thalberg, I. (1977) *Perception, Emotion and Action*, Oxford: Blackwell.

Thomas, M., Walker, A., Wilmot, A. and Bennett, N. (eds) (1998) *Living in Britain: Results from the 1996 General Household Survey*, London: Stationery Office.

Thorne, C. (ed.) (1992) *The Waltham Book of Dog and Cat Behaviour*, Oxford: Pergamon Press.

Tibbalds, F. (1992) *Making People-friendly Towns: improving the public environment in towns and cities*, Harlow: Longman.

Tibbat, D. (2000) 'Swings and roundabouts: play – community involvement', *Leisure Manager*, January: 12–14.

Tietenberg, T. (1994) *Environmental Economics and Policy*, New York: Harper Collins.

Timmerman, J.G., Ottens, J.J. and Ward, R.C. (2000) 'The information cycle as a framework for defining information goals for water-quality monitoring', *Environmental Management*, 25, 3: 229–39.

Timothy, D.J. (1998) 'Cooperative tourism planning in a developing destination', *Journal of Sustainable Tourism*, 52–68.

Timothy, D.J. and Wall, G. (1997) 'Selling to tourists: Indonesian street vendors', *Annals of Tourism Research*, 24, 2: 322–40.

Timothy Cochrane Associates (1984) *Providing for Children's Play in the Countryside*, Perth: Countryside Commission for Scotland.

Todd, S.E. and Williams, P.W. (1996) 'From white to green: a proposed environmental management system framework for ski areas', *Journal of Sustainable Tourism*, 4, 3: 147–73.

Toner, J.P. (1995) *Leisure and Ancient Rome*, Cambridge: Polity Press.

Torkildsen, G. (1986) *Leisure and Recreation Management*, second edition, London: Spon.

Tourism and Environment Forum (2000) *Operational Plan 2000 to 2003*, Inverness: Tourism and Environment Initiative.

Tourism and Environment Task Force (1993) *Going Green: guidelines for the Scottish tourism industry*, Edinburgh: Scottish Tourist Board.

Tourism and Rural Initiatives Consultancy (1994) *Enjoying the Outdoors: a summary report on responses to the consultation paper*, Edinburgh: Scottish Natural Heritage.

Tresidder, R. (1999) 'Tourism and sacred landscapes', in Crouch, D. (ed.) *Leisure/Tourism Geographies: practices and geographical knowledge*, London: Routledge, 136–48.

Tribe, J. (1995) *The Economics of Leisure and Tourism: environments, markets and impacts*, Oxford: Butterworth Heinemann.

Tribe, J., Font, X., Griffiths, N., Vickery, R. and Yale, K. (2000) *Environmental Management for Rural Tourism and Recreation*, London: Cassell.

Tuck, M. (1976) *How do we choose?*, London: Methuen.

Tull, D. and Hawkins, D. (1984) *Marketing Research: measurement and method*, third edition, New York: Collier Macmillan.

Turner, R.K. (ed.) (1993) *Sustainable Environmental Economics and Management: Principles and practice*, London: Belhaven Press.

Tyson, P. (1995) Trends: Data in Venice: Saving outdoor art. Online http://www.techreview.com/articles/oct95/TrendTyson.html (21 June 2000).

UK Sport (1999) *Major Events: guidance document*, London: UK Sport.

Ulrich, R.S., Dimberg, U. and Driver, B.L. (1991) 'Psychophysiological indicators of leisure benefits', in Driver, B.L., Brown, P.J. and Peterson, G.L. (eds) *Benefits of Leisure*, State College, Pennsylvania: Venture Publishing, 73–89.

UNEP (1999) Agenda 21, Section 1.1. Online http://www.unep.org/Documents/Default.asp?Document ID = 52 & Article ID = 49 (5 March 2001).

United Nations (1948) *Universal Declaration of Human Rights*, New York: United Nations.

Urry, J. (1990) *The Tourist Gaze: Leisure and Travel in Contemporary Society*, London: Sage.

—— (1999) 'Sensing leisure spaces', in Crouch, D. (ed.) *Leisure/Tourism Geographies: practices and geographical knowledge*, London: Routledge, 34–45.

US Department of Agriculture Forest Service (1982) *ROS Users Guide*, Washington, DC: US Department of Agriculture, Forest Service.

—— (1995) *Sustaining Forests and Communities: collaborative planning*, FS-578, Washington, DC: US Department of Agriculture, Forest Service.

US Department of Agriculture Forest Service and Bureau of Land Management, (1990) *Tread Lightly on Public and Private Lands*, Washington, DC: US Department of Agriculture, Forest Service and Bureau of Land Management.

Uzzell, D. (1989) *Heritage Interpretation II, The Visitor Experience*, London: Belhaven Press.

—— (1998) 'Planning for interpretive experiences', in Uzzell, D. and Ballantyne, R. (eds) *Contemporary Issues in Heritage and Environmental Interpretation: problems and prospects*, London: Stationery Office, 232–52.

Uzzell, D. and Ballantyne, R. (eds) (1998) *Contemporary Issues in Heritage and Environmental Interpretation: problems and prospects*, London: Stationery Office.

van der Zande, A.N. (1984) 'Outdoor Recreation and Birds: conflict or symbiosis: Impacts of outdoor recreation upon density and breeding success of birds in dune and forest areas in the Netherlands', PhD thesis, Leiden.

van der Zande, A.N. and Vos, P. (1984) 'Impact of a semi-experimental increase in recreation intensity on the densities of birds in groves and hedges on a lake shore in the Netherlands', *Biological Conservation*, 30: 237–59.

van Matre, S. and Weiler, B. (eds) (1983) *The Earth Speaks: an acclimatization journal*, Greenville, WV: Institute for Earth Education.

van Pragg, H.J. (1992) 'Industrial leadership: a practical example in the hotel industry', in Department of National Heritage (ed.) *Tourism and the Environment: challenges and choices for the '90s*, London: Department of National Heritage, 62–6.

Valentine, G. (1999) 'Consuming pleasures: Food, leisure and the negotiations of sexual relations', in Crouch, D. (ed.) *Leisure/Tourism Geographies: practices and geographical knowledge*, London: Routledge, 164–80.

Vaughan, D.R., Farr, H. and Slee, R.W. (2000) 'Estimating and interpreting the local economic benefits of visitor spending: an explanation', *Leisure Studies* 19: 95–118.

Veal, A.J. (1987) *Leisure and the Future*, London: Allen & Unwin.

—— (1992) *Research Methods for Leisure and Tourism: a practical guide*, Harlow: Longman and Institute of Leisure and Amenity Management.

—— (1994) *Leisure Policy and Planning*, London: Longman and Institute of Leisure and Amenity Management.

—— (1999) 'Forecasting leisure and recreation', in Jackson, E.L. and Burton, T.L. (eds) *Leisure Studies: prospects for the twenty-first century*, State College, PA: Venture Publishing, 385–96.

Veblen, T., (1899) *The Theory of Leisure Class*, London: Allen & Unwin.

Vickerman, R.W. (1978) 'Economics and leisure studies: themes, problems and policies in the economics of leisure', in Talbot, M. and Vickerman, R.W. (eds) *Social and Economic Costs and Benefits of Leisure*, LSA Conference Paper 8, Leeds: Leisure Studies Association, 1–26.

Victurine, R. (2000) 'Building tourism excellence at the community level: capacity building for community-based entrepreneurs in Uganda', *Journal of Travel Research*, 38, 3: 221–9.

Vistad, O. (2000) '*Allemansretten*: on Access and Recreation in Norway', in *Enjoyment and Understanding of the Natural Heritage: finding a new balance between rights and responsibilities*, Edinburgh: Scottish Natural Heritage.

Voase, R. (1999) '"Consuming" tourist sites/sights: a note on York', *Leisure Studies* 18: 289–96.

Voelkl, J.E. and Ellis, G.D. (1998) 'Measuring flow experiences in daily life: an examination of the items used to measure challenge and skill', *Journal of Leisure Research*, 30, 3: 380–89.

von Frisch, K. (1970) *The Dancing Bees: an account of the life and senses of the honey bee*, London: Methuen, 113 ff.

Vorkinn, M. (1998) 'Visitor response to management regulations: a study among recreation users in southern norway', *Environmental Management*, 22, 5: 737–46.

Walker, G.J., Hull IV, R.B. and Roggenbuck, J.W. (1998) 'On-site optimal experiences and their relationship to off-site benefits', *Journal of Leisure Research*, 30, 4: 453–71.

Walker, S. (1995) *Report of the 1993 United Kingdom Day Visits Survey*, Cardiff: Countryside Recreation Network.

Wall, G. (1996) 'One name, two destinations: planned and unplanned coastal resorts in Indonesia', in Harrison, L.C. and Husbands, W. (eds) *Practicing Responsible Tourism: international case studies in tourism planning, policy and development*, Chichester: Wiley, 41–57.

—— (1997) 'Is ecotourism sustainable?' *Environmental Management*, 21, 4: 483–91.

Wall, G. and Wright, C. (1977), *The Environmental Impact of Outdoor Recreation*, Department of Geography Publication Series 11, Waterloo, Ontario: University of Waterloo.

Walpole, M.J. and Goodwin, H.J. (2000) 'Local economic impacts of dragon tourism in Indonesia', *Annals of Tourism Research*, 27, 3: 559–76.

Walsh, R.G. (1986) *Recreation Economic Decision: comparing benefits and costs*, State College, Pennsylvania: Venture Publishing Inc.

Walton, J.K. (1983) 'Municipal government and the holiday industry in Blackpool, 1876–1914', in Walton, J.K. and Walvin, J. (eds) *Leisure in Britain, 1780–1939*, Manchester: Manchester University Press.

Wang, B. and Manning, R.E. (1999) 'Computer simulation modeling for recreation management: a study on carriage road use in Acadia National Park, Maine, USA', *Environmental Management*, 23, 2: 193–203.

Wankel, L.M. and Berger, B.G. (1991) 'Personal and social benefits of sport and physical activity', in Driver, B.L., Brown, P.J. and Peterson, G.L. (eds) *Benefits of Leisure*, State College, Pennsylvania: Venture Publishing, 121–44.

Warren, S. (1999) 'Cultural contestation at Disneyland Paris', in Crouch, D., (ed.) *Leisure/ Tourism Geographies: practices and geographical knowledge*, London: Routledge, 109–25.

Warrington, J. and White, A. (1986) *Sport and Recreation for Women and Girls: action guide for providers*, London: Sports Council, London and South East Region.

Waters, I. (1994) *Entertainment, Arts and Cultural Services*, Harlow: Longman.

Watson, D. (1997) *Going Green: a handbook for managers of tourism businesses*, Inverness: Tourism and Environment Initiative.

Wearing, S. and Wearing, B. (2000) 'Smoking as a fashion accessory in the 1990s: conspicuous consumption, identity, and adolescent women's leisure choices', *Leisure Studies*, 19: 45–58.

Welch, D. (1991) *The Management of Urban Parks*, Harlow: Longman.

Weston, S.A. (1996) *Commercial Recreation and Tourism: an introduction to business oriented recreation*, London: Brown & Benchmark.

Wheeller, B. (1992) 'Alternative tourism: a deceptive ploy', in Cooper, C.P. and Lockwood, A. (eds) *Progress in Tourism, Recreation and Hospitality Management* IV, London: Belhaven Press, 140–5.

Whinam, J. and Chilcott, N. (1999) 'Impacts of trampling in alpine environments in central Tasmania', *Journal of Environmental Management*, 57: 205–20.

Whitby, M. and Falconer, K. (1998) *Reinstatement, Renewal or Re-creation of Access Rights: the economics of a right to roam*, Working Paper Series 38, Newcastle: Centre for Rural Economy, University of Newcastle upon Tyne.

Whiteman, A. and Sinclair, J. (1994) *The Costs and Benefits of Planting Three Community Forests: Forest of Mercia, Thames Chase, and Great Northern Forest*, Edinburgh: Forestry Commission.

Wightman, A.D. (1996), *Who owns Scotland? A study in land ownership*, Edinburgh: Canongate Books.

—— (1999) *Scotland: Land and Power: the agenda for land reform*, Edinburgh: Luath Press.

Wilcox, D. (1994) *The Guide to Effective Participation*, Brighton: Partnership Books.

Williams, R. and Buncombe, A. (1999) 'Our generation of couch potato kids, stuck in their rooms and glued to TV', *Independent*, 19 March, p. 3.

Williams, S. (1995a) 'On the street: public space for popular leisure', in Leslie, D. (ed.) *Tourism and Leisure: Perspectives on Provision*, Brighton: Leisure Studies Association, 23–35.

—— (1995b) 'Urban playgrounds: rethinking the city as an environment for children's play', in Lawrence, L., Murdoch, E. and Parker, S. (eds) *Professional and Development Issues in Leisure, Sport and Education*, LSA Conference Paper 56, Brighton: Leisure Studies Association.

—— (1995c) *Outdoor Recreation and the Urban Environment*, London: Routledge.

Wilson, F. (1971) *Structure: the essence of architecture*, London: Studio Vista.

Winter, D.L. and Palucki, L.J. (1999), 'Anticipated responses to a fee program: the key is trust', *Journal of Leisure Research*, 31, 3: 207–26.

Wolanska, T. (1989) 'Lifelong education for physical recreation', in Murphy, W. (ed.) *Leisure, Labour and Lifestyles: international comparisons*, V, *Children, Schools and Education for Leisure*, LSA Conference Paper 36, Eastbourne: Leisure Studies Association.

Wondolleck, J. (1997) 'Incorporating hard-to-define values into public lands decision making: a conflict management perspective' in Driver, B.L., Dustin, D., Baltic, T., Elsner, G. and Peterson, G. (eds) *Nature and the Human Spirit: toward an expanded land management ethic*, State College, Pennsylvania: Venture Publishing, 257–62.

Wood, C., Dipper, B. and Jones, C. (2000) 'Auditing the assessment of the environmental impacts of planning projects', *Journal of Environmental Planning and Management*, 43,1: 23–47.

Wood, R. (ed.) (1994), *Environmental Economics, Sustainable Management and the Countryside*, proceedings of a workshop, Cardiff: Countryside Recreation Network.

Wooley, H. and Noor-ul-Amin (1999), 'Pakistani teenagers' use of public open space in Sheffield', *Managing Leisure*, 4: 156–67.

Woolmore, L. (ed.) (1995), *Pathways to Partnership: enabling people with disabilities to make informed choices about their discovery and enjoyment of the countryside*, Northampton: Northamptonshire County Council.

Wright, M. (ed.) (1974) *The Complete Indoor Gardener*, London: Pan.

Wright, M. and Wilsdon, J. (1999) 'When efficient ain't sufficient', *Green Futures*, 16: 26–8.

Wright, M., Tickell, O., Kinver, M., Henderson, C., Frankel, C., McQuillan, R. and Smith, A. (2000) 'The Hocketan Housing Project', *Green Futures*, 22: 31.

Young, J.Z. (1971) *An Introduction to the Study of Man*, London: Oxford University Press.

Young, M. (1999) 'Cognitive maps of nature-based tourists', *Annals of Tourism Research*, 26,4: 817–39.

Zabinksi, C.A. and Gannon, J.E. (1997) 'Effects of recreational impacts on microbial communities', *Environmental Management*, 21, 2: 233–8.

Zadek, S. (1995) 'Redefining business: the rise of social auditing', *New Economics*, autumn: 6–9.

Zalatan, A. (1998) 'Wives' involvement in tourism decision processes', *Annals of Tourism Research*, 25, 4: 890–903.

Zanger, B.R.K. (1998) 'Professional style', in Parks, J.B., Zanger, B.R.K. and Quarterman, J. (eds) *Contemporary Sport Management*, Champaign IL: Human Kinetics, 295–302.

Zeldin, T. (1994) *An Intimate History of Humanity*, London: Sinclair-Stevenson.

—— (1999) *Conversation: how talk can change your life*, London: Harvill Press.

Zietsma,C. and Vertinsky, I. (1999), 'Cognitive Framing of Corporate Environmental Response: Examples from the Forestry Industry', Edmonton, Alta: Sustainable Forest Management Network Conference.

Index